W9-BLC-072

ELECTRONIC
INSTRUMENTATION
AND
MEASUREMENT
TECHNIQUES

2nd Edition

ELECTRONIC INSTRUMENTATION AND MEASUREMENT TECHNIQUES

William David Cooper

Prentice-Hall, Inc., *Englewood Cliffs, New Jersey* *07632*

Library Congress Cataloging in Publication Data

COOPER, WILLIAM DAVID.
    Electronic instrumentation and measurement techniques.

    Includes bibliographies and index.
    1. Electric measurements.   2. Electric meters.
3. Electronic measurements.   4. Electronic instruments.
I. Title.
TK275.C63 1978   621.3815′4   77-24528
ISBN   0-13-251710-8

Printed in the United States of America

10   9   8   7   6   5   4

Prentice-Hall International, Inc., *London*
Prentice-Hall of Australia Pty. Limited, *Sydney*
Prentice-Hall of Canada, Ltd., *Toronto*
Prentice-Hall of India Private Limited, *New Delhi*
Prentice-Hall of Japan, Inc., *Tokyo*
Prentice-Hall of Southeast Asia Pte. Ltd., *Singapore*
Whitehall Books Limited, *Wellington, New Zealand*

Contents

Preface

The second edition of *Electronic Instrumentation and Measurement Techniques* is designed to serve as a text for students of electrical and electronic engineering at both two- and four-year colleges and technical institutes, and as a handbook for practicing engineers. It presents a thorough treatment of the principles that govern the operation and behavior of electrical and electronic test instruments and provides practical information on the use and limitations of instruments in typical measurement applications.

Compared to the first edition, the structure of the revised text and the style of presentation have remained largely unchanged. Numerous modifications have been made to the text where experience has shown the need for improvements, and new material has been added in areas where technological advances have been made. In response to suggestions by several readers, the mathematical content has been made more consistent throughout the text, and calculus has been avoided wherever possible.

Except for editorial improvements and some reorganization of the subject matter, Chapters 1 through 5 have remained practically unchanged from the first edition. Important additions were made to Chapter 6, where the application of potentiometers has been expanded to include the volt box and the shunt box, and to Chapter 7, where the Murray and Varley loop tests have been added to dc bridge measurements. A section on unbalance conditions in ac bridges has been added to Chapter 8. Chapter 9 has been entirely revised and updated, and now includes major new sections on vertical and horizontal deflection systems, various types of delay line, and oscillo-scope probes. Some additional material has been introduced in the remaining chapters, while less relevant material, notably the chapter on number systems in the first edition, has been deleted.

The author wishes to acknowledge the assistance given by the editorial staff at Prentice-Hall, Inc., and is grateful to the many companies for their contributions of data, photographs, and diagrams.

Washington, D.C.   WILLIAM DAVID COOPER

1

Measurement and Error

1-1 DEFINITIONS

Measurement generally involves using an instrument as a physical means of determining a quantity or variable. The instrument serves as an extension of human faculties and in many cases enables a person to determine the value of an unknown quantity which his unaided human faculties could not measure. An instrument, then, may be defined as *a device for determining the value or magnitude of a quantity or variable*. The *electronic* instrument, as its name implies, is based on electrical or electronic principles for its measurement function. An electronic instrument may be a relatively uncomplicated device of simple construction such as a basic dc current meter (see Chapter 4). As technology expands, however, the demand for more elaborate and more accurate instruments increases and produces new developments in instrument design and application. To use these instruments intelligently, one needs to understand their operating principles and to appraise their suitability for the intended application.

Measurement work employs a number of terms which should be defined here.

Instrument: a device for determining the value or magnitude of a quantity or variable.

Accuracy: closeness with which an instrument reading approaches the true value of the variable being measured.

1

Precision: a measure of the reproducibility of the measurements; i.e., given a fixed value of a variable, precision is a measure of the degree to which successive measurements differ from one another.

Sensitivity: the ratio of output signal or response of the instrument to a change of input or measured variable.

Resolution: the smallest change in measured value to which the instrument will respond.

Error: deviation from the true value of the measured variable.

Several techniques may be used to minimize the effects of errors. For example, in making precision measurements, it is advisable to record a series of observations rather than rely on one observation. Alternate methods of measurement, as well as the use of different instruments to perform the same experiment, provide a good technique for increasing accuracy. Although these techniques tend to increase the *precision* of measurement by reducing environmental or random error, they cannot account for instrumental error.*

This chapter provides an introduction to different types of error in measurement and to the methods generally used to express errors, in terms of the most reliable value of the measured variable.

1-2   ACCURACY AND PRECISION

Accuracy refers to the degree of closeness or conformity to the true value of the quantity under measurement. *Precision* refers to the degree of agreement within a group of measurements or instruments.

To illustrate the distinction between accuracy and precision, two voltmeters of the same make and model may be compared. Both meters have knife-edged pointers and mirror-backed scales to avoid parallax, and they have carefully calibrated scales. They may therefore be read to the same *precision*. If the value of the series resistance in one meter changes considerably, its readings may be in error by a fairly large amount. Therefore the *accuracy* of the two meters may be quite different. (To determine which meter is in error, a comparison measurement with a standard meter should be made.)

Precision is composed of two characteristics: *conformity* and the number of *significant figures* to which a measurement may be made. Consider, for example, that a resistor, whose true resistance is 1,384,572 Ω, is measured by an ohmmeter which consistently and repeatedly indicates 1.4 MΩ. But can the observer "read" the true value from the scale? His estimates from the scale reading consistently yield a value of 1.4 MΩ. This is as close to the true value as he can read the scale by estimation. Although there are no deviations

---

*Melville B. Stout, *Basic Electrical Measurements*, 2nd ed. (Englewood Cliffs, N.J.: Prentice-Hall, Inc., 1960), pp. 21–26.

from the observed value, the error created by the limitation of the scale reading is a *precision* error. The example illustrates that conformity is a necessary, but not sufficient, condition for precision because of the lack of significant figures obtained. Similarly, precision is a necessary, but not sufficient, condition for accuracy.

Too often the beginning student is inclined to accept instrument readings at face value. He is not aware that the accuracy of a reading is *not* necessarily guaranteed by its precision. In fact, good measurement technique demands *continuous skepticism* as to the accuracy of the results.

In critical work, good practice dictates that the observer make an independent set of measurements, using different instruments or different measurement techniques, not subject to the same systematic errors. He must also make sure that the instruments function properly and are calibrated against a known standard, and that no outside influence affects the accuracy of his measurements.

1-3   SIGNIFICANT FIGURES

An indication of the precision of the measurement is obtained from the number of *significant figures* in which the result is expressed. Significant figures convey actual information regarding the magnitude and the measurement precision of a quantity. The more significant figures, the greater the precision of measurement.

For example, if a resistor is specified as having a resistance of 68 Ω, its resistance should be closer to 68 Ω than to 67 Ω or 69 Ω. If the value of the resistor is described as 68.0 Ω, it means that its resistance is closer to 68.0 Ω than it is to 67.9 Ω or 68.1 Ω. In 68 Ω there are two significant figures; in 68.0 Ω there are three. The latter, with more significant figures, expresses a measurement of greater precision than the former.

Often, however, the total number of digits may not represent measurement precision. Frequently, large numbers with zeros before a decimal point are used for approximate populations or amounts of money. For example, the population of a city is reported in six figures as 380,000. This may imply that the true value of the population lies between 379,999 and 380,001, which is six significant figures. What is meant, however, is that the population is closer to 380,000 than to 370,000 or 390,000. Since in this case the population can be reported only to two significant figures, how can large numbers be expressed?

A more technically correct notation uses *powers of ten*, 38×10^4 or 3.8×10^5. This indicates that the population figure is only accurate to two significant figures. Uncertainty caused by zeros to the *left* of the decimal point is therefore usually resolved by *scientific notation* using powers of ten. Reference to the velocity of light as 186,000 mi/s, for example, would cause no misunderstanding to anyone with a technical background. But 1.86×10^5 mi/s leaves no confusion.

It is customary to record a measurement with all the digits of which we are sure nearest to the true value. For example, in reading a voltmeter, the voltage may be read as 117.1 V. This simply indicates that the voltage, read by the observer to best estimation, is closer to 117.1 V than to 117.0 V or 117.2 V. Another way of expressing this result indicates the *range of possible error*. The voltage may be expressed as 117.1 ± 0.05 V, indicating that the value of the voltage lies between 117.05 V and 117.15 V.

When a number of independent measurements are taken in an effort to obtain the best possible answer (closest to the true value), the result is usually expressed as the arithmetic *mean* of all the readings, with the range of possible error as the *largest deviation* from that mean. This is illustrated in Example 1-1.

Example 1-1: A set of independent voltage measurements taken by four observers was recorded as 117.02 V, 117.11 V, 117.08 V, and 117.03 V. Calculate (a) the average voltage, (b) the range of error.

SOLUTION:

(a) $$E_{av} = \frac{E_1 + E_2 + E_3 + E_4}{N}$$

$$= \frac{117.02 + 117.11 + 117.08 + 117.03}{4} = 117.06 \ V$$

(b) $$\text{Range} = E_{max} - E_{av} = 117.11 - 117.06 = 0.05 \text{ V}$$

but also $$E_{av} - E_{min} = 117.06 - 117.02 = 0.04 \text{ V}$$

The average range of error therefore equals

$$\frac{0.05 + 0.04}{2} = \pm 0.045 = \pm 0.05 \text{ V}$$

When two or more measurements with different degrees of accuracy are added, *the result is only as accurate as the least accurate measurement*. Suppose that two resistances are *added* in series as in Example 1-2.

Example 1-2: Two resistors, R_1 and R_2, are connected in series. Individual resistance measurements, using a Wheatstone bridge, give $R_1 = 18.7 \ \Omega$ and $R_2 = 3.624$ Ω. Calculate the total resistance to the appropriate number of significant figures.

SOLUTION:

$$R_1 = 18.7 \ \Omega \text{ (three significant figures)}$$

$$R_2 = 3.624 \ \Omega \text{ (four significant figures)}$$

$$R_T = R_1 + R_2 = 22.324 \ \Omega \text{ (five significant figures)} = 22.3 \ \Omega$$

The doubtful figures are written in *italics* to indicate that in the addition of R_1 and R_2 the last three digits of the sum are doubtful figures. There is no value whatsoever in retaining the last two digits (the *2* and the *4*) because one of the resistances is accurate only to three significant figures or tenths of an ohm. The result should therefore also be reduced to three significant figures or the nearest tenth, i.e., *22.3* Ω.

The number of significant figures in *multiplication* may increase rapidly, but again only the appropriate figures are retained in the answer, as shown in Example 1-3.

Example 1-3: In calculating voltage drop, a current of 3.18 A is recorded in a resistance of 35.68 Ω. Calculate the voltage drop across the resistor to the appropriate number of significant figures.

SOLUTION: $E = IR = (35.68) \times (3.18) = 113.4624 = 113$ V

Since there are three significant figures involved in the multiplication, the answer can be written only to a maximum of three significant figures.

In Example 1-3, the current, I, has three significant figures and R has four; and the result of the multiplication has only three significant figures. This illustrates that the answer cannot be known to an accuracy greater than the *least* poorly defined of the factors. Note also that if extra digits accumulate in the answer, they should be discarded or rounded off. In the usual practice, if the (least significant) digit in the first place to be discarded is less than five, it and the following digits are dropped from the answer. This was done in Example 1-3. If the digit in the first place to be discarded is five or greater, the previous digit is increased by one. For three-digit precision, therefore, 113.46 should be rounded off to 113; and 113.74 to 114.

Addition of figures with a range of doubt is illustrated in Example 1-4.

Example 1-4: Add 826 ± 5 to 628 ± 3.

SOLUTION: $N_1 = 826 \pm 5 \,(= \pm 0.605\%)$

$N_2 = 628 \pm 3 \,(= \pm 0.477\%)$

Sum $= 1,454 \pm 8 \,(= \pm 0.55\%)$

Note in Example 1-4 that the doubtful parts are *added*, since the ± sign means that one number may be high and the other low. The worst possible combination of range of doubt should be taken in the answer. The percentage doubt in the original figure N_1 and N_2 does not differ greatly from the percentage doubt in the final result.

If the same two numbers are *subtracted*, as in Example 1-5, there is an interesting comparison between addition and subtraction with respect to the range of doubt.

Example 1-5: Subtract 628 ± 3 from 826 ± 5 and express the range of doubt in the answer as a percentage.

SOLUTION: $N_1 = 826 \pm 5 (= \pm 0.605\%)$

$N_2 = 628 \pm 3 (= \pm 0.477\%)$

Difference $= 198 \pm 8 (= \pm 4.04\%)$

Again, in Example 1-5, the doubtful parts are added for the same reason as in Example 1-4. Comparing the results of addition and subtraction of the same numbers in Examples 1-4 and 1-5, note that the precision of the results, when expressed in *percentages*, differs greatly. The final result after subtraction shows a large increase in percentage doubt compared to the percentage doubt after addition. The percentage doubt increases even more when the difference between the numbers is relatively small. Consider the case illustrated in Example 1-6.

Example 1-6: Subtract 437 ± 4 from 462 ± 4 and express the range of doubt in the answer as a percentage.

SOLUTION: $N_1 = 462 \pm 4 (= \pm 0.87\%)$

$N_2 = 437 \pm 4 (= \pm 0.92\%)$

Difference $= 25 \pm 8 (= \pm 32\%)$

Example 1-6 illustrates clearly that one should avoid measurement techniques depending on subtraction of experimental results because the range of doubt in the final result may be greatly increased.

1-4 TYPES OF ERROR

No measurement can be made with perfect accuracy, but it is important to find out what the accuracy actually is and how different errors have entered into the measurement. A study of errors is a first step in finding ways to reduce them. Such a study also allows us to determine the accuracy of the final test result.

Errors may come from different sources and are usually classified under three main headings:

Gross errors: largely human errors, among them misreading of instruments, incorrect adjustment and improper application of instruments, and computational mistakes.

Systematic errors: shortcomings of the instruments, such as defective or worn parts, and effects of the environment on the equipment or the user.

Random errors: those due to causes that cannot be directly established because of random variations in the parameter or the system of measurement.

Each of these classes of errors will be discussed briefly and some methods will be suggested for their reduction or elimination.

1-4.1 Gross Errors

This class of errors mainly covers *human* mistakes in reading or using instruments and in recording and calculating measurement results. As long as human beings are involved, some gross errors will inevitably be committed. Although complete elimination of gross errors is probably impossible, one should try to anticipate and correct them. Some gross errors are easily detected; others may be very elusive. One common gross error, frequently committed by beginners in measurement work, involves the improper use of an instrument. In general, indicating instruments change conditions to some extent when connected into a complete circuit, so that the measured quantity is altered by the method employed. For example, a well-calibrated voltmeter may give a misleading reading when connected across two points in a high-resistance circuit (Example 1-7). The same voltmeter, when connected in a low-resistance circuit, may give a more dependable reading (Example 1-8). These examples illustrate that the voltmeter has a "loading effect" on the circuit, altering the original situation by the measurement process.

Example 1-7: A voltmeter, having a sensitivity of 1,000 Ω/V, reads 100 V on its 150-V scale when connected across an unknown resistor in series with a milliammeter.

When the milliammeter reads 5 mA, calculate (a) apparent resistance of the unknown resistor, (b) actual resistance of the unknown resistor, (c) error due to the loading effect of the voltmeter.

SOLUTION:

(a) The total circuit resistance equals

$$R_T = \frac{V_T}{I_T} = \frac{100 \text{ V}}{5 \text{ mA}} = 20 \text{ k}\Omega$$

Neglecting the resistance of the milliammeter, the value of the unknown resistor is $R_X = 20$ kΩ.

(b) The voltmeter resistance equals

$$R_V = 1000 \frac{\Omega}{V} \times 150 \text{ V} = 150 \text{ k}\Omega$$

Since the voltmeter is in parallel with the unknown resistance, we can write

$$R_X = \frac{R_T R_V}{R_V - R_T} = \frac{20 \times 150}{130} = 23.05 \text{ k}\Omega$$

(c) % error $= \dfrac{\text{actual} - \text{apparent}}{\text{actual}} \times 100\% = \dfrac{23.05 - 20}{23.05} \times 100\%$

$= 13.23\%$

Example 1-8: Repeat Example 1-7 if the milliammeter reads 800 mA and the voltmeter reads 40 V on its 150-V scale.

SOLUTION:

(a) $R_T = \dfrac{V_T}{I_T} = \dfrac{40 \text{ V}}{0.8 \text{ A}} = 50 \ \Omega$

(b) $R_V = 1{,}000 \dfrac{\Omega}{\text{V}} \times 150 \text{ V} = 150 \text{ k}\Omega$

$R_X = \dfrac{R_T R_V}{R_V - R_T} = \dfrac{50 \times 150}{149.95} = 50.1 \ \Omega$

(c) % error $= \dfrac{50.1 - 50}{50.1} \times 100\% = 0.2\%$

Errors caused by the loading effect of the voltmeter can be avoided by using it intelligently. For example, a low-resistance voltmeter should not be used to measure voltages in a vacuum tube amplifier. In this particular measurement, a high-input impedance voltmeter (such as a VTVM or TVM) is required.

A large number of gross errors can be attributed to carelessness or bad habits, such as improper reading of an instrument, recording the result differently from the actual reading taken, or adjusting the instrument incorrectly. Consider the case in which a multirange voltmeter uses a single set of scale markings with different number designations for the various voltage ranges. It is easy to use a scale which does not correspond to the setting of the range selector of the voltmeter. A gross error may also occur when the instrument is not set to zero before the measurement is taken; then all the readings are off.

Errors like these cannot be treated mathematically. They can be avoided only by taking care in reading and recording the measurement data. Good practice requires making more than one reading of the same quantity, preferably by a different observer. Never place complete dependence on one reading but take at least three separate readings, preferably under conditions in which instruments are switched off-on.

1-4.2 Systematic Errors

This type of error is usually divided into two different categories: (1) instrumental errors, defined as shortcomings of the instrument; (2) environmental errors, due to external conditions affecting the measurement.

Instrumental errors are errors inherent in measuring instruments because of their mechanical structure. For example, in the d'Arsonval movement friction in bearings of various moving components may cause incorrect readings. Irregular spring tension, stretching of the spring, or reduction in tension due to improper handling or overloading of the instrument will result in errors. Other instrumental errors are calibration errors, causing the instrument to read high or low along its entire scale. (Failure to set the instrument to zero before making a measurement has a similar effect.)

There are many kinds of instrumental errors, depending on the type of instrument used. The experimenter should always take precautions to insure that the instrument he is using is operating properly and does not contribute excessive errors for the purpose at hand. Faults in instruments may be detected by checking for erratic behavior, and stability and reproducibility of results. A quick and easy way to check an instrument is to compare it to another with the same characteristics or to one that is known to be more accurate.

Instrumental errors may be avoided by (1) selecting a suitable instrument for the particular measurement application; (2) applying correction factors after determining the amount of instrumental error; (3) calibrating the instrument against a standard.

Environmental errors are due to conditions external to the measuring device, including conditions in the area surrounding the instrument, such as the effects of changes in temperature, humidity, barometric pressure, or of magnetic or electrostatic fields. Thus a change in ambient temperature at which the instrument is used causes a change in the elastic properties of the spring in a moving-coil mechanism and so affects the reading of the instrument. Corrective measures to reduce these effects include air conditioning, hermetically sealing certain components in the instrument, use of magnetic shields, and the like.

Systematic errors can also be subdivided into *static* or *dynamic* errors. Static errors are caused by limitations of the measuring device or the physical laws governing its behavior. A static error is introduced in a micrometer when excessive pressure is applied in torquing the shaft. Dynamic errors are caused by the instrument's not responding fast enough to follow the changes in a measured variable.

1-4.3 Random Errors

These errors are due to unknown causes and occur even when all systematic errors have been accounted for. In well-designed experiments, few random errors usually occur, but they become important in high-accuracy work. Suppose a voltage is being monitored by a voltmeter which is read at half-hour intervals. Although the instrument is operated under ideal environmental conditions and has been accurately calibrated before the measurement, it will be found that the readings vary slightly over the period of observation. This variation cannot be corrected by any method of calibration or other known method of control and it cannot be explained without minute investigation. The only way to offset these errors is by increasing the number of readings and using statistical means to obtain the best approximation of the true value of the quantity under measurement.

1-5 STATISTICAL ANALYSIS

A statistical analysis of measurement data is common practice because it allows an analytical determination of the uncertainty of the final test result. The outcome of a certain measurement method may be predicted on the basis of sample data without having detailed information on all the disturbing factors. To make statistical methods and interpretations meaningful, a large number of measurements is usually required. Also, systematic errors should be small compared with residual or random errors, because statistical treatment of data cannot remove a fixed bias contained in all the measurements.

1-5.1 Arithmetic Mean

The most probable value of a measured variable is the artithmetic mean of the number of readings taken. The best approximation will be made when the number of readings of the same quantity is very large. Theoretically, an infinite number of readings would give the best result, although in practice, only a finite number of measurements can be made. The arithmetic mean is given by the following expression:

$$\bar{x} = \frac{x_1 + x_2 + x_3 + x_4 + \ldots + x_n}{n} = \frac{\sum x}{n} \qquad (1\text{-}1)$$

where
$$\bar{x} = \text{arithmetic mean}$$
$$x_1, x_2, x_n = \text{readings taken}$$
$$n = \text{number of readings}$$

Example 1-1 showed how the arithmetic mean is used.

1-5.2 Deviation from the Mean

Deviation is the departure of a given reading from the arithmetic mean of the group of readings. If the deviation of the first reading, x_1, is called d_1, and that of the second reading, x_2, is called d_2, and so on, then the deviations from the mean can be expressed as

$$d_1 = x_1 - \bar{x} \qquad d_2 = x_2 - \bar{x} \qquad d_n = x_n - \bar{x} \qquad (1\text{-}2)$$

Note that the deviation from the mean may have a positive or a negative value and that the algebraic sum of all the deviations must be zero.

Example 1-9 illustrates the computation of deviations.

Example 1-9: A set of independent current measurements was taken by six observers and recorded as 12.8 mA, 12.2 mA, 12.5 mA, 13.1 mA, 12.9 mA, and 12.4 mA. Calculate (a) the arithmetic mean, (b) the deviations from the mean.

SOLUTION:

(a) Using Eq. (1-1), we see that the arithmetic mean equals

$$\bar{x} = \frac{12.8 + 12.2 + 12.5 + 13.1 + 12.9 + 12.4}{6} = 12.65 \text{ mA}$$

(b) Using Eq. (1-2), we see that the deviations are

$$d_1 = 12.8 - 12.65 = 0.15 \text{ mA}$$
$$d_2 = 12.2 - 12.65 = -0.45 \text{ mA}$$
$$d_3 = 12.5 - 12.65 = -0.15 \text{ mA}$$
$$d_4 = 13.1 - 12.65 = 0.45 \text{ mA}$$
$$d_5 = 12.9 - 12.65 = 0.25 \text{ mA}$$
$$d_6 = 12.4 - 12.65 = -0.25 \text{ mA}$$

Note that the algebraic sum of all the deviations equals zero.

1-5.3   Average Deviation

The average deviation is an indication of the precision of the instruments used in making the measurements. Highly precise instruments will yield a low average deviation between readings. By definition, average deviation is the sum of the *absolute* values of the deviations divided by the number of readings. The absolute value of the deviation is the value without respect to sign. Average deviation may be expressed as

$$D = \frac{|d_1| + |d_2| + |d_3| + \ldots + |d_n|}{n} = \frac{\sum |d|}{n} \qquad \text{(1-3)}$$

Example 1-10 shows how average deviation is calculated.

Example 1-10: Calculate the average deviation for the data given in Example 1-9.

SOLUTION:

$$D = \frac{0.15 + 0.45 + 0.15 + 0.45 + 0.25 + 0.25}{6} = 0.283 \text{ mA}$$

1-5.4   Standard Deviation

In statistical analysis of random errors, the root-mean-square deviation or *standard deviation* is a very valuable aid. By definition, the standard deviation σ of an infinite number of data is the square root of the sum of *all* the individual deviations squared, divided by the number of readings. Expressed mathematically:

$$\sigma = \sqrt{\frac{d_1^2 + d_2^2 + d_3^2 + \ldots + d_n^2}{n}} = \sqrt{\frac{\sum d_i^2}{n}} \qquad (1\text{-}4)$$

In practice, of course, the possible number of observations is finite. The standard deviation of a *finite* number of data is given by

$$\sigma = \sqrt{\frac{d_1^2 + d_2^2 + d_3^2 + \ldots + d_n^2}{n-1}} = \sqrt{\frac{\sum d_i^2}{n-1}} \qquad (1\text{-}5)$$

Equation (1-5) will be used in Example 1-11.

Another expression for essentially the same quantity is the *variance* or *mean square deviation*, which is the same as the standard deviation except that the square root is not extracted. Therefore

$$\text{variance } (V) = \text{mean square deviation} = \sigma^2$$

The variance is a convenient quantity to use in many computations because variances are additive. The standard deviation, however, has the advantage of being of the same units as the variable, making it easy to compare magnitudes. Most scientific results are now stated in terms of standard deviation.

1-6 PROBABILITY OF ERRORS

1-6.1 Normal Distribution of Errors

Table 1-1 shows a tabulation of 50 voltage readings that were taken at small time intervals and recorded to the nearest 0.1 V. The nominal value of the measured voltage was 100.0 V. The result of this series of measurements

Table 1-1

TABULATION OF VOLTAGE READINGS

Voltage reading (volts)	Number of readings
99.7	1
99.8	4
99.9	12
100.0	19
100.1	10
100.2	3
100.3	1
	50

can be presented graphically in the form of a block diagram or *histogram* in which the number of observations is plotted against each observed voltage reading. The histogram of Fig. 1-1 represents the data of Table 1-1.

Figure 1-1 shows that the largest number of readings (19) occurs at the central value of 100.0 V, while the other readings are placed more or less symmetrically on either side of the central value. If more readings were taken at smaller increments, say 200 readings at 0.05-V intervals, the distribution of observations would remain approximately symmetrical about the central value and the shape of the histogram would be about the same as before. With more and more data, taken at smaller and smaller increments, the contour of the histogram would finally become a smooth curve, as indicated by the broken line in Fig. 1-1. This bell-shaped curve is known as a Gaussian curve. The sharper and narrower the curve, the more definitely an observer may state that the most probable value of the true reading is the central value or mean reading.

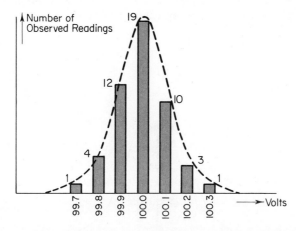

FIGURE 1-1   Histogram showing the frequency of occurrence of the 50 voltage readings of Table 1-1. The broken curve represents the limiting case of the histogram when a large number of readings at small increments are taken.

The Gaussian or Normal law of error forms the basis of the analytical study of random effects. Although the mathematical treatment of this subject is beyond the scope of this text, the following qualitative statements are based on the Normal law:

(a) All observations include small disturbing effects, called random errors.
(b) Random errors can be positive or negative.

(c) There is an equal probability of positive and negative random errors.

We can therefore expect that measurement observations include plus and minus errors in more or less equal amounts, so that the total error will be small and the mean value will be the true value of the measured variable.

The possibilities as to the form of the error distribution curve can be stated as follows:

(a) Small errors are more probable than large errors.
(b) Large errors are very improbable.
(c) There is an equal probability of plus and minus errors so that the probability of a given error will be symmetrical about the zero value.

The error distribution curve of Fig. 1-2 is based on the Normal law and shows a symmetrical distribution of errors. This normal curve may be regarded as the limiting form of the histogram of Fig. 1-1 in which the most probable value of the true voltage is the mean value of 100.0 V.

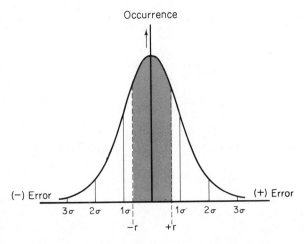

FIGURE 1-2   Curve for the Normal law. The shaded portion indicates the region of probable error, where $r = \pm 0.6745\ \sigma$.

1-6.2 Probable Error

The area under the Gaussian probability curve of Fig. 1-2, between the limits $+\infty$ and $-\infty$, represents the entire number of observations. The area under the curve between the $+\sigma$ and $-\sigma$ limits represents the cases that differ from the mean by no more than the standard deviation. Integration of the area under the curve within the $\pm\sigma$ limits gives the total number of cases within these limits. For normally dispersed data, following the Gaussian distribution, approximately 68 per cent of all the cases lie between the limits of $+\sigma$ and $-\sigma$ from the mean. Corresponding values of other deviations, expressed in terms of σ, are given in Table 1-2.

Table 1-2

AREA UNDER THE PROBABILITY CURVE

Deviation ($\pm$) (σ)	*Fraction of total area included*
0.6745	0.5000
1.0	0.6828
2.0	0.9546
3.0	0.9972

If, for example, a large number of nominally 100-Ω resistors is measured and the mean value is found to be 100.00 Ω, with a standard deviation (S.D.) of 0.20 Ω, we know that on the average 68 per cent (or roughly two-thirds) of all the resistors have values which lie between limits of $\pm 0.20\,\Omega$ of the mean. There is then approximately a two to one chance that any resistor, selected from the lot at random, will lie within these limits. If larger odds are required, the deviation may be extended to a limit of $\pm 2\sigma$, in this case $\pm 0.40\,\Omega$. According to Table 1-2, this now includes 95 per cent of all the cases, giving ten to one odds that any resistor selected at random lies within $\pm 0.40\,\Omega$ of the mean value of 100.00 Ω.

Table 1-2 also shows that half of the cases are included in the deviation limits of $\pm 0.6745\,\sigma$. The quantity r is called the *probable error* and is defined as

$$\text{probable error } r = \pm 0.6745\,\sigma \qquad \text{(1-6)}$$

This value is *probable* in the sense that there is an even chance that any one observation will have a random error no greater than $\pm r$. Probable error has been used in experimental work to some extent in the past, but standard deviation is more convenient in statistical work and is given preference.

Example 1-11: Ten measurements of the resistance of a resistor gave 101.2 Ω, 101.7 Ω, 101.3 Ω, 101.0 Ω, 101.5 Ω, 101.3 Ω, 101.2 Ω, 101.4 Ω, 101.3 Ω, and 101.1 Ω.

Assume that only random errors are present. Calculate (a) the arithmetic mean, (b) the standard deviation of the readings, (c) the probable error.

SOLUTION: With a large number of readings a simple tabulation of data is very convenient and avoids confusion and mistakes.

| Reading | Deviation | |
x	d	d^2
101.2	−0.1	0.01
101.7	0.4	0.16
101.3	0.0	0.00
101.0	−0.3	0.09
101.5	0.2	0.04
101.3	0.0	0.00
101.2	−0.1	0.01
101.4	0.1	0.01
101.3	0.0	0.00
101.1	−0.2	0.04
$\sum x = 1{,}013.0$	$\sum |d| = 1.4$	$\sum d^2 = 0.36$

(a) Arithmetic mean, $\bar{x} = \dfrac{\sum x}{n} = \dfrac{1{,}013.0}{10} = 101.3 \ \Omega$

(b) Standard deviation, $\sigma = \sqrt{\dfrac{d^2}{n-1}} = \sqrt{\dfrac{0.36}{9}} = 0.2 \ \Omega$

(c) Probable error $= 0.6745\sigma = 0.6745 \times 0.2 = 0.1349 \ \Omega$

1-7 LIMITING ERRORS

In most indicating instruments the accuracy is guaranteed to a certain percentage of full-scale reading. Circuit components (such as capacitors, resistors, etc.) are guaranteed within a certain percentage of their rated value. The limits of these deviations from the specified values are known as *limiting errors* or *guarantee errors*. For example, if the resistance of a resistor is given as 500 Ω ±10 per cent, the manufacturer guarantees that the resistance falls between the limits 450 Ω and 550 Ω. The maker is not specifying a standard deviation or a probable error, but promises that the error is no greater than the limits set.

Example 1-12: A 0–150-V voltmeter has a guaranteed accuracy of 1 per cent full-scale reading. The voltage measured by this instrument is 83 V. Calculate the limiting error in per cent.

SOLUTION: The magnitude of the limiting error is

$$0.01 \times 150 \ \text{V} = 1.5 \ \text{V}$$

The percentage error at a meter indication of 83 V is

$$\frac{1.5}{83} \times 100 \text{ per cent} = 1.81 \text{ per cent}$$

$$(\ 83)$$

It is important to note in Example 1-12 that a meter is guaranteed to have an accuracy of better than 1 per cent of the full-scale reading, but when the meter reads 83 V the limiting error increases to 1.81 per cent. Correspondingly, when a smaller voltage is measured, the limiting error will increase further. If the meter reads 60 V, the per cent limiting error is $1.5/60 \times 100 = 2.5$ per cent; if the meter reads 30 V, the limiting error is $1.5/30 \times 100 = 5$ per cent. The increase in per cent limiting error, as smaller voltages are measured, occurs because the magnitude of the limiting error is a fixed quantity based on the full-scale reading of the meter. Example 1-12 shows the importance of taking measurements *as close to full scale as possible*.

Measurements or computations, *combining* guarantee errors, are often made. Example 1-13 illustrates such a computation.

Example 1-13: Three decade boxes, each guaranteed to ± 0.1 per cent are used in a Wheatstone bridge to measure the resistance of an unknown resistor R_x. Calculate the limits on R_x imposed by the decade boxes.

SOLUTION: The equation for bridge balance shows that R_x can be determined in terms of the resistances of the three decade boxes and $R_x = R_1 R_2 / R_3$, where R_1, R_2, and R_3 are the resistances of the decade boxes, guaranteed to ± 0.1 per cent. One must recognize that the two terms in the numerator may both be positive to the full limit of 0.1 per cent and the denominator may be negative to the full 0.1 per cent, giving a resultant error of 0.3 per cent. The guarantee error is thus obtained by taking the *direct sum* of all the possible errors, adopting the algebraic signs which give the worst possible combination.

As a further example, using the relationship $P = I^2 R$, as shown in Example 1-14, consider computing the power dissipation in a resistor.

Example 1-14: The current passing through a resistor of $100 \pm 0.2 \, \Omega$ is 2.00 ± 0.01 A. Using the relationship $P = I^2 R$, calculate the limiting error in the computed value of power dissipation.

SOLUTION: Expressing the guaranteed limits of both current and resistance in percentages instead of units, we obtain

$$I = 2.00 \pm 0.01 \text{ A} = 2.00 \text{ A} \pm 0.5\%$$
$$R = 100 \pm 0.2 \, \Omega = 100 \, \Omega \pm 0.2\%$$

If the worst possible combination of errors is used, the limiting error in the power dissipation is ($P = I^2 R$):

$$(2 \times 0.5\%) + 0.2\% = 1.2\%.$$

Power dissipation should then be written as follows:

$$P = I^2 R = (2.00)^2 \times 100 = 400 \text{ W} \pm 1.2\% = 400 \pm 4.8 \text{ W}$$

REFERENCES

1. Bartholomew, Davis, *Electrical Measurements and Instrumentation*, chaps. 1, 2. Boston: Allyn and Bacon, Inc., 1963.

2. Frank, Ernest, *Electrical Measurement Analysis*, chap. 14. New York: McGraw-Hill Book Company, Inc., 1959.

3. Stout, Melville B., *Basic Electrical Measurements*, 2nd ed., chap. 2. Englewood Cliffs, N.J.: Prentice-Hall, Inc., 1960.

4. Young, Hugh D., *Statistical Treatment of Experimental Data*. New York: McGraw-Hill Book Company, Inc., 1962.

QUESTIONS

1. What is the difference between accuracy and precision?

2. List four sources of possible errors in instruments.

3. What are the three general classes of errors?

4. Define
   - (a) instrumental error,
   - (c) calibration error,
   - (e) random error,
   - (b) limiting error,
   - (d) environmental error,
   - (f) probable error.

PROBLEMS

1. A 0–100-V voltmeter has 200 scale divisions that can be read to $\frac{1}{2}$ division. Determine the resolution of the meter in volts.

2. A digital voltmeter has a read-out range from 0 to 9,999 counts. Determine the resolution of the instrument in volts when the full-scale reading is 9.999 V.

3. State the number of significant figures in each of the following:
   - (a) 542,
   - (c) 27.25,
   - (e) 40×10^6,
   - (b) 0.65,
   - (d) 0.00005,
   - (f) 20,000.

4. Four resistors are placed in series. The values of the resistors are 28.4 Ω, 4.25 Ω, 56.605 Ω, 0.75 Ω, with an uncertainty of one unit in the last digit of each number. Calculate the total series resistance. Give only significant figures in the answer.

5. A voltage drop of 112.5 V is measured across a resistor passing a current of 1.62 A. Calculate the power dissipation of the resistor. Give only significant figures in the answer.

6. A voltmeter, having a sensitivity of 10 kΩ/V, reads 75 V on its 100-V scale when connected across an unknown resistor. When the current through the

resistor is 1.5 mA, calculate (a) the apparent resistance of the unknown resistor, (b) the actual resistance of the unknown resistor, (c) the percentage error due to the loading effect of the voltmeter.

7. The voltage across a resistor is 200 V, with a probable error of ± 2 per cent, and the resistance is 42 Ω with a probable error of ± 1.5 per cent. Calculate (a) the power dissipated in the resistor, (b) the percentage error in the answer.

8. The following values were obtained from the measurements of the value of a resistor: 147.2 Ω, 147.4 Ω, 147.9 Ω, 148.1 Ω, 147.1 Ω, 147.5 Ω, 147.6 Ω, 147.4 Ω, 147.6 Ω, and 147.5 Ω. Calculate (a) the arithmetic mean, (b) the average deviation, (c) the standard deviation, (d) the probable error of the average of the ten readings.

9. Six determinations of a quantity, as entered on the data sheet and presented to you for analysis, are 12.35, 12.71, 12.48, 10.24, 12.63, and 12.58. Examine the data and on the basis of your conclusions calculate (a) the arithmetic mean, (b) the standard deviation, (c) the probable error in per cent of the average of the readings.

10. Two resistors have the following ratings:

$$R_1 = 36\,\Omega \pm 5\% \quad \text{and} \quad R_2 = 75\,\Omega \pm 5\%$$

Calculate (a) the magnitude of error in each resistor, (b) the limiting error in ohms and in per cent when the resistors are connected in series, (c) the limiting error in ohms and in per cent when the resistors are connected in parallel.

11. The resistance of an unknown resistor is determined by the Wheatstone bridge method. The solution for the unknown resistance is stated as $R_x = R_1 R_2 / R_3$,

where $R_1 = 500\,\Omega \pm 1\%$

$R_2 = 615\,\Omega \pm 1\%$

$R_3 = 100\,\Omega \pm 0.5\%$

Calculate (a) the nominal value of the unknown resistor, (b) the limiting error in ohms of the unknown resistor, (c) the limiting error in per cent of the unknown resistor.

12. A resistor is measured by the voltmeter-ammeter method. The voltmeter reading is 123.4 V on the 250-V scale and the ammeter reading is 283.5 mA on the 500-mA scale. Both meters are guaranteed to be accurate within ± 1 per cent of full-scale reading. Calculate (a) the indicated value of the resistance, (b) the limits within which you can guarantee the result.

13. In a dc circuit, the voltage across a component is 64.3 V and the current is 2.53 A. Both current and voltage are given with an uncertainty of one unit in the last place. Calculate the power dissipation to the appropriate number of significant figures.

14. A power transformer was tested to determine losses and efficiency. The input power was measured as 3,650 W and the delivered output power was 3,385 W, with each reading in doubt by ± 10 W. Calculate (a) the percentage uncertainty

in the losses of the transformer, (b) the percentage uncertainty in the efficiency of the transformer, as determined by the difference in input and output power readings.

15. The power factor and phase angle in a circuit carrying a sinusoidal current are determined by measurements of current, voltage, and power. The current is read as 2.50 A on a 5-A ammeter, the voltage as 115 V on a 250-V voltmeter, and the power as 220 W on a 500-W wattmeter. The ammeter and voltmeter are guaranteed accurate to within ± 0.5 per cent of full-scale indication and the wattmeter to within ± 1 per cent of fullscale reading. Calculate (a) the percentage accuracy to which the power factor can be guaranteed, (b) the possible error in the phase angle.

16. The arms of a Wheatstone bridge, marked in order around the bridge, are B, A, X, and R. The three known arms have the following constants:

$$A = 840 \ \Omega \ (\text{S.D.} = 1 \ \Omega)$$
$$B = 90 \ \Omega \ (\text{S.D.} = 0.5 \ \Omega)$$
$$R = 250 \ \Omega \ (\text{S.D.} = 1 \ \Omega)$$

Calculate (a) the probable value of X, (b) the standard deviation of X.

2

Systems of Units
of Measurement

2-1 FUNDAMENTAL AND DERIVED UNITS

To specify and perform calculations with physical quantities, the physical quantities must be defined both in *kind* and *magnitude*. The standard measure of each kind of physical quantity is the *unit*; the number of times the unit occurs in any given amount of the same quantity is the *number of measure*. For example, when we speak of a distance of 100 meters, we know that the meter is the unit of length and that the number of units of length is one hundred. The physical quantity, *length*, is therefore defined by the unit, *meter*. Without the unit, the number of measure has no physical meaning.

In science and engineering, two kinds of units are used: *fundamental units* and *derived units*. The fundamental units in mechanics are measures of *length*, *mass*, and *time*. The sizes of the fundamental units, whether foot or meter, pound or kilogram, second or hour, are arbitrary and can be selected to fit a certain set of circumstances. Since length, mass, and time are fundamental to most other physical quantities besides those in mechanics, they are called the *primary* fundamental units. Measures of certain physical quantities in the thermal, electrical, and illumination disciplines are also represented by fundamental units. These units are used only when these particular classes are involved, and they may therefore be defined as *auxiliary* fundamental units.

All other units which can be expressed in terms of the fundamental units are called *derived* units. Every derived unit originates from some physical law defining that unit. For example, the area (A) of a rectangle is proportional to its length (l) and breadth (b), or $A = lb$. If the meter has been chosen as the unit of length, then the area of a rectangle of 3 m by 4 m is 12 m². Note that the numbers of measure are multiplied ($3 \times 4 = 12$) as well as the units (m $\times$ m = m²). The derived unit for area (A) is then the square meter (m²).

A derived unit is recognized by its *dimensions*, which can be defined as the complete algebraic formula for the derived unit. The dimensional *symbols* for the fundamental units of length, mass, and time are L, M, and T, respectively. The dimensional symbol for the derived unit of area is L^2 and that for volume, L^3. The dimensional symbol for the unit of force is LMT^{-2}, which follows from the defining equation for force. The dimensional formulas of the derived units are particularly useful for converting units from one system to another, as is shown in Sec. 2-6.

For convenience, some derived units have been given new names. For example, the derived unit of force in the SI system is called the newton (N), instead of the dimensionally correct name kg m/s².

2-2 SYSTEMS OF UNITS

In 1790 the French government issued a directive to the French Academy of Sciences to study and to submit proposals for a single system of weights and measures to replace all other existing systems. The French scientists decided, as a first principle, that a *universal system* of weights and measures should not depend on man-made reference standards, but instead be based on permanent measures provided by nature. As the *unit of length*, therefore, they chose the *meter*, defined as the ten-millionth part of the distance from the pole to the equator along the meridian passing through Paris. As the *unit of mass* they chose the mass of a cubic centimeter of distilled water at 4°C and normal atmospheric pressure (760 mm Hg) and gave it the name *gram*. As the third unit, the *unit of time*, they decided to retain the traditional second, defining it as 1/86,400 of the mean solar day.

As a second principle, they decided that all other units should be *derived* from the aforementioned three *fundamental units* of length, mass, and time. Next—the third principle—they proposed that all multiples and submultiples of basic units be in the *decimal system*, and they devised the system of prefixes in use today. Table 2-1 lists the decimal multiples and submultiples.

The proposals of the French Academy were approved and introduced as the *metric system* of units in France in 1795. The metric system aroused considerable interest elsewhere and finally, in 1875, 17 countries signed the so-called Metre Convention, making the metric system of units the legal system. Britain and the United States, although signatories of the convention, recognized its legality only in international transactions but did not accept the metric system for their own domestic use.

Table 2-1

DECIMAL MULTIPLES AND SUBMULTIPLES

Name	Symbol	Equivalent
tera	T	10^{12}
giga	G	10^9
mega	M	10^6
kilo	k	10^3
hecto	h	10^2
deca	da	10
deci	d	10^{-1}
centi	c	10^{-2}
milli	m	10^{-3}
micro	μ	10^{-6}
nano	n	10^{-9}
pico	p	10^{-12}
femto	f	10^{-15}
atto	a	10^{-18}

Britain, in the meantime, had been working on a system of electrical units, and the British Association for the Advancement of Science decided on the centimeter and the gram as the fundamental units of length and mass. From this developed the *centimeter-gram-second* or *CGS absolute system* of units, used by physicists all over the world. Complications arose when the CGS system was extended to electric and magnetic measurements because of the need to introduce at least one more unit in the system. In fact, two parallel systems were established. In the CGS *electrostatic system*, the unit of electric charge was derived from the centimeter, gram, and second by assigning the value 1 to the permittivity of free space in Coulomb's law for the force between electric charges. In the CGS *electromagnetic system*, the basic units are the same and the unit of magnetic pole strength is derived from them by assigning the value 1 to the permeability of free space in the inverse square formula for the force between magnetic poles.

The *derived* units for electric current and electric potential in the electromagnetic system, the ampere and the volt, are used in practical measurements. These two units, and the corresponding ones, such as the coulomb, ohm, henry, farad, etc., were incorporated in a third system, called the *practical system*. Further simplification in the establishment of a truly universal system came as the result of pioneer work by the Italian engineer Giorgi, who pointed out that the practical units of current, voltage, energy, and power, used by electrical engineers, were compatible with the meter-kilogram-second system. He suggested that the metric system be expanded into a *coherent* system of units by including the practical electrical units. The Giorgi system, adopted by many countries in 1935, came to be known as the MKSA system of units in which the ampere was selected as the fourth basic unit.

A more comprehensive system was adopted in 1954 and designated in 1960 by international agreement as the Système International d'Unités (SI). In the SI system, six basic units are used, namely, the meter, kilogram, second, and ampere of the MKSA system and, in addition, the degree Kelvin and the candela as the units of temperature and luminous intensity, respectively. The SI units are replacing other systems in science and technology; they have been adopted as the legal units in France, and will become obligatory in other metric countries.

The six basic SI quantities and units of measurement, with their unit symbols, are listed in Table 2-2.

Table 2-2

BASIC SI QUANTITIES, UNITS, AND SYMBOLS

Quantity	Unit	Symbol
Length	meter	m
Mass	kilogram	kg
Time	second	s
Electric current	ampere	A
Thermodynamic temperature	degree Kelvin	°K
Luminous intensity	candela	cd

2-3 ELECTRIC AND MAGNETIC UNITS

Before listing the SI units (sometimes called the *International* MKS system of units), a brief look at the origin of the electrical and magnetic units seems appropriate. The practical electrical and magnetic units with which we are familiar, such as the volt, ampere, ohm, henry, etc., were first derived in the CGS systems of units.

The *CGS electrostatic system* (CGSe) is based on Coulomb's experimentally derived law for the force between two electric charges. Coulomb's law states that

$$F = k\frac{Q_1 Q_2}{r^2} \tag{2-1}$$

where $F =$ the force between the charges, expressed in CGSe units of force ($g\,cm/s^2 =$ dyne)

   $k =$ a proportionality constant

   $Q_{1,2} =$ electric charges, expressed in (derived) CGSe units of electric charge (statcoulomb)

   $r =$ the separation between the charges, expressed in the fundamental CGSe unit of length (centimeter)

Coulomb also found that the proportionality factor k depended on the medium, varying inversely as its permittivity ϵ. (Faraday called permittivity the *dielectric constant*.) Coulomb's law then takes the form

$$F = \frac{Q_1 Q_2}{\epsilon r^2} \qquad (2\text{-}2)$$

Since ϵ is a numerical value depending only on the medium, a value of 1 was assigned to the permittivity of free space, ϵ_0, thereby defining ϵ_0 as the *fourth fundamental unit* of the CGSe system. Coulomb's law then allowed the unit of electric charge Q to be determined in terms of these four fundamental units by the relation

$$\text{dyne} = \frac{\text{g cm}}{\text{s}^2} = \frac{Q^2}{(\epsilon_0 = 1)\,\text{cm}^2}$$

and therefore, dimensionally,

$$Q = \text{cm}^{3/2}\text{g}^{1/2}\text{s}^{-1} \qquad (2\text{-}3)$$

The CGSe unit of electric charge was given the name *statcoulomb*.

The derived unit of electric charge in the CGSe system of units allowed other electrical units to be determined by their defining equations. For example, *electric current* (symbol I) is defined as the rate of flow of electric charge and is expressed as

$$I = \frac{Q}{t} \qquad \text{(statcoulomb/sec)} \qquad (2\text{-}4)$$

The unit for electric current in the CGSe system was given the name *stat-ampere*. Electric *fieldstrength, E, potential difference, V,* and *capacitance, C,* can similarly be derived from their defining equations.

The basis of the *CGS electromagnetic system* of units (CGSm) is Coulomb's experimentally determined law for the force between two magnetic poles, which states that

$$F = k\frac{m_1 m_2}{r^2} \qquad (2\text{-}5)$$

The proportionality factor, k, was found to depend on the medium in which the poles were placed, varying inversely with the magnetic *permeability* μ of the medium. The factor k was assigned the value 1 for the permeability of free space, μ_0, so that $k = 1/\mu_0 = 1$. This established the permeability of free space, μ_0, as the *fourth fundamental unit* of the CGSm system. The derived electromagnetic unit of pole strength was then defined in terms of these four fundamental units by the relation:

$$\text{dyne} = \frac{\text{g cm}}{\text{s}^2} = \frac{m^2}{(\mu_0 = 1)\,\text{cm}^2}$$

and therefore, dimensionally,

$$m = \text{cm}^{3/2}\text{g}^{1/2}\text{s}^{-1} \qquad (2\text{-}6)$$

The derived unit of magnetic polestrength in the CGSm system led to the determination of other magnetic units, again by their defining equations. *Magnetic flux density* (symbol B), for example, is defined as the magnetic force per unit polestrength, where both force and polestrength are derived units in the CGSm system. Dimensionally, B is found to be equal to $cm^{-1/2}g^{1/2}s^{-1}$ (dyne-second/abcoulomb-centimeter) and is given the name *gauss*. Similarly, other magnetic units can be derived from defining equations and we find that the unit for *magnetic flux* (symbol Φ) is given the name *maxwell*; the unit for *magnetic fieldstrength* (symbol H), the name *oersted*; and the unit for *magnetic potential difference* or *magnetomotive force* (symbol U), the name *gilbert*.

The two CGS systems were linked together by Faraday's discovery that a moving magnet could induce an electric current in a conductor, and conversely, that electricity in motion could produce magnetic effects. Ampere's law of the magnetic field relates electric current (I) to magnetic fieldstrength (H),* quantitatively connecting the magnetic units in the CGSm system to the electric units in the CGSe system. The dimensions of the two systems did not agree exactly, and numerical conversion factors were introduced. The two systems finally formed one *practical system of electrical units* which was officially adopted by the International Electrical Congress.

These practical electrical units, derived from the CGSm system, were later defined in terms of so-called international units. It was thought at the time (1908) that the establishment of the practical units from the definitions of the CGS system would be too difficult for most laboratories and it was therefore decided (unfortunately) to define the practical units in a way which would make it fairly simple to establish them. The *ampere*, therefore, was defined in terms of the rate of deposition of silver from a silver nitrate solution by passing a current through that solution and the *ohm* as the resistance of a specified column of mercury. These units and those derived from them were called *international units*. As measurement techniques improved, it was found that small differences existed between the CGSm derived practical units and the international units, which were then specified as follows:

1 int. ohm $= 1.00049 \, \Omega$ (practical CGSm unit)
1 int. ampere $= 0.99985$ A
1 int. volt $= 1.00034$ V
1 int. coulomb $= 0.99985$ C
1 int. farad $= 0.99951$ F
1 int. henry $= 1.00049$ H
1 int. watt $= 1.00019$ W
1 int. joule $= 1.00019$ J

---

*See a textbook on electromagnetic theory.

Particulars of the electric and magnetic units, and their defining relationships, are given in Table 2-3. Multiplication factors for conversion into SI units are given in the columns headed CGSm and CGSe.

Table 2-3

ELECTRIC AND MAGNETIC UNITS

| | SI *unit* | | *Conversion factors* | |
Quantity and symbol	*Name and symbol*	*Defining equation*	CGSm	CGSe§
Electric current, I	ampere A	$F_z = 10^{-7}I^2\dfrac{dN^*}{dz}$	10	$10/c$
Electromotive force, E	volt V	$p\dagger = IE$	10^{-8}	$10^{-8}c$
Potential, V	volt V	$p\dagger = IV$	10^{-8}	$10^{-8}c$
Resistance, R	ohm Ω	$R = V/I$	10^{-9}	$10^{-9}c$
Electric charge, Q	coulomb C	$Q = It$	10	$10/c$
Capacitance, C	farad F	$C = Q/V$	10^9	$10^9/c^2$
Electric fieldstrength, E	— V/m	$E = V/l$	10^{-6}	$10^{-6}c$
Electric flux density, D	— C/m²	$D = Q/l^2$	10^5	$10^5/c$
Permittivity, ϵ	— F/m	$\epsilon = D/E$	—	$10^{11}/4\pi c^2$
Magnetic fieldstrength, H	— A/m	$\oint H\,dl = nI$	$10^{3/4}$	—
Magnetic flux, Φ	weber Wb	$E = d\Phi/dt$	10^{-8}	—
Magnetic flux density, B	tesla T	$B = \Phi/l^2\ddagger$	10^{-4}	—
Inductance, L, M	henry H	$M = \Phi/I$	10^{-9}	—
Permeability, μ	— H/m	$\mu = B/H$	$4\pi \times 10^{-7}$	—

    * N denotes Neumann's integral for two linear circuits each carrying the current I. F_z is the force between the two circuits in the direction defined by coordinate z, the circuits being in a vacuum.

    † p denotes power.

    ‡ l^2 denotes area.

    § c = velocity of light in free space in cm/s = 2.997925×10^{10}.

2-4 INTERNATIONAL SYSTEM OF UNITS

The international MKSA system of units was adopted in 1960 by the Eleventh General Conference of Weights and Measures under the name *Système International d'Unités* (SI). The SI system is replacing all other systems in the metric countries and its widespread acceptance dooms other systems to eventual obsolescence.

The six fundamental SI quantities are listed in Table 2-2. The derived units are expressed in terms of these six basic units by defining equations. Some examples of defining equations are given in Table 2-3 for the electric and magnetic quantities. Table 2-4 lists, together with the fundamental

Table 2-4

FUNDAMENTAL, SUPPLEMENTARY, AND DERIVED UNITS

Quantity	Symbol	Dimension	Unit	Unit symbol
Fundamental				
Length	l	L	meter	m
Mass	m	M	kilogram	kg
Time	t	T	second	s
Electric current	I	I	ampere	A
Thermodynamic temperature	T	Θ	degree Kelvin	°K
Luminous intensity			candela	cd
Supplementary*				
Plane angle	α, β, γ	$[\text{L}]°$	radian	rad
Solid angle	Ω	$[\text{L}^2]°$	steradian	sr
Derived				
Area	A	L^2	square meter	m²
Volume	V	L^3	cubic meter	m³
Frequency†	f	T^{-1}	hertz	Hz (1/s)
Density	ρ	L^{-3}M	kilogram per cubic meter	kg/m³
Velocity	v	LT^{-1}	meter per second	m/s
Angular velocity	ω	$[\text{L}]°\text{T}$	radian per second	rad/s
Acceleration	a	LT^{-2}	meter per second squared	m/s²
Angular acceleration	α	$[\text{L}]°\text{T}^{-2}$	radian per second squared	rad/s²
Force	F	LMT^{-2}	newton	N (kg m/s²)
Pressure, stress	p	$\text{L}^{-1}\text{MT}^{-2}$	newton per square meter	N/m²
Work, energy	W	L^2MT^{-2}	joule	J (N m)
Power	P	L^2MT^{-3}	watt	W (J/s)
Quantity of electricity	Q	TI	coulomb	C (A s)
Potential difference, electromotive force	V	$\text{L}^2\text{MT}^{-3}\text{I}^{-1}$	volt	V (W/A)
Electric fieldstrength	E, ϵ	$\text{LMT}^{-3}\text{I}^{-1}$	volt per meter	V/m
Electric resistance	R	$\text{L}^2\text{MT}^{-3}\text{I}^2$	ohm	Ω (V/A)
Electric capacitance	C	$\text{L}^{-2}\text{M}^{-1}\text{T}^4\text{I}^2$	farad	F (A s/V)
Magnetic flux	Φ	$\text{L}^2\text{MT}^{-2}\text{I}^{-1}$	weber	Wb (v s)
Magnetic fieldstrength	H	L^{-1}I	ampere per meter	A/m
Magnetic flux density†	B	$\text{MT}^{-2}\text{I}^{-1}$	tesla	T (Wb/m²)
Inductance	L	$\text{L}^2\text{MT}^{-2}\text{I}^2$	henry	H (V s/A)
Magnetomotive force	U	I	ampere	A
Luminous flux			lumen	lm (cd sr)
Luminance			candela per square meter	cd/m²
Illumination			lux	lx (lm/m²)

    *The Eleventh General Conference designated these units as *supplementary*, although it could be argued that they are derived units.

    †In some countries, frequency is expressed not in Hz but in the equivalent unit, cycle per second (c/s), and magnetic flux density, not in T, but in the equivalent weber per square meter (Wb/m²).

quantities which are repeated in this table, the supplementary and derived units in the SI which are recommended for use by the General Conference.

The first column in Table 2-4 shows the *quantities* (fundamental, supplementary, and derived). The second column gives the *equation symbol* for each quantity. The third column lists the *dimension* of each derived unit in terms of the six fundamental dimensions. The fourth column gives the name of each *unit*; the fifth, the unit *symbol*. The *unit* symbol should not be confused with the *equation* symbol; i.e., the equation symbol for resistance is R, but the unit abbreviation (symbol) for ohm is Ω.

2-5 OTHER SYSTEMS OF UNITS

The English system of units uses the *foot*(ft), the *pound-mass* (lb), and the *second* (s) as the three fundamental units of length, mass, and time, respectively. Although the measures of length and weight are legacies of the Roman occupation of Britain and therefore rather poorly defined, the *inch* (defined as one-twelfth of the foot) has since been fixed at *exactly* 25.4 mm. Similarly, the measure for the pound (lb) has been determined as *exactly* 0.45359237 kg. These two figures allow all units in the English system to be converted into SI units.

Starting with the fundamental units, foot, pound, and second, the mechanical units may be derived simply by substitution into the dimensional equations of Table 2-4. For example, the unit of density will be expressed in lb/ft^3 and the unit of acceleration in ft/s^2. The derived unit of force in the ft-lb-s system is called the *poundal* and is the force required to accelerate 1 pound-mass at the rate of 1 ft/s^2. As a result, the unit for work or energy becomes the foot-poundal (ft pdl).

Various other systems have been devised and were used in various parts of the world. The *MTS* (meter-tonne-second) system was especially designed for engineering purposes in France and provided a replica of the CGS system except that the length and mass units (meter and tonne, respectively) were more suitable in practical engineering applications. *Gravitational* systems define the second fundamental unit as the *weight* of a mass measure; i.e., as the force by which that mass is attracted to the earth by gravity. In contrast to the gravitational systems, the so-called absolute systems, as the CGS and SI, use the mass measure as the second fundamental unit, but its value is independent of gravitational attraction.

Since English measures are still extensively used, both in Britain and on the North American continent, conversion into the SI becomes necessary if we wish to work in that system. Table 2-5 lists some of the common conversion factors for English into SI units.

Table 2-5

ENGLISH INTO SI CONVERSION

	English unit	Symbol	Metric equivalent	Reciprocal
Length	1 foot	ft	30.48 cm	0.0328084
	1 inch	in.	25.4 mm	0.0393701
Area	1 square foot	ft^2	9.29030×10^2 cm^2	0.0107639×10^{-2}
	1 square inch	in.^{2}	6.4516×10^2 mm^2	0.155000×10^{-2}
Volume	1 cubic foot	ft^3	0.0283168 m^3	35.3147
Mass	1 pound (avdp)	lb	0.45359237 kg	2.20462
Density	1 pound per cubic foot	lb/ft^3	16.0185 kg/m^3	0.062428
Velocity	1 foot per second	ft/s	0.3048 m/s	3.28084
Force	1 poundal	pdl	0.138255 N	7.23301
Work, energy	1 foot-poundal	ft pdl	0.0421401 J	23.7304
Power	1 horsepower	hp	745.7 W	0.00134102
Temperature	degree F	°F	$5(t-32)/9$°C	–

2-6 CONVERSION OF UNITS

It is often necessary to convert physical quantities from one system of units into another. Section 2-1 stated that a physical quantity is expressed in both unit and number of measure: it is the unit that must be converted, not the number of measure. Dimensional equations are very convenient for converting the numerical value of a dimensional quantity, when the units are transformed from one system to the other. The technique requires a knowledge of the numerical relation between the fundamental units and some dexterity in the manipulation of multiples and submultiples of the units.

The method used in converting from one system into the other is illustrated by a number of examples of progressively increasing difficulty.

Example 2-1: The floor area of an office building is 5,000 m^2. Calculate the floor area in ft^2.

SOLUTION: To convert the unit m^2 into the new unit ft^2, we must know the relation between them. In Table 2-5 the metric equivalent of 1 ft is 30.48 cm, or 1 ft = 0.3048 m. Therefore

$$A = 5,000 \text{ m}^2 \times \left(\frac{1 \text{ ft}}{0.3048 \text{ m}}\right)^2 = 53,800 \text{ ft}^2$$

Example 2-2: The floor area of a classroom measures 30 ft by 24 ft. Calculate the floor area in m^2.

SOLUTION: Again consulting Table 2-5 we find that the reciprocal of the ft-to-cm conversion is 0.0328084. Therefore

$$1 \text{ cm} = 0.0328 \text{ ft} \quad \text{or} \quad 1 \text{ m} = 3.28 \text{ ft}.$$

$$A = 30 \text{ ft} \times 24 \text{ ft} = 720 \text{ ft}^2$$

$$\text{or } A = 720 \text{ ft}^2 \times \left(\frac{1 \text{ ft}}{3.28 \text{ ft}}\right)^2 = 67.3 \text{ m}^2$$

Example 2-3: A flux density in the CGS system is expressed as 20 maxwells/cm². Calculate the flux density in lines/in². (NOTE: 1 maxwell = 1 line.)

SOLUTION:

$$B = \frac{20 \text{ maxwells}}{\text{cm}^2} \times \left(\frac{2.54 \text{ cm}}{\text{in.}}\right)^2 \times \frac{1 \text{ line}}{1 \text{ maxwell}} = 129 \text{ lines/in}^2.$$

Example 2-4: The velocity of light in free space is given as 2.997925×10^8 m/s. Express the velocity of light in km/hr.

SOLUTION:

$$c = 2.997925 \times 10^8 \frac{\text{m}}{\text{s}} \times \frac{1 \text{ km}}{10^3 \text{ m}} \times \frac{3.6 \times 10^3 \text{s}}{1 \text{ hr}} = 10.79 \times 10^8 \text{ km/hr}$$

Example 2-5: Express the density of water, 62.5 lb/ft³, in (a) lb/in.³, (b) g/cm³.

SOLUTION:

(a) $\text{density} = \frac{62.5 \text{ lb}}{\text{ft}^3} \times \left(\frac{1 \text{ ft}}{12 \text{ in.}}\right)^3 = 3.62 \times 10^{-2} \text{ lb/in.}^3$

(b) $\text{density} = 3.62 \times 10^{-2} \frac{\text{lb}}{\text{in.}^3} \times \frac{453.6 \text{ g}}{1 \text{ lb}} \times \left(\frac{1 \text{ in.}}{2.54 \text{ cm}}\right)^3 = 1 \text{ g/cm}^3$

Example 2-6: The speed limit on a highway is 60 km/hr. Calculate the limit in (a) mi/hr, (b) ft/s.

SOLUTION:

(a) $\text{speed limit} = \frac{60 \text{ km}}{\text{hr}} \times \frac{10^3 \text{ m}}{1 \text{ km}} \times \frac{10^2 \text{ cm}}{1 \text{ m}} \times \frac{1 \text{ in.}}{2.54 \text{ cm}} \times \frac{1 \text{ ft}}{12 \text{ in.}} \times \frac{1 \text{ mi}}{5,280 \text{ ft}}$

$= 37.4 \text{ mi/hr}$

(b) $\text{speed limit} = \frac{37.4 \text{ mi}}{\text{hr}} \times \frac{5,280 \text{ ft}}{1 \text{ mi}} \times \frac{1 \text{ hr}}{3.6 \times 10^3 \text{ s}} = 54.9 \text{ ft/s}$

REFERENCES

1. Hvistendahl, H. S., *Engineering Units and Physical Quantities*. London: Macmillan and Co., Ltd., 1964.
2. Kaye, G. W. C., and T. H. Laby, *Tables of Physical and Chemical Constants*, 13th ed. London: Longmans, Green and Co., Ltd., 1966.

PROBLEMS

1. Using powers of ten, express the following in Hz:
   (a) 1,500 Hz
   (b) 20 kHz
   (c) 1,800 kHz
   (d) 0.5 MHz
   (e) 50 MHz
   (f) 1.2 GHz

2. Using powers of ten, express the following in V:
   (a) 24 mV   (b) 540 μV   (c) 4.4 kV
   (d) 1.2 MV   (e) 16 nV   (f) 0.4 mV

3. Using powers of ten, express the following in A:
   (a) 23.5 mA   (b) 45 μA   (c) 0.25 mA
   (d) 72 nA   (e) 620 μA   (f) 74.6 nA

4. Using powers of ten, express the following in μA:
   (a) 0.00036 A   (b) 0.027 A   (c) 0.250 mA
   (d) 25 pA   (e) 2.5 A   (f) 1.275 mA

5. Calculate the height in cm of a man 5 ft 11 in. tall.

6. Calculate the mass in kg of 1 yd^3 of iron when the density of iron is 7.86 g/cm^3.

7. Calculate the conversion factor to change mi/hr to ft/s.

8. An electrically charged body has an excess of 10^{15} electrons. Calculate its charge in C.

9. A train covers a distance of 220 mi in 2 hr and 45 min. Calculate the average velocity of the train in m/s.

10. Two electric charges are separated by a distance of 1 m. If one charge is $+10$ C and the other charge -6 C, calculate the force of attraction between the charges in N and in lb. Assume that the charges are placed in a vacuum.

11. The practical unit of electrical energy is the kWh. The unit of energy in the SI is the joule (J). Calculate the number of joules in 1 kWh.

12. A crane lifts a 100-kg mass a height of 20 m in 5 s. Calculate (a) the work done by the crane, in SI units, (b) the increase of potential energy of the mass, in SI units, (c) the power, or rate of doing the work, in SI units.

13. Calculate the voltage of a battery if a charge of 3×10^{-4} C residing on the positive battery terminal possesses 6×10^{-2} J of energy.

14. An electric charge of 0.035 C flows through a copper conductor in 5 min. Calculate the average current in mA.

15. An average current of 25 μA is passed through a wire for 30 s. Calculate the number of electrons transferred through the conductor.

16. The speed limit on a four-lane highway is 70 mi/hr. Calculate the speed limit in (a) km/hr, (b) ft/s.

17. The density of copper is 8.93 g/cm^3. Express the density in (a) kg/m^3, (b) lb/ft^3.

18. The melting point of magnesium is 650°C. Express the melting point in (a) °F, (b) °K.

3

Standards of Measurement

3-1 CLASSIFICATION OF STANDARDS

A standard of measurement is a physical representation of a unit of measurement. A unit is realized by reference to an arbitrary material standard or to natural phenomena including physical and atomic constants. For example, the fundamental unit of mass in the international system (SI) is the *kilogram*, defined as the mass of a cubic decimeter of water at its temperature of maximum density of 4°C (see Sec. 2-2). This unit of mass is represented by a material standard: the mass of the International Prototype Kilogram, consisting of a platinum-iridium alloy cylinder. This cylinder is preserved at the International Bureau of Weights and Measures at Sèvres, near Paris, and is the *material representation* of the kilogram. Similar standards have been developed for other units of measurement, including standards for the fundamental units as well as for some of the derived mechanical and electrical units.

Just as there are fundamental and derived units of measurement, we find different types of *standards of measurement*, classified by their function and application in the following categories:

(a) International standards
(b) Primary standards
(c) Secondary standards
(d) Working standards

The *international standards* are defined by international agreement. They represent certain units of measurement to the closest possible accuracy that production and measurement technology allow. International standards are periodically evaluated and checked by *absolute measurements* in terms of the fundamental units (see Table 2-2). These standards are maintained at the International Bureau of Weights and Measures and are not available to the ordinary user of measuring instruments for purposes of comparison or calibration.

The *primary* (basic) *standards* are maintained by national standards laboratories in different parts of the world. The National Bureau of Standards (NBS) in Washington is responsible for maintenance of the primary standards in North America. Other national laboratories include the National Physical Laboratory (NPL) in Great Britain and, the oldest in the world, the Physikalisch-Technische Reichsanstalt in Germany. The primary standards, again *representing* the fundamental units and some of the derived mechanical and electrical units, are independently calibrated by absolute measurements at each of the national laboratories. The results of these measurements are compared against each other, leading to a world average figure for the primary standard. Primary standards are not available for use outside the national laboratories. One of the main functions of primary standards is the verification and calibration of secondary standards.

Secondary standards are the basic *reference* standards used in industrial measurement laboratories. These standards are maintained by the particular involved industry and are checked locally against other reference standards in the area. The responsibility for maintenance and calibration of secondary standards rests entirely with the industrial laboratory itself. Secondary standards are generally sent to the national standards laboratories on a periodic basis for calibration and comparison against the primary standards. They are then returned to the industrial user with a *certification* of their measured value in terms of the primary standard.

Working standards are the principal tools of a measurement laboratory. They are used to check and calibrate general laboratory instruments for accuracy and performance or to perform comparison measurements in industrial applications. A manufacturer of precision resistances, for example, may use a *standard resistor* (a *working* standard) in the quality control department of his plant to check his testing equipment. In this case, he *verifies* that his measurement setup performs within the required limits of accuracy.

In electrical and electronic measurement we are concerned with the electrical and magnetic standards of measurement. These are discussed in the following sections. We have seen, however, that electrical units can be traced back to the basic units of length, mass, and time (in fact, the national laboratories perform *measurements* to relate derived electrical units to fundamental units) and they deserve some investigation here.

3-2 STANDARDS FOR MASS, LENGTH, AND VOLUME

The metric *unit of mass* was originally defined as the mass of a cubic decimeter of water at its temperature of maximum density. The *material representation* of this unit is the International Prototype Kilogram, preserved at the International Bureau of Weights and Measures near Paris. The *primary standard* of mass in North America is the United States Prototype Kilogram, preserved by the NBS to an accuracy of 1 part in 10^8 and occasionally verified against the standard at the International Bureau. *Secondary standards* of mass, kept by the industrial laboratories, generally have an accuracy of 1 ppm (part per million) and may be verified against the NBS primary standard. Commercial *working standards* are available in a wide range of values to suit almost any application. Their accuracy is in the order of 5 ppm. The working standards, in turn, are checked against the secondary laboratory standards.

The *pound* (lb), established by the Weights and Measures Act of 1963 (which actually came into effect on January 31, 1964), is defined as equal to 0.45359237 kg *exactly*. All countries which retain the pound as the basic unit of measurement have now adopted the new definition, which supersedes the former imperial standard pound made of platinum.

The metric *unit of length* (the meter), initially defined as the ten-millionth part of the meridianal quadrant through Paris (Sec. 2-2), was materially represented by the distance between two lines engraved on a platinum-iridium bar preserved at the International Bureau of Weights and Measures near Paris. In 1960 the meter was redefined more accurately in terms of an *optical standard*, namely, the orange-red radiation of a krypton atom. The internationally specified krypton-86 discharge lamp, excited and observed under well-defined conditions, emits orange light whose wavelength now constitutes the basic standard of length to 1 ppm. The meter, as the SI unit of length, is now defined as equal to 1,650,763.73 wavelengths in vacuum of the orange-red radiation of the krypton-86 atom. The optically defined standard of length represents the same fundamental unit of length as the former platinum-iridium bar, but its accuracy is an order of magnitude greater.

The *yard* is defined as 0.9144 m *exactly* (1 in. = 25.4 mm exactly) and by this definition then also depends on the krypton-86 wavelength standard. This definition of the yard supersedes the former definition in terms of the imperial standard yard. All countries which retain the yard as the basic unit of measurement have now adopted this new definition.

The most widely used industrial *working standards* of length are precision *gage blocks*, made of steel. These steel blocks have two plane parallel surfaces, a specified distance apart, with accuracy tolerances in the 0.5–0.25-micron range (1 micron = one millionth of 1 m). The development and use of precision gage blocks, low in cost and of high accuracy, have made it possible to manufacture interchangeable industrial components in a very economical application of precision measurement.

The unit of *volume* is a derived quantity and is not represented by an international standard. The NBS, however, has constructed a number of primary standards of volume, calibrated in terms of the absolute dimensions of length and mass. Secondary derived standards of volume are available and may be calibrated in terms of the NBS primary standards.

3-3 TIME AND FREQUENCY STANDARDS*

Since early times men have sought a reference standard for a uniform time scale together with means to interpolate from it a small time interval. For many centuries the time reference used was the rotation of the earth about its axis with respect to the sun. Precise astronomical observations have shown that the rotation of the earth about the sun is very irregular, owing to secular and irregular variations in the rotational speed of the earth. Since the time scale based on this apparent *solar time* does not represent a uniform time scale, other avenues were explored. *Mean solar time* was thought to give a more accurate time scale. A mean solar day is the average of all the apparent days in the year. A *mean solar second* is then equal to 1/86,400 of the mean solar day. The mean solar second, thus defined, is still inadequate as the fundamental unit of time, since it is tied to the rotation of the earth, which is now known to be nonuniform.

The system of *universal time* (UT), or mean solar time, is also based on the rotation of the earth about its axis. This system is known as UT_0 and is subject to periodic, long-term, and irregular variations. Correction of UT_0 has led to two subsequent universal time scales: UT_1 and UT_2. UT_1 recognizes the fact that the earth is subject to polar motion, and the UT_1 time scale is based on the true angular rotation of the earth, corrected for polar motion. The UT_2 time scale is UT_1 with an additional correction for *seasonal* variations in the rotation of the earth. These variations are apparently caused by seasonal displacement of matter over the earth's surface, such as changes in the amount of ice in the polar regions as the sun moves from the southern hemisphere to the northern and back again through the year. This *cyclic* redistribution of mass acts on the earth's rotation since it produces changes in its moment of inertia. The *epoch*, or *instant of time*, of UT_2 can be established to an accuracy of a few milliseconds, but it is not usually distributed to this accuracy. The epoch indicated by the standard radio time signals may differ

---

*Application note AN 52 (*Frequency and Time Standards*), published by Hewlett-Packard, Palo Alto, Calif., describes methods of frequency comparisons, time scales, worldwide standards broadcasts.

from the epoch of UT_2 by as much as 100 ms. The actual values of the differences are given in bulletins published by the national time services (NBS) and by the Bureau Internationale de l'Heure (Paris Observatory).

The search for a truly universal time unit has led astronomers to define a time unit called *ephemeris time* (ET). ET is based on astronomical observations of the motion of the moon about the earth. Since 1956 the *ephemeris second* has been defined by the International Bureau of Weights and Measures as the fraction 1/31,556,925.9747 of the tropical year for 1900 January 0 at 12 h ET, and adopted as the *fundamental invariable unit of time*. A disadvantage of the use of the ephemeris second is that it can be determined only several years in arrears and then only indirectly, by observations of the positions of the sun and the moon. For *physical measurements*, the unit of time interval has now been defined in terms of an *atomic standard*. The universal second and the ephemeris second, however, will continue to be used for navigation, geodetic surveys, and celestial mechanics.

Development and refinement of *atomic resonators* have made possible control of the frequency of an oscillator and, hence, by frequency conversion, *atomic clocks*. The transition between two energy levels, E_1 and E_2, of an atom is accompanied by the emission (or absorption) of radiation having a frequency given by $hv = E_2 - E_1$, where h is Planck's constant. Provided that the energy states are not affected by external conditions, such as magnetic fields, the frequency v is a *physical constant*, depending *only* on the internal structure of the atom. Since frequency is the inverse of time interval, such an atom provides a *constant time interval*. Atomic transitions of various metals were investigated, and the first atomic clock, based on the cesium atom, was put into operation in 1955. The time interval, provided by the cesium clock, is more accurate than that provided by a clock calibrated by astronomical measurements. The *atomic unit of time* was first related to UT but was later expressed in terms of ET. The International Committee of Weights and Measures has now defined the second in terms of the frequency of the cesium transition, assigning a value of 9,192,631,770 Hz to the hyperfine transition of the cesium atom unperturbed by external fields.

The *atomic definition* of the second realizes an accuracy much greater than that achieved by astronomical observations, resulting in a more uniform and much more convenient time base. Determinations of time intervals can now be made in a few minutes to greater accuracy than was possible before in astronomical measurements that took many years to complete. An atomic clock with a precision exceeding 1 μs per day is in operation as a primary frequency standard at the NBS. An atomic time scale, designated NBS-A, is maintained with this clock.

NBS disseminates its time and frequency standards by broadcasts from several radio stations, operating at different transmission frequencies from various parts of the continental United States and Hawaii. Complete information regarding broadcast schedules and changes in station operation may be obtained upon request from the NBS.

3-4 ELECTRICAL STANDARDS

3-4.1 The Absolute Ampere

The international system of units (SI) defines the *ampere* (the fundamental unit of electric current) as the constant current which, if maintained in two straight parallel conductors of infinite length and negligible circular cross section placed 1 m apart in a vacuum, will produce between these conductors a force equal to 2×10^{-7} newton per meter length. Early measurements of the absolute value of the ampere were made with a *current balance* which measured the force between two parallel conductors. These measurements were rather crude and the need was felt to produce a more practical and reproducible standard for the national laboratories. By international agreement, the value of the *International Ampere* was based on the electrolytic deposition of silver from a silver nitrate solution. The International Ampere was then defined as that current which deposits silver at the rate of 1.118 mg/s from a standard silver-nitrate solution. Difficulties were encountered in the exact measurement of the deposited silver and slight discrepancies existed between measurements made independently by the various national standards laboratories.

In 1948 the International Ampere was superseded by the *Absolute Ampere*. The determination of the Absolute Ampere is again made by means of a current balance, which *weighs* the *force* exerted between two current-carrying coils. Improvement in the techniques of force measurement yields a value for the ampere far superior to the early measurements. The relationship between the force and the current which produces the force can be calculated from fundamental electromagnetic theory concepts and reduces to a simple computation involving the geometric dimensions of the coils. The Absolute Ampere is now the *fundamental unit of electric current* in the SI and is universally accepted by international agreement.

Instruments manufactured before 1948 are calibrated in terms of the International Ampere but newer instruments are using the Absolute Ampere as the basis for calibration. Since both types of instrument may be found side by side in one laboratory, the NBS has established conversion factors to relate both units. These factors are given in Sec. 2-3.

Voltage, current, and resistance are related by Ohm's law of constant proportionality ($E = IR$). The specification of any two quantities automatically sets the third. Two types of material standards form a combination which conveniently serves to maintain the ampere with high precision over long periods of time: the *standard resistor* and the *standard cell* (for voltage). Each of these is descibed below.

3-4.2   Resistance Standards

The absolute value of the ohm in the SI system is defined in terms of the fundamental units of length, mass, and time. The *absolute measurement* of the ohm is carried out by the International Bureau of Weights and Measures in Sèvres and also by the national standards laboratories, which preserve a group of *primary* resistance standards. The NBS maintains a group of those primary standards (1-Ω standard resistors) which are periodically checked against each other and are occasionally verified by absolute measurements. The standard resistor is a coil of wire of some alloy like *manganin* which has a high electrical resistivity and a low temperature coefficient of resistance (almost constant temperature-resistance relationship). The resistance coil is mounted in a double-walled sealed container (Fig. 3-1) to prevent changes in resistance due to moisture conditions in the atmosphere. With a set of four or five 1-Ω resistors of this type, the unit of resistance can be represented with a precision of a few parts in 10^7 over several years.

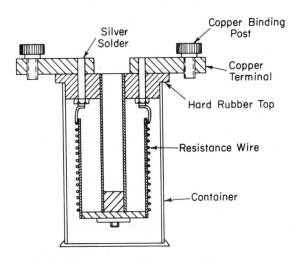

FIGURE 3-1   Cross-sectional view of a double-walled resistance standard (courtesy Hewlett-Packard Co.).

Secondary standards and *working* standards are available from some instrument manufacturers in a wide range of values, usually in multiples of 10 Ω. These standard resistors are made of alloy resistance wire, such as manganin or Evanohm. Figure 3-2 is a photograph of a laboratory secondary standard, sometimes referred to as a *transfer resistor*. The resistance coil of the transfer resistor is supported between polyester film to reduce stresses on the wire and to improve the stability of the resistor. The coil is immersed in moisture-free oil and placed in a sealed can. The connections to the coil are silver soldered, and the terminal hooks are made of nickel-plated oxygen-free

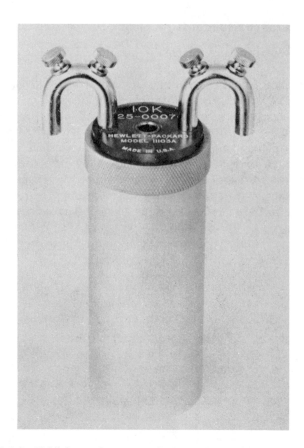

FIGURE 3-2   10-kilohm resistance standard (courtesy Hewlett-Packard Co.).

copper. The transfer resistor is checked for stability and temperature characteristics at its rated power and a specified operating temperature (usually
25°C). A *calibration report* accompanying the resistor specifies its traceability
to NBS standards and includes the α and β temperature coefficients. Although
the selected resistance wire provides almost constant resistance over a fairly
wide temperature range, the exact value of the resistance at any temperature
can be calculated from the formula

$$R_t = R_{25°C} + \alpha(t - 25) + \beta(t - 25)^2 \tag{3-1}$$

where R_t = resistance at the ambient temperature t

$R_{25°C}$ = resistance at 25°C

α and β = temperature coefficients

Temperature coefficient α is usually less than 10×10^{-6}, and coefficient β
lies between -3×10^{-7} and -6×10^{-7}. This means that a change in temperature of 10°C from the specified reference temperature of 25°C may cause

a change in resistance of 30 to 60 ppm (parts per million) from the nominal value.

Transfer resistors find application in industrial, research, standards, and calibration laboratories. In typical applications, the transfer resistor may be used for resistance and ratio determinations or in the construction of ultra-linear decade dividers which can then be used for the calibration of universal ratio sets, voltboxes, and Kelvin-Varley dividers.

3-4.3 Voltage Standards

The *primary voltage standard*, selected by the NBS for the maintenance of the volt, is the *normal*, or *saturated*, *Weston cell*. The Weston cell has a positive electrode of mercury and a negative electrode of cadmium amalgam (10 per cent Cd). The electrolyte is a solution of cadmium sulphate. These components are placed in an H-shaped glass container, as shown in Fig. 3-3.

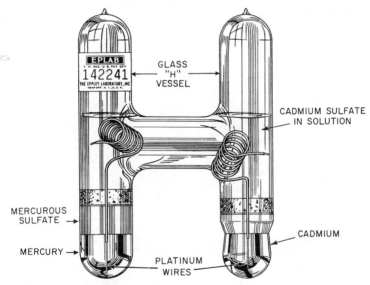

FIGURE 3-3   Construction details of a saturated Weston cadmium cell (courtesy The Eppley Laboratory, Inc.).

There are two types of Weston cell: the *saturated* cell, in which the electrolyte is saturated at all temperatures by cadmium sulphate crystals covering the electrodes, and the *unsaturated* cell, in which the concentration of cadmium sulphate is such that it produces saturation at 4°C. The unsaturated cell has a negligible temperature coefficient of voltage at normal room temperatures. The saturated cell has a voltage variation of approximately -40 μV per 1°C rise, but is better reproducible and more stable than the unsaturated cell.

National standards laboratories, such as the NBS, maintain a number of *saturated* cells as the *primary standard* for voltage. The cells are kept in an oil bath to control their temperature to within 0.01°C. The voltage of the Weston saturated cell at 20°C is 1.01858 V (absolute), and the emf at other temperatures is given by the formula

$$e_t = e_{20^\circ C} - 0.000046(t - 20) - 0.00000095(t - 20)^2 + 0.00000001(t - 20)^3$$

$$(3\text{-}2)$$

Saturated Weston cells remain satisfactory as voltage standards for periods of 10 to 20 years, provided that they are carefully treated. Their drift in voltage is on the order of 1 μV per year. Since saturated cells are temperature sensitive, they are unsuited for general laboratory use as secondary or working standards.

More rugged portable *secondary* and *working standards* are found in the *unsaturated* Weston cell. These cells are very similar in construction to the normal cell but they do not require exact temperature control. The emf of an unsaturated cell lies in the range of 1.0180 V to 1.0200 V and varies less than 0.01 per cent from 10°C to 40°C. The voltage of the cell is usually indicated on the cell housing, as shown in Fig. 3-4 (1.0193 abs. V). The internal resistance of Weston cells range from 500 Ω to 800 Ω. The current drawn from these

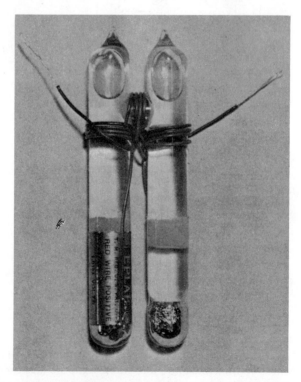

FIGURE 3-4   An unsaturated Weston cadmium cell: emf 1.0193 V, accuracy 0.1 % (courtesy The Eppley Laboratory, Inc.).

cells should therefore not exceed 100 μA, because the nominal voltage would be affected by the internal voltage drop.

Versatile *laboratory working* standards have been developed with accuracies on the order of standard cell accuracy. Figure 3-5 is a photograph

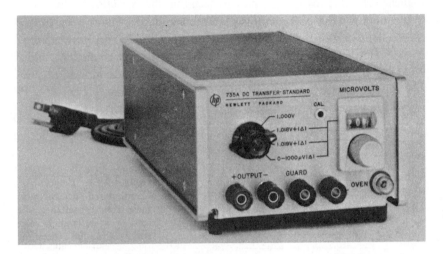

FIGURE 3-5   A dc transfer standard that can be used as a 1.000-V reference source, a standard cell comparison instrument, and a 0–1,000 μV dc source (courtesy Hewlett-Packard Co.).

of a multipurpose laboratory voltage standard, called a *transfer standard*, based on the operation of a Zener diode as the voltage reference element. The instrument basically consists of a Zener-controlled voltage source placed in a temperature-controlled environment to improve its long-term stability, and a precision output voltage divider. The temperature-controlled oven is held to within $\pm 0.03°$C over an ambient temperature range of 0°C to 50°C, providing an output stability on the order of 10 ppm/month. The four available outputs are (a) a 0–1,000-μV source with 1-μV resolution, called (Δ); (b) a 1.000-V reference for voltbox potentiometric measurements; (c) a 1.018 + (Δ) reference for saturated cell comparisons; (d) a 1.0190 + (Δ) reference for unsaturated cell comparisons. The dc transfer standard can be used as a transfer instrument and can be moved to the piece of equipment to be calibrated, since it can easily be disconnected from the power line at one location and set up at a different location where it will recover to within ± 1 ppm in approximately 30 minutes warm-up time.

3-4.4   Capacitance Standards

Since the unit of resistance is represented by the standard resistor, and the unit of voltage by the standard Weston cell, many electrical and magnetic units may be expressed in terms of these standards. The unit of *capacitance* (the farad) can be measured with a Maxwell dc commutated bridge, where the capacitance is computed from the resistive bridge arms and the frequency of the dc commutation. This bridge is shown in Fig. 3-6. Although the exact derivation of the expression for capacitance in terms of the resistances and the frequency is rather involved, it may be seen that the capacitor could be measured by this method. Since both resistance and frequency can be determined very accurately, the value of the capacitance can be measured with great accuracy. *Standard capacitors* are usually constructed from interleaved metal plates with air as the dielectric material. The area of the plates and the distance between them must be known very accurately, and the capacitance of the air capacitor can be determined from these basic dimensions. The NBS maintains a bank of air capacitors as standards and uses them to calibrate the secondary and working standards of measurement laboratories and industrial users.

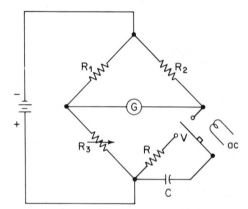

FIGURE 3-6 Commutated dc method for measuring capacitance. Capacitor C is alternately charged and discharged through the commutating contact and resistor R. Bridge balance is obtained by adjustment of R_3, allowing exact determination of the capacitance value in terms of bridge-arm constants and frequency of commutation.

Capacitance working standards can be obtained in a range of suitable values. Smaller values are usually air capacitors, whereas the larger capacitors use solid dielectric materials. The high dielectric constant and the very thin dielectric layer accounts for the compactness of these standards. Silver-mica capacitors make excellent working standards; they are very stable and have a very low dissipation factor (Sec. 8-5), a very small temperature coefficient, and little or no aging effect. Mica capacitors are available in decade mounting, but decade capacitors are usually not guaranteed better than 1 per cent. Fixed standards are generally used where accuracy is important.

3-4.5 Inductance Standards

The *primary inductance standard* is derived from the ohm and the farad, rather than from the large geometrically constructed inductors used in the determination of the absolute value of the ohm. The NBS selected a *Campbell standard* of mutual inductance as the primary standard for both mutual and self-inductance. Inductance *working standards* are commercially available in a wide range of practical values, both fixed and variable. A typical set of fixed inductance standards includes values from approximately 100 μH to 10 H, with a guaranteed accuracy of 0.1 per cent at a specified operating frequency. Variable inductors are also available. Typical mutual inductance accuracy is on the order of 2.5 per cent and inductance values range from 0 to 200 mH. *Distributed capacitance* exists between the windings of these inductors, and the errors they introduce must be taken into account. These considerations are usually specified with commercial equipment.

3-5 MAGNETIC STANDARDS

3-5.1 Ballistic Measurements

Measurements of magnetic flux generally involve the use of a *ballistic galvanometer*. The ballistic galvanometer is essentially a d'Arsonval movement, specifically designed for long-period operation (20 s to 30 s) and of high sensitivity. In ballistic measurements, the coil receives a *momentary* impulse of current, which causes it to swing to one side and then return to rest in an oscillatory motion, governed by the circuit damping (Sec. 4-2.3). When the current impulse is of sufficiently short duration, the initial deflection from its rest position is directly proportional to the quantity of electrical discharge through the coil. The relative magnitude of the current impulse is measured in terms of the *initial* angular deflection of the coil and we can write

$$Q = K\theta \tag{3-3}$$

where Q = charge in coulombs

       K = galvanometer sensitivity, in coulomb/radian of deflection

       θ = angular deflection of the coil, in radians

The sensitivity, K, depends on the damping and should be obtained experimentally by a *calibration* check conducted under the *actual* conditions of use.

Several procedures can be used to calibrate the ballistic galvanometer, these include the capacitor method, the solenoid method, and the mutual inductor method. The last method is shown in Fig. 3-7 in which a current source in the primary circuit is coupled to the ballistic galvanometer under test by means of a mutual inductance. Reversal of the known primary current I causes a galvanometer deflection θ, proportional to the circuit constants and the galvanometer sensitivity.

It can be shown that the total charge in the circuit, caused by the change in current from $+I$ to $-I$, equals

$$Q = \frac{2MI}{R} \text{ (coulombs)} \tag{3-4}$$

where M = mutual inductance, in henries

       R = total resistance in the secondary circuit

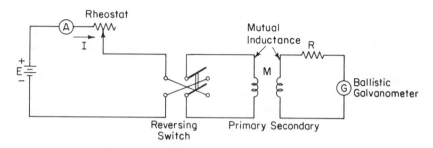

FIGURE 3-7   Calibration of a ballistic galvanometer by the mutual inductance method.

Substituting Eq. (3-4) into Eq. (3-3) yields a value for the galvanometer sensitivity and

$$K = \frac{2MI}{R\theta} \tag{3-5}$$

Once calibrated, the ballistic galvanometer may be used to measure the flux produced by a variety of permanent magnets. The method is shown in Fig. 3-8. A *search coil*, surrounding a permanent magnet whose flux is to be measured, is connected in series with a ballistic galvanometer and a variable resistor. The variable resistor is generally adjusted to give critical damping to the galvanometer. If the permanent magnet is withdrawn *quickly* from the search coil, an impulse of current is produced and the galvanometer deflects. The quantity of charge through the ballistic galvanometer is directly proportional to the total flux ϕ of the permanent magnet and the number of turns N

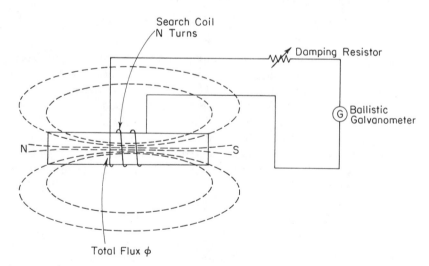

FIGURE 3-8   Flux measurement by ballistic galvanometer.

of the search coil, and inversely proportional to the total circuit resistance R, so that

$$Q = \frac{N\phi}{R} \text{ (coulombs)} \tag{3-6}$$

The galvanometer deflection is, from Eq. (3-3),

$$\theta = \frac{Q}{K} = \frac{N\phi}{KR} \text{ (radians)} \tag{3-7}$$

Rearranging Eq. (3-7) provides an expression for the flux being measured and

$$\phi = \frac{KR\theta}{N} \text{ (webers)} \tag{3-8}$$

It should be stressed that the sensitivity factor K must be evaluated for the circuit resistance used in the measurement setup.

3-5.2 Magnetic Flux Standards

The measurement method described in Fig. 3-8 is used to measure the *standard flux* produced by a variety of permanent magnets. These permanent magnets are then maintained as *magnetic flux standards*.

It is often useful to have a standard flux source independent of external exciting current. The *Hibbert magnetic standard* (Fig. 3-9) is such a device. A permanent magnet is enclosed in a soft iron container which has a narrow circular air gap. A brass cylinder is suspended within the air gap. On this cylinder is wound an insulated winding of conducting material such as copper. At the release of a catch, the brass cylinder and the winding assembly drop through the flux in the air gap. The resulting electrical current induced in the

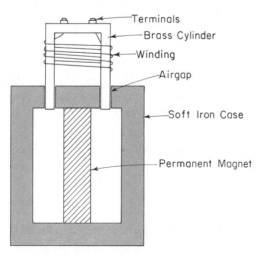

FIGURE 3-9   Principle of construction of the Hibbert magnetic standard.

winding is proportional to the rate at which the magnetic flux was cut by the moving winding. Since the local gravitational field is the only force acting on the winding, the rate at which the flux was cut is constant. Therefore, the induced current is directly proportional to the flux in the air gap. The Hibbert standard is a *secondary* standard and must be calibrated against the mutual inductance method described earlier.

3-6 STANDARDS OF TEMPERATURE AND LUMINOUS INTENSITY

Thermodynamic temperature is one of the basic SI quantities and its unit is the degree Kelvin (Sec. 2-2). The thermodynamic Kelvin scale is recognized as the *fundamental scale* to which all temperatures should be referred. The temperatures on this scale are designated as °K and denoted by the symbol T. The magnitude of the degree Kelvin has been fixed by defining the thermodynamic temperature of the *triple point* of water at *exactly* 273.16°K. The triple point of water is the temperature of equilibrium between ice, liquid water, and its vapor.

Since temperature measurements on the thermodynamic scale are inherently difficult, the Seventh General Conference of Weights and Measures adopted in 1927 a *practical scale* which has been modified several times and is now called the *International Practical Scale of Temperature*. The temperatures on this scale are designated as °C (degree Celsius) and denoted by the symbol t. The Celsius scale has two *fundamental* fixed points: the boiling point of water as 100°C and the triple point of water as 0.01°C, both points established at atmospheric pressure. A number of *primary* fixed points have been established above and below the two fundamental points. These are the boiling point of oxygen (-182.97°C), the boiling point of sulphur (444.6°C), the freezing point of silver (960.8°C), and the freezing point of gold (1,063°C). The numerical values of all these points are reproducible quantities at atmospheric pressure. The conversion between the Kelvin scale and the Celsius scale follows the relationship:

$$t(°C) = T(°K) - T_0 \qquad (3-9)$$

where $T_0 = 273.15$ degrees.

The primary *standard thermometer* is a platinum resistance thermometer of special construction so that the platinum wire is not subjected to strain. Interpolated values between the fundamental and primary fixed points on the scale are calculated by formulas based on the properties of the platinum resistance wire.

The primary *standard of luminous intensity* is a full radiator (black body or Planckian radiator) at the temperature of solidification of platinum (2,042°K approx.). The *candela* is then defined as one-sixtieth of the luminous intensity per cm² of the full radiator. Secondary standards of luminous intensity are special tungsten filament lamps, operated at a temperature whereby their spectral power distribution in the visible region matches that of the basic standard. These secondary standards are recalibrated against the basic standard at periodic intervals.

REFERENCES

1. Kaye, G. W. C., and T. H. Laby, *Tables of Physical and Chemical Constants*, 13th ed. London: Longmans, Green and Co., Ltd., 1966.

2. Philco Technological Center, *Electronic Precision Measurement Techniques and Experiments*. Englewood Cliffs, N.J.: Prentice-Hall, Inc., 1964.

3. Stout, Melville B., *Basic Electrical Measurements*, 2nd ed. Englewood Cliffs, N.J.: Prentice-Hall, Inc., 1960.

QUESTIONS

1. Describe briefly the differences between primary and secondary standards in terms of their accuracy and their use.

2. What is meant by the "atomic time scale"? How is this time scale related to UT_2?

3. NBS radio stations WWV and WWVB transmit standard time signals which may be used in the calibration of laboratory equipment such as clocks and counters. Briefly describe the kind of services offered by these radio stations and indicate how the transmitted signal can be traced back to the primary time standard.

4. Name several precautions that should be taken when using a standard Weston cell.

5. What is the emf of a normal Weston cell at 20°C and how much does the emf change when the cell is used at 0°C?

6. You are asked to determine the internal resistance of an unsaturated Weston cell. Describe a method that would supply the correct answer.

7. You suspect that the emf of one of the standard cells in the calibration laboratory may be in error by a fairly large amount. You would like to check this but realize that an ordinary voltmeter would draw too much current and would most certainly damage the cell. What circuit arrangement can you think of to perform this measurement?

8. A time code generator contains a precision oscillator that must be checked daily against a standard frequency transmission of radio station WWV. With the aid of a block diagram, explain how this could be done.

9. Describe briefly the construction of the primary standards for the absolute ohm and the henry.

4

Direct-current Indicating Instruments

4-1 SUSPENSION GALVANOMETER

Early measurements of direct current required a suspension galvanometer. This instrument was the forerunner of the moving-coil instrument, basic to most dc indicating movements currently used. Figure 4-1 shows the construction of a *suspension* galvanometer.

A *coil* of fine wire is suspended in a magnetic field produced by a *permanent magnet*. According to the fundamental law of electromagnetic force, the coil will rotate in the magnetic field when it carries an electric current. The fine filament suspension of the coil serves to carry current to and from it, and the elasticity of the filament sets up a moderate torque in opposition to the rotation of the coil. The coil will continue to deflect until its electromagnetic torque balances the mechanical countertorque of the suspension. The coil deflection therefore is a measure of the magnitude of the current carried by the coil. A *mirror* attached to the coil deflects a beam of light, causing a magnified light spot to move on a *scale* at some distance from the instrument. The optical effect is that of a pointer of great length but zero mass.

With modern refinements, the suspension galvanometer is still used in certain high-sensitivity laboratory measurements when the delicacy of the instrument is not objectionable and portability is not required.

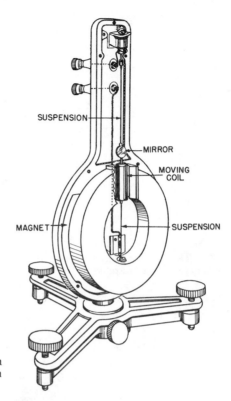

FIGURE 4-1 An early suspension galvanometer (courtesy Weston Instruments, Inc.).

4-2 TORQUE AND DEFLECTION OF THE GALVANOMETER

4-2.1 Steady-state Deflection

Although the suspension galvanometer is neither a practical nor portable instrument, the principles governing its operation apply equally to its more modern version, the *permanent-magnet moving-coil mechanism* (PMMC). Figure 4-2 shows the construction of the PMMC mechanism. The different parts of the instrument are identified alongside the figure.

Here again we have a coil, suspended in the magnetic field of a permanent magnet, this time in the shape of a horseshoe. The coil is suspended so that it can rotate freely in the magnetic field. When current flows in the coil, the developed electromagnetic (EM) torque causes the coil to rotate. The EM torque is counterbalanced by the mechanical torque of control springs attached to the movable coil. The balance of torques, and therefore the

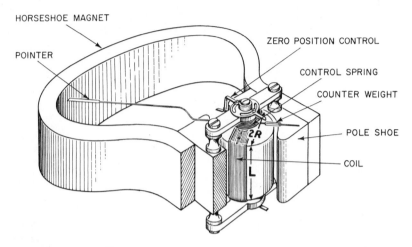

FIGURE 4-2 Construction details of the external magnet PMMC movement (courtesy Weston Instruments, Inc.).

angular position of the movable coil, is indicated by a pointer against a fixed reference, called a *scale*.

The equation for the developed torque, derived from the basic law for electromagnetic torque, is

$$T = B \times A \times I \times N \tag{4-1}$$

where $T =$ torque, newton-meter (N-m)

$B =$ flux density in the air gap, webers/square meter (Wb/m²)

$A =$ effective coil area, square meters (m²)

$I =$ current in the movable coil, amperes (A)

$N =$ turns of wire on the coil

Equation (4-1) shows that the developed torque is directly proportional to the flux density of the field in which the coil rotates, the current in the coil, and the coil constants (area and turns). Since both flux density and coil area are *fixed* parameters for a given instrument, the developed torque is a direct indication of the current in the coil. This torque causes the pointer to deflect to a *steady-state* position where it is balanced by the opposing control-spring torque.

Equation (4-1) also shows that the designer may vary only the value of the control torque and the number of turns on the moving coil to measure a given full-scale current. The practical coil area generally ranges from approximately 0.5 to 2.5 cm². Flux densities for modern instruments usually range from 1,500 to 5,000 gauss (0.15 to 0.5 Wb/m²). Thus a wide choice of mechanisms is available to the designer to meet many different measurement applications.

A typical panel PMMC instrument, with a $3\frac{1}{2}$-in. case, a 1-mA range, and full-scale deflection of 100 degrees of arc, would have the following characteristics:*

$$A = 1.75 \text{ cm}^2$$
$$B = 2,000 \text{ G } (0.2 \text{ Wb/m}^2)$$
$$N = 84 \text{ turns}$$
$$T = 2.92 \times 10^{-6} \text{ N-m}$$
$$\text{Coil resistance} = 88 \text{ }\Omega$$
$$\text{Power dissipation} = 88 \text{ }\mu\text{W}$$

4-2.2   Dynamic Behavior

In Sec. 4-2.1 we considered the galvanometer as a simple indicating instrument in which the deflection of the pointer is directly proportional to the magnitude of the current applied to the coil. This is perfectly satisfactory when we are dealing with a steady-state condition in which we are mainly interested in obtaining a reliable reading of a direct current. In some applications, however, the *dynamic* behavior of the galvanometer (such as speed of response, damping, overshoot) can be important. For example, when an alternating or varying current is applied to a recording galvanometer, the written record produced by the motion of the moving coil includes the response characteristics of the moving element itself and it is therefore important to consider its dynamic behavior.

The dynamic behavior of the galvanometer can be observed by suddenly interrupting the applied current, so that the coil swings back from its deflected position toward the zero position. It will be seen that as a result of inertia of the moving system the pointer swings past the zero mark in the opposite direction, and then oscillates back and forth around zero. These oscillations gradually die down as a result of the damping of the moving element, and the pointer will finally come to rest at zero.

The motion of a moving coil in a magnetic field is characterized by three quantities:

(a) The moment of inertia (J) of the moving coil about its axis of rotation
(b) The opposing torque (S) developed by the coil suspension
(c) The damping constant (D)

Data Sheets, Weston Instruments, Inc., Newark, N.J.

The differential equation that relates these three factors yields three possible solutions, each of which describes the dynamic behavior of the coil in terms of its deflection angle θ. The three types of behavior are shown in the curves of Fig. 4-3 and are known as *overdamped*, *underdamped*, and *critically damped*.

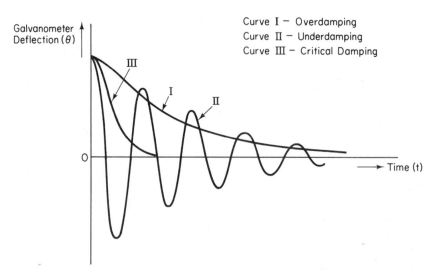

FIGURE 4-3   Dynamic behavior of a galvanometer.

Curve I of Fig. 4-3 shows the overdamped case in which the coil returns slowly to its rest position, without overshoot or oscillations. The pointer seems to approach the steady-state position in a sluggish manner. This case is of minor interest because we prefer to operate under the conditions of curve II or curve III for most applications. Curve II of Fig. 4-3 shows the under-damped case in which the motion of the coil is subject to damped sinusoidal oscillations. The rate at which these oscillations die away is determined by the damping constant (D), the moment of inertia (J), and the counter torque (S) produced by the coil suspension. Curve III of Fig. 4-3 shows the critically damped case in which the pointer returns promptly to its steady-state position, without oscillations.

Ideally, the galvanometer response should be such that the pointer travels to its final position without overshoot; hence, the movement should be critically damped. In practice, the galvanometer is usually slightly under-damped, causing the pointer to overshoot a little before coming to rest. This method is perhaps less direct than critical damping, but it assures the user that the movement has not been damaged because of rough handling, and it compensates for any additional friction that may develop in time because of dust or wear.

4-2.3   Damping Mechanisms

Galvanometer damping is provided by two mechanisms: mechanical and electromagnetic. *Mechanical* damping is caused mainly by the motion of the coil through the air surrounding it; it is independent of any electrical current through the coil. Friction of the movement in its bearings and flexing of the suspension springs caused by the rotating coil also contribute to the mechanical damping effects. *Electromagnetic* damping is caused by induced effects in the moving coil as it rotates in the magnetic field, provided that the coil forms part of a closed electrical circuit.

PMMC instruments are generally constructed to produce as little viscous damping as possible and the required degree of damping is added. One of the simplest damping mechanisms is provided by an aluminum vane, attached to the shaft of the moving coil. As the coil rotates, the vane moves in an air chamber. The amount of clearance between the chamber walls and the air vane effectively controls the degree of damping.

Some instruments use the principle of electromagnetic damping (Lenz's law), where the movable coil is wound on a light aluminum frame. The rotation of the coil in the magnetic field sets up circulating currents in the conductive metal frame, causing a retarding torque that opposes the motion of the coil. Indeed, the same principle is often used to protect PMMC instruments during shipment by placing a metal shorting strap across the coil terminals to reduce deflection.

A galvanometer may also be damped by connecting a resistor across the coil. When the coil rotates in the magnetic field, a voltage is generated in the coil which circulates a current through the coil and the external resistor. This produces an opposing, or retarding, torque that damps the motion of the movement. For any galvanometer, a value for the external resistor can be found that produces critical damping. This resistance is called the *Critical Damping Resistance External* (CDRX); it is an important galvanometer constant. The dynamic damping torque produced by the CDRX depends on the total circuit resistance: the smaller the total circuit resistance, the larger the damping torque.

One way to determine the CDRX consists of observing the galvanometer swing when a current is applied or removed from the coil. Beginning with the oscillating condition, decreasing values of external resistances are tried until a value is found for which the overshoot just disappears. A determination like this is not very precise, but it is adequate for most practical purposes. The value of the CDRX may also be computed from known galvanometer constants.

4-3   PERMANENT-MAGNET MOVING-COIL MECHANISM (PMMC)

4-3.1   D'Arsonval Movement

The basic PMMC movement of Fig. 4-2 is often called the *d'Arsonval* movement, after its inventor. This design offers the largest magnet in a given space and is used when maximum flux in the air gap is required. It provides an instrument with very low power consumption and low current required for *full-scale deflection* (fsd). Figure 4-4 shows a *phantom view* of the d'Arsonval movement.

Inspection of the photograph of Fig. 4-4 shows a *permanent magnet* of

FIGURE 4-4   Phantom photograph of the external moving-coil mechanism shows details of the coil construction, the external horseshoe magnet, and the indicating pointer (courtesy Weston Instruments, Inc.)

horseshoe form, with soft iron *pole pieces* attached to it. Between the pole pieces is a *cylinder* of soft iron, which serves to provide a uniform magnetic field in the air gap between the pole pieces and the cylinder. The *coil* is wound on a light metal *frame* and is mounted so that it can rotate freely in the air gap. The *pointer*, attached to the coil, moves over a graduated *scale* and indicates the angular deflection of the coil and therefore the current through the coil.

The Y-shaped member is the *zero adjust* control and is connected to the fixed end of the *front control-spring.* An *eccentric pin* through the instrument cover engages the Y-shaped member so that the zero position of the pointer can be adjusted from outside the case. Two phosphor-bronze *conductive springs*, normally equal in strength, provide the calibrated force opposing the moving-coil torque. Constancy of spring performance is essential to maintain instrument accuracy. The spring thickness is accurately controlled in manufacture to avoid permanent set of the springs. Current is conducted to and from the coil by the control springs.

The entire moving system is statically balanced for all deflection positions by three *balance weights*, as shown in Fig. 4-5. The pointer, springs, and

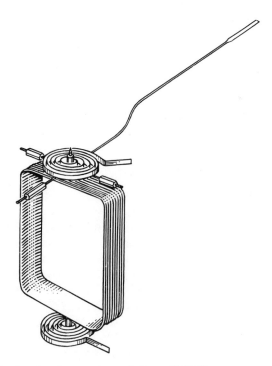

FIGURE 4-5   Details of a moving coil for a PMMC movement, showing the control springs and the indicator with its counterbalance weights (courtesy Weston Instruments, Inc.).

pivots are assembled to the coil structure by means of pivot bases, and the entire movable-coil element is supported by jewel bearings. Different bearing systems are shown in Fig. 4-6.

(a)

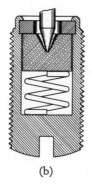

(b)

FIGURE 4-6   Details of instrument bearings. (a) V-jewel bearing. (b) Spring-back jewel bearing (courtesy Weston Instruments, Inc.).

The V-jewel, shown in Fig. 4-6(a), is almost universally used in instrument bearings. The pivot, bearing in the pit in the jewel, may have a radius at its tip from 0.01 mm to 0.02 mm, depending on the weight of the mechanism and the vibration the instrument will encounter. The radius of the pit in the jewel is slightly larger than the pivot radius, so that the contact area is circular, only a few microns across. The V-jewel design of Fig. 4-6(a) has the least friction of any practical type of instrument bearing. Although the moving elements of instruments are designed to have the smallest possible weight, the extremely minute area of contact between pivot and jewel results in stresses on the order of 10 kg/mm². If the weight of the moving element is further increased, the contact area does not increase in proportion so that the stress is even greater. Stresses set up by relatively moderate accelerations (like jarring or dropping an instrument) may consequently cause pivot damage. Specially protected (*ruggedized*) instruments use the spring-back (incabloc) jewel bearing, whose construction is shown in Fig. 4-6(b). It is located in its normal position by the spring and is free to move axially when the shock to the mechanism becomes severe.

The scale markings of the basic dc PMMC instrument are usually linearly spaced because the torque (and hence the pointer deflection) is directly proportional to the coil current. [See Eq. (4-1) for the developed torque.] The basic PMMC instrument is therefore a *linear-reading dc device*. The power requirements of the d'Arsonval movement are surprisingly small: typical values range from 25 μW to 200 μW. Accuracy of the instrument is generally on the order of 2 to 5 per cent of full-scale reading.

If low-frequency alternating current is applied to the movable coil, the deflection of the pointer would be up-scale for one half-cycle of the input waveform and down-scale (in the opposite direction) for the next half-cycle. At powerline frequencies (60 Hz) and above, the pointer could not follow the rapid variations in direction and would quiver slightly around the zero mark, seeking the *average* value of the alternating current (which equals zero). The PMMC instrument is therefore unsuitable for ac measurements, unless the current is rectified before application to the coil (Sec. 5-4).

4-3.2   Core-magnet Construction

In recent years, with the development of Alnico and other improved magnetic materials, it has become feasible to design a magnetic system in which the magnet itself serves as the core. These magnets have the obvious advantage of being relatively unaffected by external magnetic fields, eliminating the magnetic shunting effects in steel panel construction, where several meters operating side by side may affect each other's readings. The need for magnetic shielding, in the form of iron cases, is also eliminated by the *core-magnet* construction. Details of the core-magnet self-shielding movement are shown in Fig. 4-7.

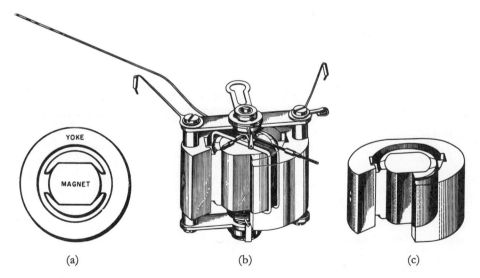

FIGURE 4-7   Construction details of the core-magnet moving-coil mechanism. (a) The magnet with its poleshoes is surrounded by the yoke, which acts as a magnetic shield. (b) The assembled movement. (c) A cut-away view of the yoke, the core, and the poleshoes (courtesy Weston Instruments, Inc.).

Self-shielding makes the core-magnet mechanism particularly useful in aircraft and aerospace applications, where a multiplicity of instruments must be mounted in close proximity to each other. An example of this type of mounting may be found in the *cross-pointer indicator*, where as many as five mechanisms are mounted in one case to form a unified display. Obviously, the elimination of iron cases and the corresponding weight reduction are of great advantage in aircraft and aerospace instruments.

4-3.3   Taut-band Suspension

The *suspension-type* galvanometer mechanism has been known for many years. Until recently the device was used only in the laboratory where high sensitivities were required and the torque was extremely low (because of small currents). It was desirable in such instruments to eliminate even the low friction of pivots and jewels. The suspension galvanometer (Fig. 4-1) had to be used in the upright position, because sag in the low-torque ligaments caused the moving system to come in contact with stationary members of the mechanism in any other position. This increase in friction caused errors.

The *taut-band* instrument of Fig. 4-8 has the advantage of eliminating the friction of the jewel-pivot suspension. The movable coil is suspended by means of two *torsion ribbons*. The ribbons are placed under sufficient tension

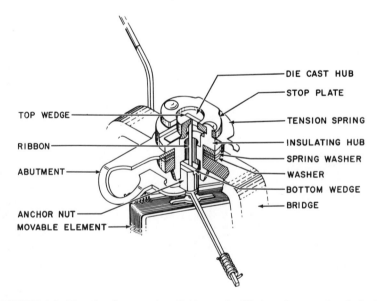

FIGURE 4-8   Taut-band suspension eliminates the friction of conventional pivot-and-jewel type suspensions. This figure shows some construction details, in particular the torsion ribbon with its tension-spring mechanism (courtesy Weston Instruments, Inc.).

to eliminate any sag, as was the case in the suspension galvanometer of Fig. 4-1. This tension is provided by a tension spring, so that the instrument can be used in any position. Generally speaking, taut-band suspension instruments can be made with higher sensitivities than those using pivots and jewels, and they can be used in almost every application presently served by pivoted instruments. Furthermore, taut-band instruments are relatively insensitive to shock and temperature and are capable of withstanding greater overloads than previous types described.

4-3.4   Temperature Compensation

The PMMC basic movement is not inherently insensitive to temperature, but it may be *temperature-compensated* by the appropriate use of series- and shunt-resistors of copper and manganin. Both the magnetic fieldstrength and spring-tension decrease with an increase in temperature. The coil resistance increases with an increase in temperature. These changes tend to make the pointer read low for a given current with respect to magnetic fieldstrength and coil resistance. The spring change, conversely, tends to cause the pointer to read high with an increase in temperature. The effects are not identical, however; hence an *uncompensated meter tends to read low* by approximately 0.2 per cent per °C rise in temperature. For purposes of instrument specification, the movement is considered to be compensated when the change in accuracy, due to a 10°C-change in temperature, is not more than one-fourth of the total allowable error.*

Compensation may be accomplished by using *swamping resistors* in series with the movable coil, as shown in Fig. 4-9(a). The swamping resistor is made of manganin (which has a temperature coefficient of practically zero) combined with copper in the ratio of 20/1 to 30/1. The total resistance of coil and swamping resistor increases slightly with a rise in temperature, but only just enough to counteract the change of springs and magnet, so that the overall temperature effect is zero.

A more complete cancellation of temperature effects is obtained with the arrangement of Fig. 4-9(b). Here the *total* circuit resistance increases slightly with a rise in temperature, owing to the presence of the copper coil and the copper shunt resistor. For a fixed applied voltage, therefore, the total current decreases slightly with a rise in temperature. The resistance of the copper shunt resistor increases more than the series combination of coil and manganin resistor; hence a larger fraction of the total current passes through the coil circuit. By correct proportioning of the copper and manganin parts in the circuit, complete cancellation of temperature effects may be accomplished. One *disadvantage* of the use of swamping resistors is a reduction in the full-

---

PMMC Data Sheets, Weston Instruments, Inc., Newark, N.J.

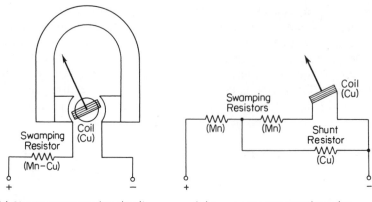

(a) Simple compensation circuit. (b) Improved compensation using
                                      series and shunt resistors.

FIGURE 4-9   Placement of swamping resistors for temperature compensation
of a meter movement.

scale sensitivity of the movement, because a higher applied voltage is necessary to sustain the full-scale current.

4-4   GALVANOMETER SENSITIVITY

Three sensitivity definitions are generally used in specifying the sensitivity of a galvanometer: (a) current sensitivity, (b) voltage sensitivity, (c)
megohm sensitivity.

Current sensitivity may be defined as the ratio of the deflection of the
galvanometer to the current producing this deflection. The current is usually
expressed in microamperes and the deflection in millimeters. For galvanometers that do not have a scale calibrated in millimeters, the deflection may be
given in scale divisions. The current sensitivity is

$$S_I = \frac{d}{I} \frac{\text{mm}}{\mu\text{A}}$$ (4-2)

where d = deflection of the galvanometer in scale divisions or in mm

I = galvanometer current in μA

Voltage sensitivity may be defined as the ratio of the galvanometer
deflection to the voltage producing this deflection. Therefore

$$S_V = \frac{d}{V} \frac{\text{mm}}{\text{mV}}$$ (4-3)

where d = deflection of the galvanometer in scale divisions or in mm

V = voltage applied to the galvanometer in mV

It is customary to consider the galvanometer together with its critical damping resistance (CDRX), and most manufacturers specify the voltage sensitivity of a galvanometer in mm/mV across the CDRX.

Megohm sensitivity may be defined as the number of megohms required in series with the (CDRX shunted) galvanometer to produce one scale division deflection when 1 V is applied to the circuit. Since the equivalent resistance of the shunted galvanometer is negligible compared with the number of megohms in series with it, the applied current practically equals $1/R$ μA and produces one division deflection. Numerically, the megohm sensitivity is equal to the current sensitivity and therefore

$$S_R = \frac{d}{I} = S_I \frac{\text{mm}}{\mu\text{A}} \qquad (4\text{-}4)$$

where d = deflection of the galvanometer in scale divisions or in mm

I = galvanometer current in μA

A fourth sensitivity figure is used with ballistic galvanometers. It is called the *ballistic sensitivity* and is defined as the ratio of the maximum deflection, d_m, of a galvanometer to the quantity Q of electric charge in a single pulse which produces this deflection. Then

$$S_Q = \frac{d_m}{Q} \frac{\text{mm}}{\mu\text{C}} \qquad (4\text{-}5)$$

where d_m = maximum galvanometer deflection in scale divisions or in mm

Q = quantity of electricity in μC

Example 4-1 illustrates a procedure for testing a galvanometer.

Example 4-1: A galvanometer is tested in the circuit of Fig. 4-10,

where $E = 1.5$ V

$R_1 = 1.0 \,\Omega$

$R_2 = 2,500 \,\Omega$

R_3 is variable

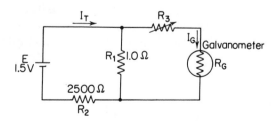

FIGURE 4-10   Circuit for testing a galvanometer.

With R_3 set at 450 Ω, the galvanometer deflection is 150 mm, and with R_3 set at 950 Ω, the deflection is reduced to 75 mm. Calculate (a) the resistance of the galvanometer, (b) the current sensitivity of the galvanometer.

SOLUTION:

(a) The fraction of the total current I_T taken by the galvanometer equals

$$I_G = \frac{R_1}{R_1 + R_3 + R_G} \times I_T$$

Since for $R_3 = 450 \Omega$, the deflection is 150 mm and for $R_3 = 950 \Omega$, the deflection is 75 mm, the galvanometer current I_G in the second case is one-half the galvanometer current in the first case. Therefore, we can write

$$I_{G_1} = 2I_{G_2} \quad \text{or} \quad \frac{1.0}{1.0 + 450 + R_G} = 2\frac{1.0}{1.0 + 950 + R_G}$$

and solving for R_G yields $R_G = 49 \Omega$.

(b) Inspection of the circuit of Fig. 4-10 indicates that the total circuit resistance, R_T, is

$$R_T = R_2 + \frac{R_1(R_3 + R_G)}{R_1 + R_3 + R_G} \approx 2,500 \Omega$$

and therefore

$$I_T = \frac{1.5 \text{ V}}{2,500 \Omega} = 0.6 \text{ mA}$$

For $R_3 = 450 \Omega$, the galvanometer current I_G is

$$I_{G_1} = \frac{R_1}{R_1 + R_3 + R_G}I_T$$

$$= \frac{1.0}{1.0 + 450 + 49} \times 0.6 \text{ mA} = 1.2 \text{ } \mu A$$

and $$S_I = \frac{150 \text{ mm}}{1.2 \text{ } \mu A} = 125 \text{ mm}/\mu A$$

4-5 DC AMMETERS

4-5.1 Shunt Resistor

The basic movement of a dc ammeter is a PMMC galvanometer. Since the coil winding of a basic movement is small and light, it can carry only very small currents. When large currents are to be measured, it is necessary to bypass the major part of the current through a resistance, called a *shunt*, as shown in Fig. 4-11.

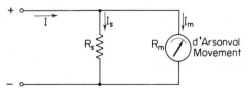

FIGURE 4-11 Basic dc ammeter circuit.

The resistance of the shunt can be calculated by applying conventional circuit analysis to Fig. 4-11,

where R_m = internal resistance of the movement (the coil)

R_s = resistance of the shunt

I_m = full-scale deflection current of the movement

I_s = shunt current

I = full-scale current of the ammeter including the shunt

Since the shunt resistance is in parallel with the meter movement, the voltage drops across the shunt and movement must be the same and we can write

$$V_{\text{shunt}} = V_{\text{movement}}$$

or $$I_s R_s = I_m R_m \quad \text{and} \quad R_s = \frac{I_m R_m}{I_s} \qquad \text{(4-6)}$$

Since $I_s = I - I_m$, we can write

$$R_s = \frac{I_m R_m}{I - I_m} \qquad \text{(4-7)}$$

For each required value of full-scale meter current we can then solve for the value of the shunt resistance required.

Example 4-2: A 1-mA meter movement with an internal resistance of 100 Ω is to be converted into a 0–100-mA ammeter. Calculate the value of the shunt resistance required.

SOLUTION: $I_s = I - I_m = 100 - 1 = 99 \text{ mA}$

$$R_s = \frac{I_m R_m}{I_s} = \frac{1 \text{ mA} \times 100 \text{ }\Omega}{99 \text{ mA}} = 1.01 \text{ }\Omega$$

The shunt resistance used with a basic movement may consist of a length of constant-temperature resistance wire within the case of the instrument or it may be an external (manganin or constantan) shunt having a very low resistance. Figure 4-12 shows an external shunt. It consists of evenly

FIGURE 4-12   High-current shunt for a switchboard instrument (courtesy Weston Instruments, Inc.)

spaced sheets of resistive material welded into a large block of heavy copper on each end of the sheets. The resistance material has a very low temperature coefficient, and a low thermoelectric effect exists between the resistance material and the copper. External shunts of this type are normally used for measuring very large currents.

4-5.2 Ayrton Shunt

The current range of the dc ammeter may be further extended by a number of shunts, selected by a *range switch*. Such a meter is called a *multi-range* ammeter. Figure 4-13 shows the schematic diagram of a multirange

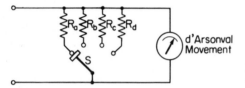

FIGURE 4-13 Schematic diagram of a simple multirange ammeter.

ammeter. The circuit has four shunts, R_a, R_b, R_c, and R_d, which can be placed in parallel with the movement to give four different current ranges. Switch S is a multiposition, *make-before-break* type switch, so that the movement will not be damaged, unprotected in the circuit, without a shunt as the range is changed.

The *universal*, or *Ayrton*, *shunt* of Fig. 4-14 eliminates the possibility of

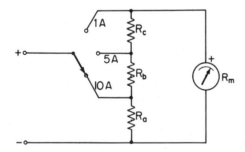

FIGURE 4-14 Universal or Ayrton shunt.

having the meter in the circuit without a shunt. This advantage is gained at the price of a slightly higher overall meter resistance. The Ayrton shunt provides an excellent opportunity to apply basic network theory to a practical circuit.

Example 4-3: Design an Ayrton shunt to provide an ammeter with current ranges of 1 A, 5 A, and 10 A. A d'Arsonval movement with an internal resistance $R_m = 50\ \Omega$ and full-scale deflection current of 1 mA is used in the configuration of Fig. 4-14.

SOLUTION: *On the 1-A range:* $R_a + R_b + R_c$ are in parallel with the 50-Ω movement. Since the movement requires 1 mA for full-scale deflection, the shunt will be required to pass a current of $1\,A - 1\,mA = 999\,mA$. Using Eq. (4-6), we get

$$R_a + R_b + R_c = \frac{1 \times 50}{999} = 0.05005\,\Omega \tag{I}$$

On the 5-A range: $R_a + R_b$ are in parallel with $R_c + R_m$ (50 Ω). In this case there will be a 1-mA current through the movement and R_c in series, and 4,999 mA through $R_a + R_b$. Again using Eq. (4-6), we get

$$R_a + R_b = \frac{1 \times (R_c + 50\,\Omega)}{4,999} \tag{II}$$

On the 10-A range: R_a now serves as the shunt and $R_b + R_c$ are in series with the movement. The current through the movement again is 1 mA, and the shunt passes the remaining 9,999 mA. Using Eq. (4-6) again, we get

$$R_a = \frac{1 \times (R_b + R_c + 50\,\Omega)}{9,999} \tag{III}$$

Solving the three simultaneous equations (I, II, and III), we obtain

$$4,999 \times \text{(I)}: \quad 4,999\,R_a + 4,999\,R_b + 4,999\,R_c = 250.2$$
$$\text{(II)}: \quad 4,999\,R_a + 4,999\,R_b - \qquad R_c = 50$$

Subtracting (II) from (I), we obtain

$$5,000\,R_c = 200.2$$
$$R_c = 0.04004\,\Omega$$

Similarly,

$$9,999 \times \text{(I)}: \quad 9,999\,R_a + 9,999\,R_b + 9,999\,R_c = 500.45$$
$$\text{(III)}: \quad 9,999\,R_a - \qquad R_b - \qquad R_c = 50$$

Subtracting (III) from (I), we obtain

$$10,000\,R_b + 10,000\,R_c = 450.45$$

Substituting the previously calculated value for R_c into this expression yields

$$10,000\,R_b = 450.45 - 400.4$$
$$R_b = 0.005005\,\Omega$$

and

$$R_a = 0.005005\,\Omega$$

This calculation indicates that for larger currents the value of the shunt resistor may become very small.

Direct-current ammeters are commercially available in a large number of ranges, from 20 μA to 50 A full-scale for a self-contained meter and to

500 A for a meter with external shunt. Laboratory-type precision ammeters are provided with a calibration chart, so that the user may correct his readings for any scale errors.

The following precautions should be observed when using an ammeter in measurement work:

(a) *Never* connect an ammeter *across* a source of emf. Because of its low resistance it would draw damaging high currents and destroy the delicate movement. *Always* connect an ammeter in series with a load capable of limiting the current.
(b) Observe the correct *polarity*. Reverse polarity causes the meter to deflect against the mechanical stop and this may damage the pointer.
(c) When using a multirange meter, first use the highest current range; then decrease the current range until substantial deflection is obtained. To increase accuracy of the observation (see Chapter 1), use the range that will give a reading as near to full-scale as possible.

4-6   DC VOLTMETERS

4-6.1   Multiplier Resistor

The addition of a series resistor, or *multiplier*, converts the basic d'Arsonval movement into a *dc voltmeter*, as shown in Fig. 4-15. The multi-

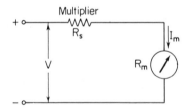

FIGURE 4-15   Basic dc voltmeter circuit.

plier limits the current through the movement so as not to exceed the value of the full-scale deflection current (I_{fsd}). A dc voltmeter measures the potential difference between two points in a dc circuit and is therefore connected *across* a source of emf or a circuit component. The meter terminals are generally marked "pos" and "neg," since polarity must be observed.

The value of a multiplier, required to extend the voltage range, is calculated from Fig. 4-15,

where I_m = deflection current of the movement (I_{fsd})

R_m = internal resistance of the movement

R_s = multiplier resistance

V = full-range voltage of the instrument

For the circuit of Fig. 4-15,

$$V = I_m(R_s + R_m)$$

Solving for R_s gives

$$R_s = \frac{V - I_m R_m}{I_m} = \frac{V}{I_m} - R_m \qquad (4\text{-}8)$$

The multiplier is usually mounted inside the case of the voltmeter for moderate ranges up to 500 V. For higher voltages, the multiplier may be mounted separately outside the case on a pair of binding posts to avoid excessive heating inside the case.

4-6.2 Multirange Voltmeter

The addition of a number of multipliers, together with a *range switch*, provides the instrument with a workable number of voltage ranges. Figure 4-16 shows a *multirange* voltmeter using a four-position switch and four

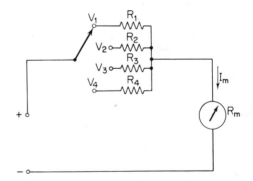

FIGURE 4-16   Multirange volt-meter.

multipliers, R_1, R_2, R_3, and R_4, for the voltage ranges V_1, V_2, V_3, and V_4, respectively. The values of the multipliers can be calculated using the method shown earlier or, alternatively, by the *sensitivity method*. The sensitivity method is illustrated by Example 4-5 in Sec. 4-7, where sensitivity is discussed.

A variation of the circuit of Fig. 4-16 is shown in Fig. 4-17, where the multipliers are connected in a series string and the range selector switches the appropriate amount of resistance in series with the movement. This system has the advantage that all multipliers except the first have standard resistance values and can be obtained commercially in precision tolerances. The low-

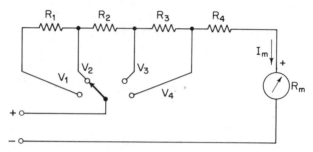

FIGURE 4-17   A more practical arrangement of multiplier resistors in the multi-range voltmeter.

range multiplier, R_4, is the only special resistor that must be manufactured to meet the specific circuit requirements.

Example 4-4: A basic d'Arsonval movement with internal resistance, $R_m = 100\ \Omega$, and full-scale current, $I_{fsd} = 1$ mA, is to be converted into a multirange dc voltmeter with voltage ranges of 0–10 V, 0–50 V, 0–250 V, and 0–500 V.

The circuit arrangement of Fig. 4-17 is to be used for this voltmeter.

SOLUTION: For the 10-V range (V_4 position of range switch), the total circuit resistance is

$$R_T = \frac{10\ \text{V}}{1\ \text{mA}} = 10\ \text{k}\,\Omega$$

$$R_4 = R_T - R_m = 10\ \text{k}\Omega - 100\ \Omega = 9{,}900\ \Omega$$

For the 50-V range (V_3 position of range switch),

$$R_T = \frac{50\ \text{V}}{1\ \text{mA}} = 50\ \text{k}\Omega$$

$$R_3 = R_T - (R_4 + R_m) = 50\ \text{k}\Omega - 10\ \text{k}\Omega = 40\ \text{k}\Omega$$

For the 250-V range (V_2 position of range switch),

$$R_T = \frac{250\ \text{V}}{1\ \text{mA}} = 250\ \text{k}\Omega$$

$$R_2 = R_T - (R_3 + R_4 + R_m) = 250\ \text{k}\Omega - 50\ \text{k}\Omega = 200\ \text{k}\Omega$$

For the 500-V range (V_1 position of range switch),

$$R_T = \frac{500\ \text{V}}{1\ \text{mA}} = 500\ \text{k}\Omega$$

$$R_1 = R_T - (R_2 + R_3 + R_4 + R_m) = 500\ \text{k}\Omega - 250\ \text{k}\Omega = 250\ \text{k}\Omega$$

Notice in Example 4-4 that only the low-range multiplier R_4 has a non-standard value.

4-7 VOLTMETER SENSITIVITY

4-7.1 Ohms-per-Volt Rating

In Sec. 4-6 it was shown that the full-scale deflection current I_{fsd} was reached on all voltage ranges when the corresponding full-scale voltage was applied. As shown in Example 4-4, a current of 1 mA is obtained for voltages of 10 V, 50 V, 250 V, and 500 V across the meter terminals. For each voltage range, the quotient of the total circuit resistance R_T and the range voltage V is always 1,000 Ω/V. This figure is often referred to as the *sensitivity*, or the *ohms-per-volt rating*, of the voltmeter. Note that the sensitivity, S, is essentially the *reciprocal* of the full-scale deflection current of the basic movement, or

$$S = \frac{1}{I_{fsd}} \frac{\Omega}{V} \qquad (4\text{-}9)$$

The sensitivity S of the voltmeter may be used to advantage in the *sensitivity method* of calculating the resistance of the multiplier in a dc voltmeter. Consider the circuit of Fig. 4-17,

where S = sensitivity of the voltmeter, in Ω/V

          V = the voltage range, as set by the range switch

      R_m = internal resistance of the movement (plus the previous series resistors)

      R_s = resistance of the multiplier

For the circuit of Fig. 4-17,

$$R_T = S \times V$$

and $R_s = (S \times V) - R_m$ **(4-10)**

Use of the sensitivity method is illustrated in Example 4-5.

Example 4-5: Repeat Example 4-4, now using the sensitivity method for calculating the multiplier resistances.

SOLUTION:

$$S = \frac{1}{I_{fsd}} = \frac{1}{0.001\ A} = 1{,}000\frac{\Omega}{V}$$

$$R_4 = (S \times V) - R_m = \frac{1{,}000\ \Omega}{V} \times 10\ V - 100\ \Omega = 9{,}900\ \Omega$$

$$R_3 = (S \times V) - R_m = \frac{1{,}000\ \Omega}{V} \times 50\ V - 10{,}000\ \Omega = 40\ k\Omega$$

$$R_2 = (S \times V) - R_m = \frac{1{,}000\ \Omega}{V} \times 250\ V - 50\ k\Omega = 200\ k\Omega$$

$$R_1 = (S \times V) - R_m = \frac{1{,}000\ \Omega}{V} \times 500\ V - 250\ k\Omega = 250\ k\Omega$$

4-7.2 Loading Effect

The sensitivity of a dc voltmeter is an important factor when selecting a meter for a certain voltage measurement. A low-sensitivity meter may give correct readings when measuring voltages in low-resistance circuits, but it is certain to produce very unreliable readings in high-resistance circuits. A voltmeter, when connected across two points in a highly resistive circuit, acts as a shunt for that portion of the circuit and thus reduces the equivalent resistance in that portion of the circuit. The meter will then give a lower indication of the voltage drop than actually existed before the meter was connected. This effect is called the *loading effect* of an instrument; it is caused principally by *low-sensitivity* instruments. The loading effect of a voltmeter is illustrated in Example 4-6.

Example 4-6: It is desired to measure the voltage across the 50-kΩ resistor in the circuit of Fig. 4-18. Two voltmeters are available for this measurement: voltmeter 1 with a sensitivity of 1,000 Ω/V and voltmeter 2 with a sensitivity of 20,000 Ω/V. Both meters are used on their 50-V range. Calculate (a) the reading of each meter; (b) the error in each reading, expressed as a percentage of the true value.

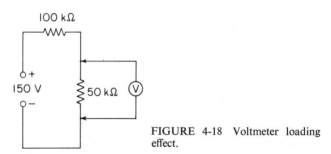

FIGURE 4-18 Voltmeter loading effect.

SOLUTION: Inspection of the circuit indicates that the voltage across the 50-kΩ resistor is

$$\frac{50 \text{ k}\Omega}{150 \text{ k}\Omega} \times 150 \text{ V} = 50 \text{ V}$$

This is the *true* value of voltage across the 50-kΩ resistor.

(a) *Voltmeter 1* ($S = 1,000 \ \Omega/V$) has a resistance of 50 V $\times$ 1,000 Ω/V $= 50$ kΩ on its 50-V range. Connecting the meter across the 50-kΩ resistor causes the equivalent parallel resistance to be decreased to 25 kΩ and the total circuit resistance to 125 kΩ. The potential difference across the combination of meter and 50-kΩ resistor is

$$V_1 = \frac{25 \text{ k}\Omega}{125 \text{ k}\Omega} \times 150 \text{ V} = 30 \text{ V}$$

Hence the voltmeter indicates a voltage of 30 V. *Voltmeter 2* ($S = 20 \text{ k}\Omega/V$)

has a resistance of 50 V $\times$ 20 kΩ/V = 1 mΩ on its 50-V range. When this meter is connected across the 50-kΩ resistor, the equivalent parallel resistance equals 47.6 kΩ. This combination produces a voltage of

$$V_2 = \frac{47.6 \text{ k}\Omega}{147.6 \text{ k}\Omega} \times 150 \text{ V} = 48.36 \text{ V}$$

which is indicated on the voltmeter.

(b) The error in the reading of voltmeter 1 is

$$\% \text{ error} = \frac{\text{true voltage} - \text{apparent voltage}}{\text{true voltage}} \times 100\%$$

$$= \frac{50 \text{ V} - 30 \text{ V}}{50 \text{ V}} \times 100\% = 40\%$$

The error in the reading of voltmeter 2 is

$$\% \text{ error} = \frac{50 \text{ V} - 48.36 \text{ V}}{50 \text{ V}} \times 100\% = 3.28\%$$

The calculation of Example 4-6 indicates that the meter with the higher sensitivity or ohms-per-volt rating gives the most reliable result. It is important to realize the factor of sensitivity, particularly when voltage measurements are made in high-resistance circuits.

The matter of reliability and accuracy of the test result raises an interesting point. When an *insensitive, yet highly accurate,* dc voltmeter is placed across the terminals of a high resistance, the meter accurately reflects the voltage condition produced by loading. The error is a human or gross error (Sec. 1-4), because the proper instrument was not selected. The meter "disturbs" the circuit, and the ideal of instrumentation, at all times, is to measure a condition without affecting it in any way. The human investigator has the responsibility to select an instrument which is precise, reliable, and sufficiently sensitive not to disturb that which is being measured. The fault lies not with the highly accurate instrument but with the investigator, who is using it incorrectly. In fact, the sophisticated instrument user could calculate the true voltage by using an insensitive yet accurate meter. Therefore *accuracy* is always required in instruments; *sensitivity* is needed only in special applications where loading disturbs that which is being measured. Example 4-7 illustrates how an insensitive yet accurate instrument is used to perform an entirely valid measurement.

Example 4-7: The only voltmeter available in a laboratory has a sensitivity of 100 Ω/V and three scales, 50 V, 150 V, and 300 V. When connected in the circuit of Fig. 4-19, the meter reads 4.65 V on its lowest (50-V) scale. Calculate the value of R_x.

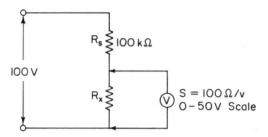

FIGURE 4-19   Use of an accurate but insensitive voltmeter to determine the resistance of R_x.

SOLUTION: The equivalent resistance of the voltmeter on its 50-V scale is

$$R_V = 100\frac{\Omega}{V} \times 10\ V = 5\ k\Omega$$

Let R_p = the parallel resistance of R_x and R_V.

$$R_p = \frac{V_p}{V_s} \times R_s = \frac{4.65}{95.35} \times 100\ k\Omega = 4.878\ k\Omega$$

Then

$$R_x = \frac{R_p \times R_V}{R_V - R_p} = \frac{4.878\ k\Omega \times 5\ k\Omega}{0.122\ k\Omega} = 200\ k\Omega$$

Example 4-7 shows that when the instrument user is aware of the limitations of his instrument, he can still make allowances provided that the voltmeter is accurate.

The following general precautions should be observed when using a voltmeter:

(a) Observe the correct polarity. Wrong polarity causes the meter to deflect against the mechanical stop and this may damage the pointer.
(b) Place the voltmeter *across* the circuit or component whose voltage is to be measured.
(c) When using a multirange voltmeter, always use the highest voltage range and then decrease the range until a good up-scale reading is obtained.
(d) Always be aware of the loading effect. The effect can be minimized by using as high a voltage range (and highest sensitivity) as possible. The precision of measurement decreases if the indication is at the low end of the scale (Sec. 1-4).

4-8   VOLTMETER-AMMETER METHOD

A popular type of resistance measurement involves the *voltmeter-ammeter method,* since the instruments required are usrally available in most laboratories. If the voltage V across the resistor and the current I through the resistor are measured, the unknown resistance R_x can be calculated by Ohm's law:

$$R_x = \frac{V}{I} \tag{4-11}$$

Equation (4-11) implies that the ammeter resistance is zero and the voltmeter resistance infinite, so that the conditions in the circuit are not disturbed.

In Fig. 4-20(a) the *true* current supplied to the load is measured by the ammeter, but the voltmeter measures the supply voltage rather than the actual load voltage. To find the true voltage across the load, the voltage drop across the ammeter must be subtracted from the voltmeter reading. If the voltmeter is placed directly across the resistor, as in Fig. 4-20(b), it measures the true load voltage, but the ammeter is in error by the amount of current drawn by the voltmeter. In either situation, an *error* is introduced in the measurement of R_x. The correct method of connecting the voltmeter depends on the value of R_x and the resistances of the voltmeter and the ammeter. In general, the ammeter resistance is low and the voltmeter resistance is high.

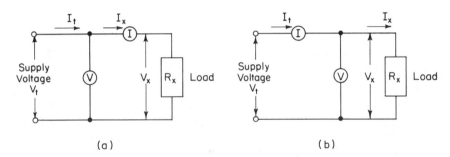

(a)   (b)

FIGURE 4-20   Effect of voltmeter and ammeter positions in voltmeter-ammeter measurements.

In Fig. 4-20(a) the ammeter reads the true value of the load current (I_x), and the voltmeter measures the supply voltage (V_t). If R_x is large compared to the internal resistance of the ammeter, the error introduced by neglecting the voltage drop across the ammeter is negligible and V_t is very close to the true load voltage V_x. The connection of Fig. 4-20(a) is therefore the best circuit when measuring *high-resistance values*.

In Fig. 4-20(b) the voltmeter reads the true value of the load voltage (V_x) and the ammeter reads the supply current (I_t). If R_x is small compared to the internal resistance of the voltmeter, the current drawn by the voltmeter does not appreciably affect the total supply current and I_t is very close to the true value of the load current I_x. The connection of Fig. 4-20(b) therefore is the best circuit when measuring *low-resistance values*.

Given an unknown value of R_x, how then can we determine if the voltmeter is connected in the right place? Consider the circuit of Fig. 4-21 in

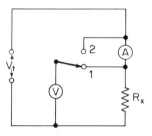

FIGURE 4-21   Effect of the volt-meter position in a voltmeter-am-meter measurement.

which the voltmeter and the ammeter may be connected in two different ways for simultaneous readings. The procedure is as follows:

(a) Connect the voltmeter across R_x with the switch in position 1, and observe the ammeter reading.
(b) Now switch the voltmeter to position 2. If the ammeter reading does not change, restore the voltmeter to position 1. The symptoms indicate a low-resistance measurement. Record both current and voltage readings and calculate R_x from Eq. (4-11).
(c) If the ammeter reading decreases when the voltmeter is changed from position 1 to position 2, leave the voltmeter at position 2. The symptoms indicate a high-resistance measurement. Record both current and voltage readings and calculate R_x from Eq. (4-11).

Voltage measurements in electronic circuits are generally made with multirange voltmeters or multimeters, with sensitivities in the range of 20 kΩ/V to 50 kΩ/V. In power measurements, where the current is usually large, voltmeters may have sensitivities as low as 100 Ω/V. Ammeter resistances depend on the design of the coil and are generally larger for low values of full-scale currents. A few typical values of ammeter resistance are given in Table 4-1.

Table 4-1

TYPICAL VALUES OF AMMETER RESISTANCE*

Full-scale current	Resistance (ohms)	
(µA)	Pivot and jewel	Taut-band
50	2,000–5,000	1,000–2,000
500	200–1,000	100–250
1,000	50–120	30–90
10,000	2–4	1–3

NOTE: Current ranges in excess of 30 mA are usually shunted.
*Data Sheets, Weston Instruments, Inc., Newark, N.J.

4-9 SERIES-TYPE OHMMETER

The series-type ohmmeter essentially consists of a d'Arsonval movement connected in series with a resistance and a battery to a pair of terminals to which the unknown is connected. The current through the movement then depends on the magnitude of the unknown resistor, and the meter indication is proportional to the value of the unknown, provided that calibration problems are taken into account. Figure 4-22 shows the elements of a simple single-range series ohmmeter.

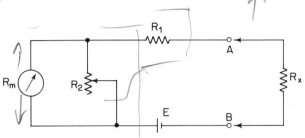

FIGURE 4-22   Series-type ohmmeter.

In Fig. 4-22, R_1 = current limiting resistor

$\qquad\qquad R_2$ = zero adjust resistor

$\qquad\qquad E$ = internal battery

$\qquad\qquad R_m$ = internal resistance of the d'Arsonval movement

$\qquad\qquad R_x$ = unknown resistor

When the unknown resistor $R_x = 0$ (terminals A and B shorted), maximum current flows in the circuit. Under this condition, shunt resistor R_2 is adjusted until the movement indicates full-scale current (I_{fsd}). The full-scale current position of the pointer is marked "0 Ω" on the scale. Similarly, when $R_x = \infty$ (terminals A and B open), the current in the circuit drops to zero and the movement indicates zero current, which is then marked "∞" on the scale. Intermediate markings may be placed on the scale by connecting different known values of R_x to the instrument. The accuracy of these scale markings depends on the repeating accuracy of the movement and the tolerances of the calibrating resistors.

Although the series-type ohmmeter is a popular design and is used extensively in portable instruments for general-service work, it has certain disadvantages. Important among these is the internal battery whose voltage decreases gradually with time and age, so that the full-scale current drops and the meter does not read "0" when A and B are shorted. The variable shunt resistor R_2 in Fig. 4-22 provides an adjustment to counteract the effect of battery change. Without R_2, it would be possible to bring the pointer back to full scale by adjusting R_1, but this would change the calibration all along the scale. Adjustment by R_2 is a superior solution, since the parallel resistance of R_2 and the coil R_m is always low compared to R_1 and therefore the change in R_2 needed for adjustment does not change the calibration very much. The circuit of Fig. 4-22 does not compensate completely for aging of the battery, but it does a reasonably good job within the expected limits of accuracy of the instrument.

A convenient quantity to use in the design of a series-type ohmmeter is the value of R_x which causes half-scale deflection of the meter. At this position, the resistance across terminals A and B is defined as the half-scale position resistance R_h. Given the full-scale current I_{fsd} and the internal resistance of the movement R_m, the battery voltage E, and the desired value of the half-scale resistance R_h, the circuit can be analyzed; i.e., values can be found for R_1 and R_2.

The design can be approached by recognizing that, if introducing R_h reduces the meter current to $\frac{1}{2}I_{fsd}$, the unknown resistance must be equal to the total internal resistance of the ohmmeter. Therefore

$$R_h = R_1 + \frac{R_2 R_m}{R_2 + R_m} \tag{4-12}$$

The total resistance presented to the battery then equals $2R_h$, and the battery current needed to supply the half-scale deflection is

$$I_h = \frac{E}{2R_h} \tag{4-13}$$

To produce full-scale deflection, the battery current must be doubled, and therefore

$$I_t = 2I_h = \frac{E}{R_h} \tag{4-14}$$

The shunt current through R_2 is

$$I_2 = I_t - I_{fsd} \tag{4-15}$$

The voltage across the shunt (E_{sh}) is equal to the voltage across the movement and

$$E_{sh} = E_m \quad \text{or} \quad I_2 R_2 = I_{fsd} R_m$$

and

$$R_2 = \frac{I_{fsd} R_m}{I_2} \tag{4-16}$$

Substituting Eq. (4-15) into Eq. (4-16), we obtain

$$R_2 = \frac{I_{fsd} R_m}{I_t - I_{fsd}} = \frac{I_{fsd} R_m R_h}{E - I_{fsd} R_h} \tag{4-17}$$

Solving Eq. (4-12) for R_1 gives

$$R_1 = R_h - \frac{R_2 R_m}{R_2 + R_m} \tag{4-18}$$

Substituting Eq. (4-17) into Eq. (4-18) and solving for R_1 yields

$$R_1 = R_h - \frac{I_{fsd} R_m R_h}{E} \qquad (4\text{-}19)$$

A typical calculation for the series-type ohmmeter is given in Example 4-8.

Example 4-8: The ohmmeter of Fig. 4-22 uses a 50-Ω basic movement requiring a full-scale current of 1 mA. The internal battery voltage is 3 V. The desired scale marking for half-scale deflection is 2,000 Ω. Calculate (a) the values of R_1 and R_2; (b) the maximum value of R_2 to compensate for a 10% drop in battery voltage; (c) the scale error at the half-scale mark (2,000 Ω) when R_2 is set as in (b).

SOLUTION:

(a) The total battery current at full-scale deflection is

$$I_t = \frac{E}{R_h} = \frac{3\text{ V}}{2{,}000\ \Omega} = 1.5\text{ mA} \qquad (4\text{-}14)$$

The current through the zero-adjust resistor R_2 then is

$$I_2 = I_t - I_{fsd} = 1.5\text{ mA} - 1\text{ mA} = 0.5\text{ mA} \qquad (4\text{-}15)$$

The value of the zero-adjust resistor R_2 is

$$R_2 = \frac{I_{fsd} R_m}{I_2} = \frac{1\text{ mA} \times 50\ \Omega}{0.5\text{ mA}} = 100\ \Omega \qquad (4\text{-}16)$$

The parallel resistance of the movement and the shunt (R_p) is

$$R_p = \frac{R_2 R_m}{R_2 + R_m} = \frac{50 \times 100}{150} = 33.3\ \Omega$$

The value of the current-limiting resistor R_1 is

$$R_1 = R_h - R_p = 2{,}000 - 33.3 = 1{,}966.7\ \Omega$$

(b) At a 10% drop in battery voltage

$$E = 3\text{ V} - 0.3\text{ V} = 2.7\text{ V}$$

The total battery current I_t then becomes

$$I_t = \frac{E}{R_h} = \frac{2.7\text{ V}}{2{,}000\ \Omega} = 1.35\text{ mA}$$

The shunt current I_2 is

$$I_2 = I_t - I_{fsd} = 1.35\text{ mA} - 1\text{ mA} = 0.35\text{ mA}$$

and the zero-adjust resistor R_2 equals

$$R_2 = \frac{I_{fsd} R_m}{I_2} = \frac{1\text{ mA} \times 50\ \Omega}{0.35\text{ mA}} = 143\ \Omega$$

(c) The parallel resistance of the meter movement and the new value of R_2 becomes

$$R_p = \frac{R_2 R_m}{R_2 + R_m} = \frac{50 \times 143}{193} = 37\ \Omega$$

Since the half-scale resistance R_h is equal to the total internal circuit resistance, R_h will increase to

$$R_h = R_1 + R_p = 1,966.7 \ \Omega + 37 \ \Omega = 2,003.7 \ \Omega$$

Therefore the true value of the half-scale mark on the meter is $2,003.7 \ \Omega$ whereas the actual scale mark is $2,000 \ \Omega$. The percentage error is then

$$\% \text{ error} = \frac{2,000 - 2,003.7}{2,003.7} \times 100\% = -0.185\%$$

The negative sign indicates that the meter reading is low.

The ohmmeter of Example 4-8 could be designed for other values of R_h, within limits. If $R_h = 3,000 \ \Omega$, the battery current would be 1 mA, which is required for the full-scale deflection current. If the battery voltage would decrease owing to aging, the total battery current would fall below 1 mA and there would then be no provision for adjustment.

4-10 SHUNT-TYPE OHMMETER

The circuit diagram of a *shunt-type ohmmeter* is shown in Fig. 4-23. It consists of a battery in series with an adjustable resistor R_1 and a d'Arsonval movement. The unknown resistance is connected across terminals A and B, in

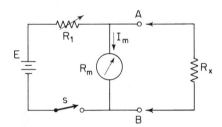

FIGURE 4-23 Shunt-type ohmmeter.

parallel with the meter. In this circuit it is necessary to have an *off-on switch* to disconnect the battery from the circuit when the instrument is not used. When the unknown resistor $R_x = 0 \ \Omega$ (A and B shorted), the meter current is zero. If the unknown resistor $R_x = \infty$ (A and B open), the current finds a path only through the meter, and by appropriate selection of the value of R_1, the pointer can be made to read full scale. This ohmmeter therefore has the "zero" mark at the left-hand side of the scale (no current) and the "infinite" mark at the right-hand side of the scale (full-scale deflection current).

The shunt-type ohmmeter is particularly suited to the measurement of *low-value resistors*. It is not a commonly used test instrument, but it is found in laboratories or for special low-resistance applications.

The analysis of the shunt-type ohmmeter is similar to that of the series-type ohmmeter (Sec. 4-9). In Fig. 4-23, when $R_x = \infty$, the full-scale meter current will be

$$I_{\text{fsd}} = \frac{E}{R_1 + R_m} \tag{4-20}$$

where $E =$ internal battery voltage

$R_1 =$ current-limiting resistor

$R_m =$ internal resistance of the movement

Solving for R_1, we find

$$R_1 = \frac{E}{I_{\text{fsd}}} - R_m \tag{4-21}$$

For any value of R_x connected across the meter terminals, the meter current decreases and is given by

$$I_m = \left\{ \frac{E}{R_1 + [R_m R_x/(R_m + R_x)]} \right\} \times \frac{R_x}{R_m + R_x}$$

or

$$I_m = \frac{ER_x}{R_1 R_m + R_x(R_1 + R_m)} \tag{4-22}$$

The meter current for any value of R_x, expressed as a fraction of the full-scale current, is

$$s = \frac{I_m}{I_{\text{fsd}}} = \frac{R_x(R_1 + R_m)}{R_1(R_m + R_x) + R_m R_x}$$

or

$$s = \frac{R_x(R_1 + R_m)}{R_x(R_1 + R_m) + R_1 R_m} \tag{4-23}$$

Defining

$$\frac{R_1 R_m}{R_1 + R_m} = R_p \tag{4-24}$$

and substituting Eq. (4-24) into Eq. (4-23), we obtain

$$s = \frac{R_x}{R_x + R_p} \tag{4-25}$$

If Eq. (4-25) is used, the meter can be calibrated by calculating s in terms of R_x and R_p.

At half-scale reading of the meter ($I_m = 0.5\, I_{\text{fsd}}$), Eq. (4-22) reduces to

$$0.5\, I_{\text{fsd}} = \frac{ER_h}{R_1 R_m + R_h(R_1 + R_m)} \tag{4-26}$$

where $R_h =$ external resistance causing half-scale deflection. To determine the relative scale values for a given value of R_1, the half-scale reading may be found by dividing Eq. (4-20) by Eq. (4-26) and solving for R_h,

$$R_h = \frac{R_1 R_m}{R_1 + R_m} \tag{4-27}$$

The analysis shows that the half-scale resistance is determined by limiting resistor R_1 and the internal resistance of the movement, R_m. The limiting resistance, R_1, is in turn determined by the meter resistance, R_m, and the full-scale deflection current, I_{fsd}.

To illustrate that the shunt-type ohmmeter is particularly suited to the measurement of very low resistances, consider Example 4-9.

Example 4-9: The circuit of Fig. 4-23 uses a 10-mA basic d'Arsonval movement with an internal resistance of 5 Ω. The battery voltage $E = 3$ V. It is desired to modify the circuit by adding an appropriate resistor R_{sh} across the movement, so that the instrument will indicate 0.5 Ω at the midpoint on its scale. Calculate (a) the value of the shunt resistor, R_{sh}; (b) the value of the current-limiting resistor, R_1.

SOLUTION:
(a) For half-scale deflection of the movement,

$$I_m = 0.5\, I_{fsd} = 5 \text{ mA}$$

The voltage across the movement is

$$E_m = 5 \text{ mA} \times 5\,\Omega = 25 \text{ mA}$$

Since this voltage also appears across the unknown resistor, R_x, the current through R_x is

$$I_x = \frac{25 \text{ mV}}{0.5\,\Omega} = 50 \text{ mA}$$

The current through the movement (I_m) plus the current through the shunt (I_{sh}) must be equal to the current through the unknown (I_x). Therefore

$$I_{sh} = I_x - I_m = 50 \text{ mA} - 5 \text{ mA} = 45 \text{ mA}$$

The shunt resistance then is

$$R_{sh} = \frac{E_m}{I_{sh}} = \frac{25 \text{ mV}}{45 \text{ mA}} = \frac{5}{9}\,\Omega$$

(b) The total battery current is

$$I_t = I_m + I_{sh} + I_x = 5 \text{ mA} + 45 \text{ mA} + 50 \text{ mA} = 100 \text{ mA}$$

The voltage drop across limiting resistor R_1 equals 3 V $-$ 25 mV $=$ 2.975 V. Therefore

$$R_1 = \frac{2.975 \text{ V}}{100 \text{ mA}} = 29.75\,\Omega$$

4-11 MULTIMETER OR VOM

The ammeter, the voltmeter, and the ohmmeter all use a d'Arsonval movement. The difference between these instruments is the circuit in which the basic movement is used. It is therefore obvious that a single instrument can be designed to perform the three measurement functions. This instrument,

which contains a *function switch* to connect the appropriate circuits to the
d'Arsonval movement, is often called a *multimeter* or *volt-ohm-milliammeter*
(VOM).

A representative example of a commercial multimeter is shown in Fig.
4-24. The circuit diagram of this meter is given in Fig. 4-25. The meter is a

FIGURE 4-24   General-purpose multimeter: Simpson Model 260 (courtesy Simpson Electric Company).

combination of a dc milliammeter, a dc voltmeter, an ac voltmeter, a multirange ohmmeter, and an output meter. (The circuits of the ac voltmeter and
the output meter are discussed in Sec. 5-4.)

Figure 4-26 shows the circuit for the dc voltmeter section, where the
common input terminals are used for voltage ranges of 0–1.5 V to 0–1,000 V.
An external voltage jack, marked "DC 5,000 V," is used for dc voltage measurements to 5,000 V. The operation of this circuit is similar to the circuit of
Fig. 4-15, which was discussed in Sec. 4-6.

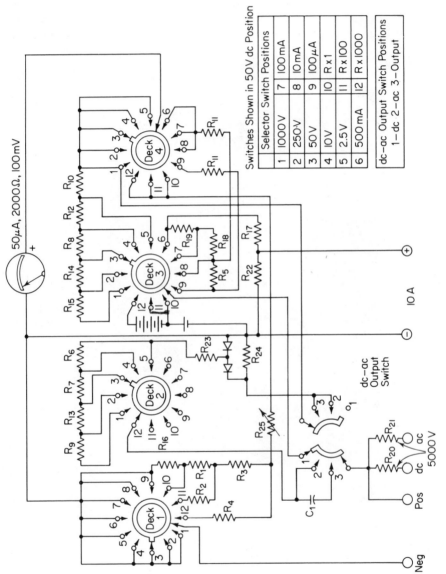

FIGURE 4-25 Schematic diagram of the Simpson Model 260 multimeter (courtesy Simpson Electric Company).

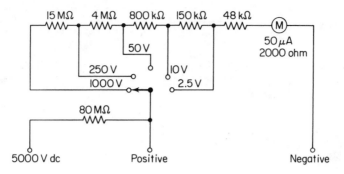

FIGURE 4-26   dc voltmeter section of the Simpson Model 260 multimeter (courtesy Simpson Electric Company).

The basic movement of the multimeter of Fig. 4-24 has a full-scale current of 50 μA and an internal resistance of 2,000 Ω. The values of the multipliers are given in Fig. 4-26. Notice that on the 5,000-V range, the range switch should be set to the 1,000-V position, but the test lead should be connected to the external jack marked "DC 5,000 V." The normal precautions for measuring voltage should be taken. Because of its fairly high sensitivity (20 kΩ/V), the instrument is suited to general-service work in the electronics field.

The circuit for measuring dc milliamperes and amperes is given in Fig. 4-27 and again the circuit is self-explanatory. The "common" (+) and

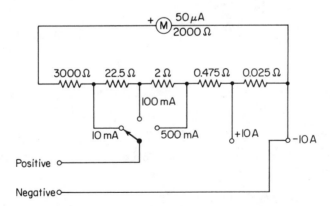

FIGURE 4-27   dc ammeter section of the Simpson Model 260 multimeter (courtesy Simpson Electric Company).

"negative" (−) terminals are used for current measurements up to 500 mA and the jacks marked "+10 A" and "−10 A" are used for the 0–10-A range.

Details of the ohmmeter section of the VOM are shown in Fig. 4-28. The circuit in Fig. 4-28(a) gives the ohmmeter circuit for a scale multiplication of

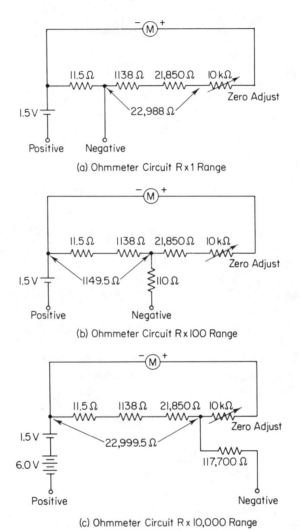

(a) Ohmmeter Circuit R x 1 Range

(b) Ohmmeter Circuit R x 100 Range

(c) Ohmmeter Circuit R x 10,000 Range

FIGURE 4-28   Ohmmeter section of the Simpson Model 260 multimeter (courtesy Simpson Electric Company).

1. Before any measurement is made, the instrument is short-circuited and the "zero-adjust" control is varied until the meter reads zero resistance (full-scale current). Notice that the circuit takes the form of a variation of the shunt-type ohmmeter. Scale multiplications of 100 and 10,000 are shown in Fig. 4-28(b), (c).

The ac voltmeter section of the meter is selected by setting the "ac-dc" switch to the "ac" position. The operation of this circuit is discussed in Sec. 5-4.

4-12   CALIBRATION OF DC INSTRUMENTS

Although detailed calibration techniques are beyond the scope of this chapter, some general procedures for the calibration of basic dc instruments are given.

Calibration of a *dc ammeter* can most easily be carried out by the arrangement of Fig. 4-29. The value of the current through the ammeter to be

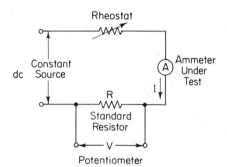

FIGURE   4-29   Potentiometer
method of calibrating a dc ammeter.

calibrated is determined by measuring the potential difference across a standard resistor by the *potentiometer method* and then calculating the current by Ohm's law. The result of this calculation is compared to the actual reading of the ammeter under calibration and inserted in the circuit. (Voltage measurements by the potentiometer method are discussed in some detail in Sec. 6-6.) A good source of constant current is required and is usually provided by storage cells or a precision power supply. A rheostat is placed in the circuit to control the current to any desired value, so that different points on the meter scale can be calibrated.

A simple method of calibrating a *dc voltmeter* is shown in Fig. 4-30,

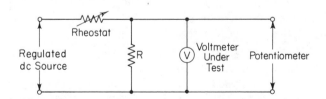

FIGURE 4-30   Potentiometer method of calibrating a dc voltmeter.

where the voltage across dropping resistor R is accurately measured with a potentiometer. The meter to be calibrated is connected across the same two points as the potentiometer and should therefore indicate the same voltage. A rheostat is placed in the circuit to control the amount of current and therefore the drop across the resistor, R, so that several points on the voltmeter scale can be calibrated. Voltmeters tested with the method of Fig. 4-30 can be calibrated with an accuracy of ± 0.01 per cent, which is well beyond the usual accuracy of a d'Arsonval movement.

The *ohmmeter* is generally considered to be an instrument of moderate accuracy and low precision. A rough calibration may be done by measuring a standard resistance and noting the reading of the ohmmeter. Doing this for several points on the ohmmeter scale and on several ranges allows one to obtain an indication of the correct operation of the instrument. Precision measurements of resistance are normally carried out by one of the bridge methods, discussed in some detail in Chapter 7.

REFERENCES

1. Bartholomew, Davis, *Electrical Measurements and Instrumentation*, chap. 4. Boston: Allyn and Bacon, Inc., 1963.

2. Frank, Ernest, *Electrical Measurement Analysis*, chap. 8. New York: McGraw-Hill Book Company, Inc., 1959.

3. Stout, Melville B., *Basic Electrical Measurements*, 2nd ed., chaps. 4, 5, 17. Englewood Cliffs, N.J.: Prentice-Hall, Inc., 1960.

4. *The Instrument Sketchbook*. Weston Instruments, Inc., Newark, N.J., 1966.

PROBLEMS

1. Determine the full-scale voltage indicated by a 500-μA meter movement with an internal resistance of 250 Ω if no multiplier is used.

2. Design a multirange dc ammeter with ranges of 0–10 mA, 0–50 mA, 0–100 mA, and 0–500 mA. A d'Arsonval movement is used with an internal resistance $R_m = 50\ \Omega$ and a full-scale deflection current $I_{fsd} = 1$ mA.
   (a) Calculate the values of the required shunts.
   (b) Draw the complete circuit diagram.

3. A simple shunted dc ammeter, using a basic meter movement with an internal resistance, $R_m = 1{,}800\ \Omega$, and a full-scale deflection current, $I_{fsd} = 100\ \mu$A, is connected into a circuit and gives a reading of 3.5 mA on its 5-mA range. This reading is checked by a recently calibrated dc ammeter which gives a reading of 4.1 mA. The implication is that the first ammeter has a faulty shunt on its 5-mA range. Calculate (a) the actual value of the faulty shunt; (b) the correct shunt for the 5-mA range.

4. Design an Ayrton shunt for a meter movement with internal resistance $R_m = 2{,}500\ \Omega$ and full-scale deflection current $I_{fsd} = 50\ \mu$A to provide current ranges of 50 μA, 100 μA, 500 μA, 10 mA, and 100 mA.

(a) Calculate the resistances of the Ayrton shunt components.

(b) Draw the schematic diagram, including the switching arrangement, for this multirange dc ammeter.

5. It is desired to convert a 50-μA dc movement with internal resistance of 1,000 Ω to a 0–2,500-V dc voltmeter. Calculate (a) the resistance of the multiplier; (b) the sensitivity of the instrument.

6. An existing 0–200-V dc voltmeter has a sensitivity of 1,000 Ω/V. Determine the value of additional series resistance required to convert this voltmeter to an instrument with a range of 0–1,000 V dc.

7. Using a 50-μA movement with internal resistance of 1,500 Ω, design a multirange voltmeter with ranges of 0–5 V, 0–10 V, 0–50 V, and 0–100 V. Calculate (a) the values of the multipliers; (b) the sensitivity of the instrument. Draw the circuit diagram of the completed design.

8. A dc microammeter with an internal resistance of 250 Ω and a full-scale deflection current of 500 μA indicates a current of 300 μA when connected into a circuit consisting of a 1.5-V dry cell and an unknown resistance. Determine the value of the unknown resistance.

9. Design a series-type ohmmeter, similar to the circuit of Fig. 4-22. The movement to be used requires 0.5 mA for full-scale deflection and has an internal resistance of 50 Ω. The internal battery has a voltage of 3.0 V. The desired value of half-scale resistance is 3,000 Ω. Calculate (a) the value of resistors R_1 and R_2; (b) the range of values of R_2, if the battery voltage may vary from 2.7 V to 3.1 V. Use the value of R_1 as calculated in (a).

10. A series-type ohmmeter, designed to operate with a 6-V battery, has a circuit diagram as shown in Fig. 4-22. The meter movement has an internal resistance of 2,000 Ω and requires a current of 100 μA for full-scale deflection. The value of resistor R_1 is 49 kΩ.

(a) Assuming that the battery voltage has fallen to 5.9 V, calculate the value of R_2 required to zero the ohmmeter.

(b) Under the conditions mentioned in (a), an unknown resistor R_x is connected to the meter, causing a 60 per cent meter deflection. Calculate the value of the unknown resistance R_x.

11. The movement of the multirange voltmeter of Fig. 4-17 has a full-scale current of 50 μA and an internal resistance of 2,000 Ω. The full-scale meter reading is 150 V with the range switch set in position V_1, 50 V with the switch in position V_2, 10 V with the switch in position V_3, and 1 V with the switch in position V_4. Calculate (a) the resistance of multipliers R_1, R_2, R_3, and R_4; (b) the sensitivity of the voltmeter.

12. A dc voltmeter is rated with a sensitivity of 10 kΩ/V and is used on its 0–150-V range to measure the voltage across the 100-kΩ resistor in Fig. 4-18. Determine the percentage error of the meter indication.

13. Design a volt-ohm-milliammeter with the following characteristics:
   (a) Voltage ranges: 0–5, 0–25, 0–100, and 0–500 V dc.
   (b) Current ranges: 0–10, 0–100, 0–500, and 0–1,000 mA dc.
   (c) Resistance ranges: 20 Ω, 2,000 Ω, and 200 kΩ at half scale.
   The movement used in this instrument is a d'Arsonval mechanism with internal resistance of 1,500 Ω and full-scale current of 50 μA. (Refer to the circuit diagrams and description of the multimeter of Fig. 4-24 for information about circuit arrangements.)

14. The dc voltmeter of Fig. 4-20(b) has a sensitivity of 1,000 Ω/V and a full-scale reading of 100 V. The meter indicates 84 V as the voltage across the load. Calculate the error in measuring the power dissipation of the load by the voltmeter-ammeter method when the ammeter indicates a current of (a) 50 mA; (b) 1 A; (c) 10 A.

5

Alternating-current Indicating Instruments

5-1 INTRODUCTION

The d'Arsonval movement responds to the *average* or *dc* value of the current through the moving coil. If the movement carries an alternating current with positive and negative half cycles, the driving torque would be in one direction for the positive alternation and in the other direction for the negative alternation. If the frequency of the ac is very low, the pointer would swing back and forth around the zero point on the meter scale. At higher frequencies, the inertia of the coil is so great that the pointer cannot follow the rapid reversals of the driving torque and hovers around the zero mark, vibrating slightly.

To measure ac on a d'Arsonval movement, some means must be devised to obtain a *unidirectional* torque that does not reverse each half cycle. One method involves rectification of the ac, so that the rectified current deflects the coil. Other methods use the heating effect of the alternating current to produce an indication of its magnitude. Some of these methods are discussed in this chapter.

5-2 ELECTRODYNAMOMETER

One of the most important ac movements is the *electrodynamometer*. It is often used in accurate ac voltmeters and ammeters, not only at the powerline frequency but also in the lower audiofrequency range. With some slight modifications, the electrodynamometer can be used as a wattmeter, a VARmeter, a power-factor meter, or a frequency meter. The electrodynamometer movement may also serve as a *transfer* instrument, because it can be calibrated on dc and then used directly on ac, establishing a direct means of equating ac and dc measurements of voltage and current.

Where the d'Arsonval movement uses a permanent magnet to provide the magnetic field in which the movable coil rotates, the electrodynamometer uses the current under measurement to produce the necessary field flux. Figure 5-1 shows a schematic arrangement of the parts of this movement.

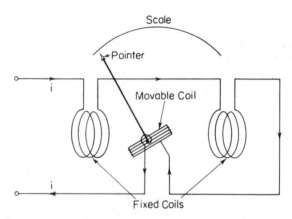

FIGURE 5-1   Schematic diagram of an electrodynamometer movement.

A fixed coil, split into two equal halves, provides the magnetic field in which the movable coil rotates. The two coil halves are connected in series with the moving coil and are fed by the current under measurement. The fixed coils are spaced far enough apart to allow passage of the shaft of the movable coil. The movable coil carries a pointer, which is balanced by counterweights. Its rotation is controlled by springs, similar to the d'Arsonval movement construction. The complete assembly is surrounded by a laminated shield to protect the instrument from stray magnetic fields which may affect its operation. Damping is provided by aluminum air vanes, moving in sector-shaped chambers. The entire movement is very solid and rigidly constructed in order to keep its mechanical dimensions stable and its calibration intact. A cutaway view of the electrodynamometer is shown in Fig. 5-2.

The operation of the instrument may be understood by returning to the expression for the torque developed by a coil suspended in a magnetic field. We previously stated, Eq. (4-1), that

$$T = B \times A \times I \times N$$

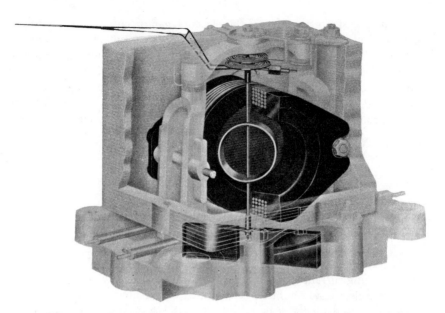

FIGURE 5-2 Phantom photograph of an electrodynamometer, showing the arrangement of fixed and movable coils. The rigidly constructed mechanism is surrounded by a laminated shield to minimize the effect of external magnetic fields on the meter indication (courtesy Weston Instruments, Inc.).

indicating that the torque, which deflects the movable coil, is directly proportional to the coil constants (A and N), the strength of the magnetic field in which the coil moves (B), and the current through the coil (I). In the electrodynamometer the flux density (B) depends on the current through the fixed coil and is therefore directly proportional to the deflection current (I). Since the coil dimensions and the number of turns on the coil frame are fixed quantities for any given meter, the developed torque becomes a function of the current squared (I^2).

If the electrodynamometer is exclusively designed for dc use, its square-law scale is easily noticed, with crowded scale markings at the very low current values, progressively spreading out at the higher current values. For ac use, the developed torque at any instant is proportional to the *instantaneous* current squared (i^2). The instantaneous value of i^2 is always positive and torque pulsations are therefore produced. The movement, however, cannot follow the rapid variations of the torque and takes up a position in which the *average* torque is balanced by the torque of the control springs. The meter deflection is therefore a function of the *mean* of the *squared current*. The scale of the electrodynamometer is usually calibrated in terms of the square root of the average current squared, and the meter therefore reads the *rms* or *effective value* of the ac.

The transfer properties of the electrodynamometer become apparent when we compare the effective value of alternating current and direct current in terms of their heating effect or transfer of power. An alternating current that produces heat in a given resistance at the same average rate as a direct current (I) has, by definition, a value of I amperes. The average rate of producing heat by a dc of I amperes in a resistance R is I^2R watts. The average rate of producing heat by an ac of i amperes during one cycle in the same resistance R is $\frac{1}{T} \int_0^T i^2R \, dt$. By definition, therefore,

$$I^2R = \frac{1}{T} \int_0^T i^2R \, dt$$

and
$$I = \sqrt{\frac{1}{T} \int_0^T i^2 \, dt} = \sqrt{\text{average } i^2}$$

This current, I, is then called the root-mean-square (rms) or effective value of the alternating current and is often referred to as the *equivalent* dc value.

If the electrodynamometer is calibrated with a direct current of 1 A and a mark is placed on the scale to indicate this 1-A dc value, then that alternating current which causes the pointer to deflect to the same mark on the scale must have an rms value of 1 A. We can therefore "transfer" a reading made with dc to its corresponding ac value and have thereby established a direct connection between ac and dc. The electrodynamometer then becomes very useful as a *calibration* instrument and is often used for this purpose because of its inherent accuracy.

The electrodynamometer, however, has certain disadvantages. One of these is its high power consumption, a direct result of its construction. The current under measurement must not only pass through the movable coil, but it must also provide the field flux. To get a sufficiently strong magnetic field, a high mmf is required and the source must supply a high current and power. In spite of this high power consumption, the magnetic field is very much weaker than that of a comparable d'Arsonval movement because there is no iron in the circuit, i.e., the entire flux path consists of air. Some instruments have been designed using special laminated steel for part of the flux path, but the presence of metal introduces calibration problems caused by frequency and waveform effects. Typical values of electrodynamometer flux density are in the range of approximately 60 gauss. This compares very unfavorably with the high flux densities (1,000–4,000 gauss) of a good d'Arsonval movement. The low flux density of the electrodynamometer immediately affects the developed torque and therefore the sensitivity of the instrument is typically very low.

The addition of a series resistor converts the electrodynamometer into a voltmeter, which again can be used to measure dc and ac voltages. For reasons previously mentioned, the sensitivity of the electrodynamometer voltmeter is low, approximately 10 to 30 Ω/V (compare this to the 20 kΩ/V of a d'Arsonval meter). The reactance and resistance of the coils also increase with increasing frequency, limiting the application of the electrodynamometer voltmeter to the lower frequency ranges. It is, however, very accurate at the powerline frequencies and is therefore often used as a *secondary* standard.

The electrodynamometer movement (even unshunted) may be regarded as an ammeter, but it becomes rather difficult to design a moving coil which can carry more than approximately 100 mA. Larger current would have to be carried to the moving coil through heavy lead-in wires, which would lose their flexibility. A shunt, when used, is usually placed across the movable coil only. The fixed coils are then made of heavy wire which can carry the large total current and it is feasible to build ammeters for currents up to 20 A. Larger values of ac currents are usually measured by using a current transformer and a standard 5-A ac ammeter (Sec. 5-11).

5-3   MOVING-IRON INSTRUMENTS

Moving-iron instruments can be classified into two types: *attraction* and *repulsion* instruments. The latter are among the more commonly used instruments. A *radial-vane* repulsion movement is shown in diagrammatic form in Fig. 5-3.

The movement consists of a stationary coil of many turns, which carries the current to be measured. Two iron vanes are placed inside the coil. One vane is rigidly attached to the coil frame; the other vane is connected to the instrument shaft which can rotate freely. The current through the coil magnetizes both vanes with the same polarity, regardless of the instantaneous direction of current. The two magnetized vanes experience a repelling force, and since only one vane can move, its displacement is an analog of the magnitude of coil current. The repelling force is proportional to the current squared, but the effects of frequency and hysteresis tend to produce a pointer deflection that is not linear and that does not have a perfect square-law relationship.

The radial-vane repulsion instrument is the most sensitive of the moving-iron mechanisms and has the most linear scale. A good design and high-quality magnetic vanes are required for instruments of good grade. Note the aluminum vane attached to the shaft, just under the pointer, which rotates in a closely fitting chamber to bring the pointer to rest quickly.

A variation of the radial-vane instrument is the *concentric-vane* repulsion movement, which is shown in the phantom photograph of Fig. 5-4. This instrument has two concentric vanes. One vane is rigidly attached to the coil frame; the other can rotate coaxially inside the stationary vane. Both vanes

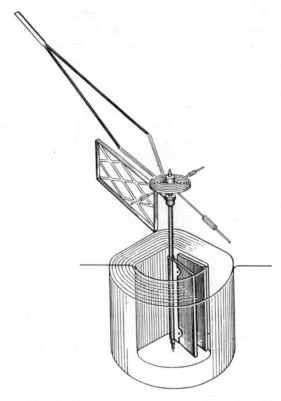

FIGURE 5-3 Radial-vane moving-iron mechanism. The aluminum damping vane, attached to the shaft just below the pointer, rotates in a closely fitting chamber to bring the pointer to rest quickly.

are magnetized by the current in the coil to the *same* polarity, causing the vanes to slip laterally under repulsion. Because the moving vane is attached to a pivoted shaft, this repulsion results in a rotational force that is a function of the current in the coil. Controlled by a spring as in other mechanisms, the final pointer position is a measure of the coil current. Since this movement, like all iron-vane instruments, does not distinguish polarity, it may be used on dc or ac; it is most commonly used for ac measurements.

Damping is obtained by a light aluminum *damping vane*, strengthened by flanges on all sides, rotating with small clearance in a closed air chamber. When used on ac, the actual operating torque is pulsating and this may cause vibration of the pointer tip. The rigid trussed pointer construction effectively eliminates such vibration over a wide frequency range and serves to prevent bending of the pointer on heavy overloads.

The concentric-vane moving-iron instrument is only moderately sensitive and has square-law scale characteristics. It is possible to modify the shape

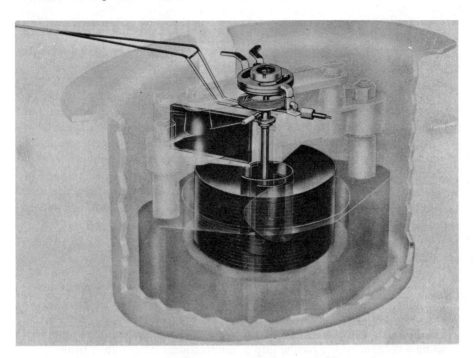

FIGURE 5-4 Phantom photograph of a concentric-vane moving-iron instrument. The figure shows details of the indicator with its counterweight, control spring, and damping vane. The moving vane may be seen as distinguished from the fixed vane in its brass retainer, and is indicated by the lightly shaded area (courtesy Weston Instruments, Inc.).

of the vanes to secure special scale characteristics, thereby "opening the scale" when desired.

   The *accuracy* of moving-iron instruments is limited mainly by the nonlinearity of the magnetization curve of the iron vanes. At low current values, the peak of the ac produces a greater displacement per unit current than the average value, resulting in an ac reading that may be appreciably higher than the equivalent dc reading at the lower end of the scale. Similarly, at the high end of the scale, the knee of the magnetization curve is approached, and the peak value of the ac will produce less deflection per unit current than the average value, so that the ac reading will be lower than the equivalent dc value.

   Hysteresis in the iron, and eddy currents in the vanes and other metal parts of the instrument further affect the accuracy of the reading. The flux density, even at full-scale values of current, is very small, so that the instrument has a rather low current sensitivity. There are no current-carrying parts in the moving system; hence the iron-vane meter is extremely rugged and reliable. It is not easily damaged, even under severe overload conditions.

Adding a suitable multiplier will convert the iron-vane movement into a *voltmeter*; adding a shunt, similarly, will produce different *current ranges*. When the iron-vane movement is used as an ac voltmeter, the frequency increases the impedance of the instrument circuit and therefore tends to give a lower reading for a given applied voltage. An iron-vane voltmeter should therefore always be *calibrated* at the frequency at which it is to be used. The usual commercial instrument may be used within its accuracy tolerance from 25 to 125 Hz. Special compensating circuits may improve the performance of the meter at higher frequencies although the upper frequency limit is not easily extended beyond about 2,500 Hz. Although these instruments will respond to dc, they cannot be used as transfer instruments. Nevertheless they are very popular because they are inexpensive and rugged, and perform adequately within their stated limitations.

5-4   RECTIFIER-TYPE INSTRUMENTS

5-4.1   Rectifier Circuits

One obvious answer to the question of ac measurement is found by using a rectifier to convert ac into a unidirectional dc and then to use a dc movement to indicate the value of the rectified ac. This method is very attractive, because a dc movement generally has a higher sensitivity than either the electrodynamometer or the moving-iron instrument.

Rectifier-type instruments generally use a PMMC movement in combination with some rectifier arrangement. The rectifier element usually consists of a germanium or a silicon diode. Copper oxide and selenium rectifiers have become obsolete, because they have small inverse voltage ratings and can handle only limited amounts of current. Germanium diodes have a peak inverse voltage (PIV) on the order of 300 V and a current rating of approximately 100 mA. Low-current silicon diode rectifiers have a PIV of up to 1,000 V and a current rating on the order of 500 mA.

Rectifiers for instrument work sometimes consist of four diodes in a bridge configuration, providing full-wave rectification. Figure 5-5 shows an ac voltmeter circuit consisting of a multiplier, a bridge rectifier, and a PMMC movement.

The bridge rectifier produces a pulsating unidirectional current through the meter movement over the complete cycle of the input voltage. Because of the inertia of the moving coil, the meter will indicate a steady deflection proportional to the *average* value of the current. Since alternating currents and voltages are usually expressed in *rms* values, the meter scale is *calibrated* in terms of the rms value of a sinusoidal waveform.

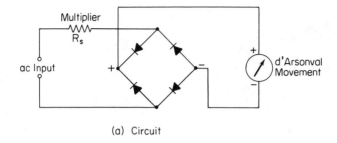

(a) Circuit

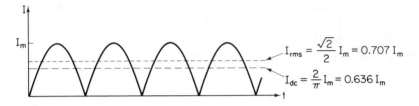

(b) Rectified Current Through Meter Movement

FIGURE 5-5 Full-wave rectifier ac voltmeter.

Example 5-1: An experimental ac voltmeter uses the circuit of Fig. 5-5(a), where the PMMC movement has an internal resistance of 50 Ω and requires a dc current of 1 mA for full-scale deflection. Assuming ideal diodes (zero forward resistance and infinite reverse resistance), calculate the value of the multiplier R_s to obtain full-scale meter deflection with 10 V ac (rms) applied to the input terminals.

SOLUTION: For full-wave rectification,

$$E_{dc} = \frac{2}{\pi} E_m = \frac{2\sqrt{2}}{\pi} E_{rms} = 0.9 \, E_{rms}$$

and $\qquad\qquad E_{dc} = 0.9 \times 10 \text{ V} = 9 \text{ V}$

The total circuit resistance, neglecting the forward diode resistance, is

$$R_t = R_s + R_m = \frac{9 \text{ V}}{1 \text{ mA}} = 9 \text{ k}\Omega$$

$$R_s = 9,000 \, \Omega - 50 \, \Omega = 8,950 \, \Omega$$

A nonsinusoidal waveform has an average value that may differ considerably from the average value of a pure sine wave (for which the meter is calibrated) and the indicated reading may be very erroneous. The *form factor* relates the average value and the rms value of time varying voltages and currents:

$$\text{form factor} = \frac{\text{effective value of the ac wave}}{\text{average value of the ac wave}}$$

For a sinusoidal waveform:

$$\text{form factor} = \frac{E_{\text{rms}}}{E_{\text{av}}} = \frac{(\sqrt{2}/2)E_m}{(2/\pi)E_m} = 1.11 \qquad \textbf{(5-1)}$$

Note that the voltmeter of Example 5-1 has a scale suitable only for sinusoidal ac measurements. The form factor of Eq. (5-1) is therefore also the factor by which the actual (average) dc current is multiplied to obtain the equivalent rms scale markings.

The ideal rectifier element should have zero forward and infinite reverse resistance. In practice, however, the rectifier is a nonlinear device, indicated by the characteristic curves of Fig. 5-6. At low values of forward current, the

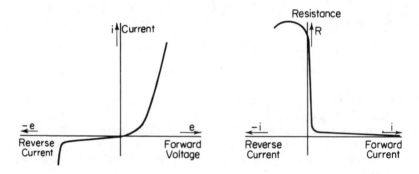

FIGURE 5-6   Characteristic curves of a solid-state rectifier.

rectifier operates in an extremely nonlinear part of its characteristic curve, and the resistance is large as compared to the resistance at higher current values. The lower part of the ac scale of a low-range voltmeter is therefore often crowded, and most manufacturers provide a separate low-voltage scale, calibrated especially for this purpose. The high resistance in the early part of the rectifier characteristics also sets a limit on the sensitivity which can be obtained in microammeters and voltmeters.

The resistance of the rectifying element changes with varying temperature, one of the major drawbacks of rectifier-type ac instruments. The meter accuracy is usually satisfactory under normal operating conditions at room temperature and is generally on the order of ± 5 per cent of full-scale reading for sinusoidal waveforms. At very much higher or lower temperatures, the resistance of the rectifier changes the total resistance of the measuring circuit sufficiently to cause the meter to be gravely in error. If large temperature variations are expected, the meter should be enclosed in a temperature-controlled box.

Frequency also affects the operation of the rectifier elements. The rectifier exhibits capacitive properties and tends to bypass the higher frequencies. Meter readings may be in error by as much as 0.5 per cent decrease for every 1-kHz rise in frequency.

5-4.2   Typical Multimeter Circuits

General rectifier-type ac voltmeters often use the arrangement shown in Fig. 5-7. Two diodes are used in this circuit, forming a full-wave rectifier with the movement so connected that it receives only half of the rectified current.

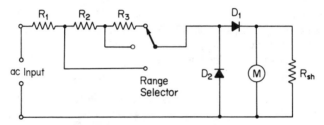

FIGURE 5-7   Typical ac voltmeter section of a commercial multimeter.

Diode D_1 conducts during the positive half-cycle of the input waveform and causes the meter to deflect according to the average value of this half-cycle. The meter movement is shunted by a resistance R_{sh}, in order to draw more current through the diode D_1 and move its operating point into the linear portion of the characteristic curve. In the absence of diode D_2, the negative half-cycle of the input voltage would apply a reverse voltage across diode D_1, causing a small leakage current in the reverse direction. The average value of the complete cycle would therefore be lower than it should be for half-wave rectification. Diode D_2 deals with this problem. On the negative half-cycle, D_2 conducts heavily, and the current through the measuring circuit, which is now in the opposite direction, bypasses the meter movement.

The commercial multimeter often uses the same scale markings for both its dc and ac voltage ranges. Since the dc component of a sine wave for half-wave rectification equals 0.45 times the rms value, a problem arises immediately. In order to obtain the same deflection on corresponding dc and ac voltage ranges, the multiplier for the ac range must be lowered proportionately. The circuit of Fig. 5-8 illustrates a solution to the problem and is discussed in some detail in Example 5-2.

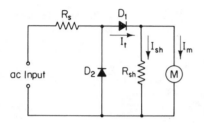

FIGURE 5-8 Computation of the multiplier resistor and the ac voltmeter sensitivity.

Example 5-2: A meter movement has an internal resistance of 100 Ω and requires 1 mA dc for full-scale deflection. Shunting resistor R_{sh}, placed across the movement, has a value of 100 Ω. Diodes D_1 and D_2 have an average forward resistance of 400 Ω each and are assumed to have infinite resistance in the reverse direction. For a 10-V ac range, calculate (a) the value of multiplier R_s, (b) the voltmeter sensitivity on the ac range.

SOLUTION:

(a) Since R_m and R_{sh} are both 100 Ω, the total current the source must supply for full-scale deflection is $I_t = 2$ mA. For half-wave rectification the equivalent dc value of the rectified ac voltage will be

$$E_{dc} = 0.45\ E_{rms} = 0.45 \times 10\ \text{V} = 4.5\ \text{V}$$

The total resistance of the instrument circuit then is

$$R_t = \frac{E_{dc}}{I_t} = \frac{4.5\ \text{V}}{2\ \text{mA}} = 2,250\ \Omega$$

This total resistance is made up of several parts. Since we are interested only in the resistance of the circuit during the half-cycle that the movement receives current, we can eliminate the infinite resistance of reverse-biased diode D_2 from the circuit. Therefore

$$R_t = R_s + R_{D_1} + \frac{R_m R_{sh}}{R_m + R_{sh}}$$

and

$$R_t = R_s + 400 + \frac{100 \times 100}{200} = R_s + 450\ \Omega$$

The value of the multiplier therefore is

$$R_s = 2,250 - 450 = 1,800\ \Omega$$

(b) The sensitivity of the voltmeter on this 10-V ac range is

$$S = \frac{2,250\ \Omega}{10\ \text{V}} = 225\ \Omega/\text{V}$$

The same movement, used in a dc voltmeter, would have given a sensitivity figure of 1,000 Ω/V.

Section 4-11 dealt with the dc circuitry of a typical multimeter, using the simplified circuit diagram of Fig. 4-25. The circuit for measuring ac volts

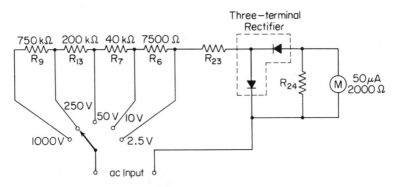

FIGURE 5-9 Multirange ac voltmeter circuit of the Simpson Model 260 multimeter (courtesy of the Simpson Electric Company).

(subtracted from Fig. 4-25), is reproduced in Fig. 5-9. Resistances R_9, R_{13}, R_7, and R_6 form a chain of multipliers for the voltage ranges of 1,000 V, 250 V, 50 V, and 10 V, respectively, and their values are indicated in the diagram of Fig. 5-9. On the 2.5-V ac range, resistor R_{23} acts as the multiplier and corresponds to the multiplier R_s of Example 5-2 shown in Fig. 5-8. Resistor R_{24} is the meter shunt and again acts to improve the rectifier operation. Both values are unspecified in the diagram and are factory selected. A little thought, however, will convince us that the shunt resistance could be 2,000 Ω, equal to the meter resistance. If the average forward resistance of the rectifier elements is 500 Ω (a reasonable assumption), then resistance R_{23} must have a value of 1,000 Ω. This follows because the meter sensitivity on the ac ranges is given as 1,000 Ω/V; on the 2.5-V ac range, the circuit must therefore have a total resistance of 2,500 Ω. This value is made up of the sum of R_{23}, the diode forward resistance, and the combination of movement and shunt-resistance, as shown in Example 5-2.

5-4.3 Decibel Measurements

Almost all VOMs and some electronic multimeters are provided with a decibel, or dB, scale. The decibel (one-tenth of a bel) represents an electric or acoustic power ratio, based on a logarithmic scale (base-10). The number of decibels corresponding to the ratio of two powers P_1 and P_2 is expressed as

$$dB = 10 \log \frac{P_1}{P_2}$$

where, generally, P_1 is the unknown power and P_2 is the so-called *reference*, or zero-level, power.

Since voltage and current are related to power by impedance, the decibel can also be used to express current and voltage ratios, provided care is taken to account for the impedances associated with them. When two voltages E_1 and E_2 or two currents I_1 and I_2 operate in *identical* impedances, the dB ratio can be expressed as

$$dB = 20 \log\frac{E_1}{E_2} \quad \text{and} \quad dB = 20 \log\frac{I_1}{I_2}$$

Conversions can be made in either direction between the number of decibels and the corresponding power, voltage, and current ratios by using standard conversion tables (see Appendix 2).

The reference level of power, generally used in communications work, equals 1 mW of power dissipated in a 600-Ω resistive load. This figure also represents a voltage of 0.775 V rms across the 600-Ω load.

For dB measurements, the ac voltage circuit of the VOM or multimeter is used in the normal manner, except that any dc in the circuit under measurement must be blocked, for example, by connecting the test lead to the "output" terminal of the VOM, and the readout is made on the dB scale. The dB scale is usually related to the lowest ac scale of the VOM, and the range selector must be set to that range if readings are to be taken directly from the dB scale. If another ac range is selected, a certain dB value must be added to the indicated dB reading.

In the VOM of Fig. 4-24 the dB scale is directly related to the 2.5-V ac scale; in fact, 0 dB (reference level) is aligned with the 0.775-V scale mark. Decibel measurements are made with the range selector set to the 2.5-V ac range. With the selector on 10 V or 50 V ac, it would be necessary to add 12 dB or 26 dB, respectively, to the actual reading. These decibel correction values are usually printed on the meter face or else are given in the instruction manual.

Note that the dB scale of any VOM or multimeter is only accurate for sinusoidal waveforms and for resistive loads of 600 Ω. If the waveform or load conditions are different from those stated, correction factors must be applied.

In a typical application the power gain of an audio amplifier is measured by taking the ratio of output power to input power, in decibels. Two separate measurements must be made, one at the input and one at the output. If both readings are made under identical conditions (input impedance equals output impedance), the arithmetic difference between the two readings is the amplifier gain. For example, if the input measurement is 3 dB (3 dB above the reference level of 1 mW into 600 Ω) and if the output reading is 16 dB, the amplifier gain is 13 dB. Referring to the conversion tables of Appendix 2, we see that this can also be expressed as a direct power ratio, and we find that 13 dB corresponds to a power ratio of 19.95. If measurements are made with unequal impedances, appropriate corrections must be made. The technique is shown in Appendix 2.

5-5 THERMOINSTRUMENTS

5-5.1 Hot-wire Mechanism

The historical forerunner of the thermoinstruments is the *hot-wire mechanism*, shown schematically in Fig. 5-10. The current under measurement passes through a fine wire tightly stretched between two terminals. A second wire is attached to the fine wire at one end and, at the other, to a spring,

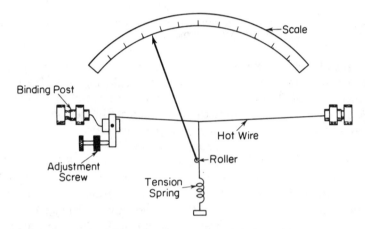

FIGURE 5-10   Schematic representation of the hot-wire ammeter.

which exerts a downward pull on the fine wire. This second wire passes over a roller to which the pointer is connected. The current under measurement causes the fine wire to heat and thus expand approximately in proportion to the heating current squared. The change in wire length drives the pointer, which indicates the magnitude of the current. Instability due to wire stretch, sluggishness in response, and lack of ambient temperature compensation have made this mechanism commercially unsatisfactory. Hot-wire mechanisms are now obsolete and have been replaced by the more sensitive, more accurate, and better-compensated combination of thermoelectric heating element and PMMC movement.

5-5.2 Thermocouple Instrument

Figure 5-11 shows a combination of a thermocouple and a PMMC movement that can be used to measure both ac and dc. This combination is called a *thermocouple instrument*, since its operation is based on the action of the thermocouple element. When two dissimilar metals are mutually in contact, a voltage is generated at the junction of the two dissimilar metals. This voltage rises in proportion to the temperature of the junction. In Fig.

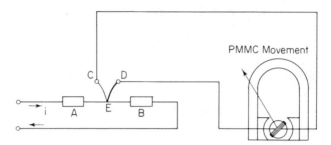

FIGURE 5-11   Schematic representation of a basic thermocouple instrument using thermocouple CDE and a PMMC movement.

5-11, *CE* and *DE* represent the two dissimilar metals, joined at point *E*, and are drawn as a light and a heavy line, to indicate dissimilarity. The potential difference between *C* and *D* depends on the temperature of the so-called *cold junction, E*. A rise in temperature causes an increase in the voltage and this is used to advantage in the thermocouple. Heating element *AB*, which is in mechanical contact with the junction of the two metals at point *E*, forms part of the circuit in which the current is to be measured. *AEB* is called the *hot junction*. Heat energy generated by the current in the heating element raises the temperature of the cold junction and causes an increase in the voltage generated across terminals *C* and *D*. This potential difference causes a dc current through the PMMC-indicating instrument. The heat generated by the current is directly proportional to the current squared (I^2R), and the temperature rise (and hence the generated dc voltage) is proportional to the square of the rms current. The deflection of the indicating instrument will therefore follow a square-law relationship, causing crowding at the lower end of the scale and spreading at the high end. The arrangement of Fig. 5-11 does not provide compensation for ambient temperature changes.

The *compensated thermoelement*, shown schematically in Fig. 5-12, produces a thermoelectric voltage in thermocouple *CED*, which is directly proportional to the current through circuit *AB*. Since the developed couple voltage is a function of the temperature difference between its hot and cold ends, this temperature difference must be caused only by the current being measured. Therefore, for accurate measurements, points *C* and *D* must be at the mean temperature of points *A* and *B*. This is accomplished by attaching

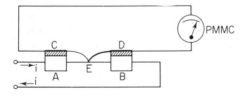

FIGURE 5-12   Compensated thermocouple to measure the thermovoltage produced by current *i* alone. Couple terminals *C* and *D* are in thermal contact with heater terminals *A* and *B*, but are electrically insulated from them.

couple ends C and D to the center of separate copper strips, whose ends are in thermal contact with A and B, but electrically insulated from them.

Self-contained thermoelectric instruments of the compensated type are available in the 0.5–20-A range. Higher current ranges are available, but in this case the heating element is external to the indicator. Thermoelements used for current ranges over 60 A are generally provided with air cooling fins.

Current measurements in the lower ranges, from approximately 0.1–0.75 A, use a *bridge-type thermoelement*, shown schematically in Fig. 5-13. This

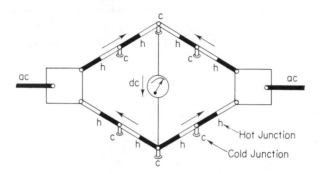

FIGURE 5-13   Bridge-type thermocouple instrument.

arrangement does not use a separate heater: the current to be measured passes directly through the thermoelements and raises their temperature in proportion to I^2R. The cold junctions (marked c) are at the pins which are embedded in the insulating frame, and the hot junctions (marked h) are at splices midway between the pins. The couples are arranged as shown in Fig. 5-13, and the resultant thermal voltage generates a dc potential difference across the indicating instrument. Since the bridge arms have equal resistances, the ac voltage across the meter is 0 V, and no ac passes through the meter. The use of several thermocouples in series provides a greater output voltage and deflection than is possible with a single element, resulting in an instrument with increased sensitivity.

Thermoinstruments may be converted into voltmeters using low-current thermocouples and suitable series resistors. Thermocouple voltmeters are available in ranges of up to 500 V and sensitivities of approximately 100 to 500 Ω/V.

A major advantage of a thermocouple instrument is that its accuracy can be as high as 1 per cent, up to frequencies of approximately 50 MHz. For this reason, it is classified as an *RF instrument*. Above 50 MHz, the skin effect tends to force the current to the outer surface of the conductor, increasing the effective resistance of the heating wire and reducing instrument accuracy. For small currents (up to 3 A), the heating wire is solid and very thin. Above 3 A the heating element is made of a tubular design to reduce the errors due to skin effect at higher frequencies.

5-5.3 Thermal Watt Converter

A thermocouple arrangement, related to the bridge-type heating element, is used in the *thermal watt converter*. This device permits measurement of ac and dc *power* by thermoelectric means. From basic ac theory we know that power is measured in watts and is expressed as $W = EI \cos \theta$, where E and I represent the phasor quantities of voltage and current, respectively, and θ represents the phase angle between them. Referring to the phasor diagram of Fig. 5-14, where a voltage phasor, E, and a current phasor, I, have been

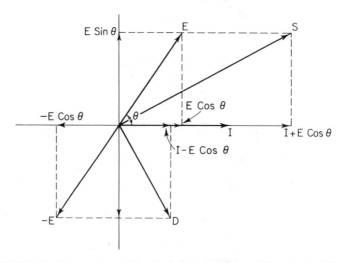

FIGURE 5-14   Geometric relationship between the sum (S) and the difference (D) of two vectors E and I at a phase angle θ.

placed at a phase angle θ, we see that the sum, S, of the two phasors can be found by the relationship

$$S^2 = E^2 + I^2 + 2\,EI \cos \theta \qquad (5\text{-}2)$$

where S represents the sum of the E phasor and the I phasor. Similarly, the difference, D, between the two phasors is found by the expression

$$D^2 = E^2 + I^2 - 2\,EI \cos \theta \qquad (5\text{-}3)$$

Subtracting Eq. (5-3) from Eq. (5-2), we obtain

$$S^2 - D^2 = 4\,EI \cos \theta \qquad (5\text{-}4)$$

where $EI \cos \theta$ is the power developed by the two phasor quantities in an electric circuit.

A circuit arrangement capable of measuring the quantity $S^2 - D^2$ can also measure a quantity proportional to $EI \cos \theta$, representing power. A thermoinstrument capable of measuring power is called a *thermal watt converter*.

Figure 5-15 shows the schematic diagram of the elementary circuit of

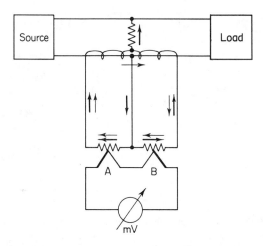

FIGURE 5-15   Elementary circuit of a thermal converter.

the thermal watt converter. For a given instant of time, the plain arrow shows the instantaneous direction of current from the current transformer. The flagged arrows show the corresponding instantaneous direction of current in the potential circuit. The heating element of thermocouple *A* receives the *sum* of the currents produced by the current transformer and the potential circuit. The heating element of thermocouple *B*, however, receives the *difference* of these currents. Through proper design the heat generated in the thermo-couples, and hence the developed emf, is proportional to the square of the current in the heater. Therefore thermocouple *A* develops an emf propor-tional to S^2, and thermocouple *B* develops an emf proportional to D^2. The voltage outputs of the thermocouples are connected so that they oppose each other. The total emf measured by the meter is proportional to $S^2 - D^2$, which was shown, by Eq. (5-4), to represent power.

In practice, *chains* of thermocouples are used instead of single couples to obtain greater developed voltages. The couples are *self-heating*, similar to those of the bridge-type element discussed earlier. This results in the practical circuit arrangement shown in Fig. 5-16.

Thermal watt converters are extremely reliable instruments and are widely used to measure power in different circuits, their outputs being summed and applied to a recording potentiometer which will draw a graph of

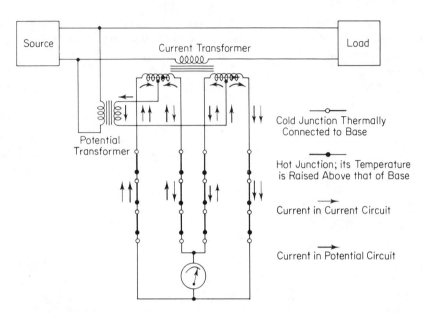

FIGURE 5-16 Circuit diagram of a thermal watt converter (courtesy Weston Instruments, Inc.).

the total power consumed by the circuits. They are also used for dc and ac instrument calibration and for instrumentation process monitoring.

5-6 ELECTROSTATIC VOLTMETER

The electrostatic voltmeter, or electrometer, is the only instrument that measures voltage *directly* rather than by the effect of the current it produces. This instrument has one distinguishing characteristic: It *consumes no power* (except during the brief transient period of initial connection to the circuit) and it therefore represents an *infinite impedance* to the circuit under measurement. Its action depends on the reaction between two electrically charged bodies (Coulomb's law). The electrostatic mechanism resembles a variable capacitor, where the force existing between the two parallel plates is a function of the potential difference applied to them. Figure 5-17 illustrates the principle of this instrument.

Plates X and Y constitute a capacitor whose capacitance increases as pointer P moves to the right. The motion of the pointer is opposed by a coil spring that also serves to provide electrical contact between connecting terminal A and plate X. When terminals X and Y are connected to points of opposite potential, the plates possess opposite charges, and the force of attraction between the two bodies of equal but opposite charge forces the

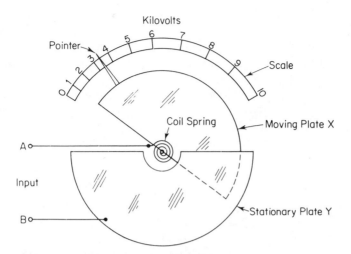

FIGURE 5-17   Schematic representation of an electrostatic voltmeter.

pointer to move to the right. The pointer will come to rest when the torque caused by the electrical attraction between the plates equals the opposing torque of the coil spring.

Analysis of the energy stored in the electric field between the capacitor plates allows us to determine an expression for the developed torque in terms of the applied voltage. The instantaneous voltage, e, across the capacitor is $e = q/C$, neglecting the leakage resistance of the air capacitor. The instantaneous energy stored in the electric field is

$$W = \frac{1}{2}\frac{q^2}{c} = \frac{1}{2}Ce^2 \qquad (5\text{-}5)$$

The instantaneous torque may be found by keeping e constant and permitting the movable plates to undergo a small angular displacement, $d\theta$. The developed torque then is

$$T_\theta = \frac{\partial W}{\partial \theta} = \frac{\partial}{\partial \theta}\left(\frac{1}{2}Ce^2\right) = \frac{1}{2}e^2\frac{\partial c}{\partial \theta} \qquad (5\text{-}6)$$

Equation (5-6) indicates that the instantaneous torque is proportional to the square of the instantaneous voltage and also depends on the manner in which C changes with θ. The average torque over an entire period T of the alternating voltage is

$$T_{\text{av}} = \frac{1}{T}\int_0^T T_\theta\, dt = \frac{1}{T}\int_0^T \frac{1}{2}e^2\frac{\partial C}{\partial \theta}\, dt = KE_{\text{rms}}^2 \qquad (5\text{-}7)$$

The deflecting torque, expressed in Eq. (5-7), is directly proportional to the square of the applied voltage, whatever its waveform, and the deflection of the electrometer may be calibrated directly in rms volts.

The electrometer can be used on *either* dc or ac and over a fairly large range of frequencies. The instrument may be calibrated with dc and the calibration is valid for ac since the deflection is independent of the waveform of the applied voltage. Since the electrometer is a "square-law" instrument, there will be no waveform error as found in the rectifier-type voltmeter. When the electrometer is first connected to a source, it draws a momentary charging current that decays exponentially. Once the plates are charged, no more current is drawn from the circuit and the meter represents infinite impedance.

The use of the instrument is limited to certain special applications, particularly in ac circuits of relatively *high voltage*, where the current taken by other instruments would result in erroneous indications. A protective resistor is generally used in series with the instrument to limit the current in case of a short circuit between the plates.

An interesting application of the same principle of electrostatic attraction or repulsion between two parallel plates is found in the *disk electrometer*. This instrument consists of two very large parallel plates, mounted in a shielded case and using quartz support pillars. The force of attraction between the parallel plates caused by the application of a potential difference is measured, and by using the exact dimensions of the plates, and their separation, the voltage between the plates is calculated. The National Bureau of Standards (NBS) uses an instrument of this type for voltage standardization up to 300,000 V. With this high-voltage electrometer the transformation ratio of high-voltage potential transformers can be directly checked by an independent method.

5-7 ELECTRODYNAMOMETERS IN POWER MEASUREMENTS

5-7.1 Single-phase Wattmeter

The electrodynamometer movement is used extensively in measuring power. It may be used to indicate *both* dc and ac power for any waveform of voltage and current and it is not restricted to sinusoidal waveforms. As described in Sec. 5-2, the electrodynamometer used as a voltmeter or an ammeter has the fixed coils and the movable coil connected in series, thereby reacting to the effect of the current squared. When used as a *single-phase power meter*, the coils are connected in a different arrangement (see Fig. 5-18).

The fixed coils, or *field coils*, shown here as two separate elements, are connected in series and carry the total line current (i_c). The movable coil, located in the magnetic field of the fixed coils, is connected in series with a current-limiting resistor across the power line and carries a small current (i_p). The instantaneous value of the current in the movable coil is $i_p = e/R_p$, where e is the instantaneous voltage across the power line, and R_p is the total resis-

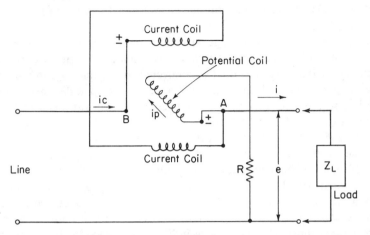

FIGURE 5-18   Diagram of an electrodynamometer wattmeter, connected to measure the power of a single-phase load.

tance of the movable coil and its series resistor. The deflection of the movable coil is proportional to the product of these two currents, i_c and i_p, and we can write for the average deflection over one period:

$$\theta_{av} = K\frac{1}{T} \int_0^T i_c i_p \, dt \tag{5-8}$$

where θ_{av} = average angular deflection of the coil

   K = instrument constant

   i_c = instantaneous current in the field coils

   i_p = instantaneous current in the potential coil

Assuming for the moment that i_c is equal to the load current, i (actually, $i_c = i_p + i$), and using the value for $i_p = e/R_p$, we see that Eq. (5-8) reduces to

$$\theta_{av} = K\frac{1}{T} \int_0^T i \frac{e}{R_p} \, dt = K_2 \frac{1}{T} \int_0^T ei \, dt \tag{5-9}$$

By definition, the average power in a circuit is

$$P_{av} = \frac{1}{T} \int_0^T ei \, dt \tag{5-10}$$

which indicates that the electrodynamometer movement, connected in the configuration of Fig. 5-18, has a deflection proportional to the average power. If e and i are sinusoidally varying quantities of the form $e = E_m \sin \omega t$ and $i = I_m \sin (\omega t \pm \theta)$, Eq. (5-9) reduces to

$$\theta_{av} = K_3 EI \cos \theta \tag{5-11}$$

where E and I represent the rms values of the voltage and the current, and θ represents the phase angle between voltage and current. Equations (5-9) and

(5-10) show that the electrodynamometer indicates the average power delivered to the load.

Wattmeters have one voltage terminal and one current terminal marked "±." When the marked current terminal is connected to the incoming line, and the marked voltage terminal is connected to the line side in which the current coil is connected, the meter will always read up-scale when power is connected to the load. If for any reason (as in the two-wattmeter method of measuring three-phase power), the meter should read backward, the *current* connections (not the voltage connections) should be reversed.

The electrodynamometer wattmeter consumes some power for maintenance of its magnetic field, but this is usually so small, compared to the load power, that it may be neglected. If a correct reading of the load power is required, the current coil should carry exactly the load current, and the potential coil should be connected across the load terminals. With the potential coil connected to point A, as in Fig. 5-18, the load voltage is properly metered, but the current through the field coils is greater by the amount i_p. The wattmeter therefore reads high by the amount of additional power loss in the potential circuit. If, however, the potential coil is connected to point B in Fig. 5-18, the field coils meter the correct load current, but the voltage across the potential coil is higher by the amount of the drop across the field coils. The wattmeter will again read high, but now by the amount of the I^2R losses in the field windings. Choice of the correct connection depends on the situation. Generally, connection of the potential coil at point A is preferred for high-current, low-voltage loads; connection at B is preferred for low-current, high-voltage loads.

The difficulty in placing the connection of the potential coil is overcome in the *compensated* wattmeter, shown schematically in Fig. 5-19. The current coil

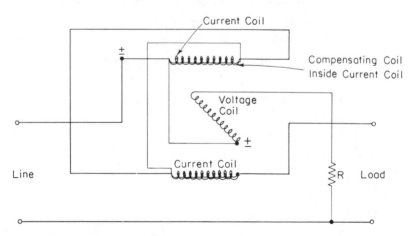

FIGURE 5-19   Diagram of a compensated wattmeter in which the effect of the current in the potential coil is canceled by the current in the compensating winding.

consists of two windings, each winding having the same number of turns. One winding uses heavy wire that carries the load current plus the current for the potential coil. The other winding uses thin wire and carries only the current to the voltage coil. This current, however, is in a direction opposite to the current in the heavy winding, causing a flux that opposes the main flux. The effect of i_p is therefore canceled out, and the wattmeter indicates the correct power.

5-7.2 Polyphase Wattmeter

Power measurements in a *polyphase* system require the use of two or more wattmeters. The total real power is then found by algebraically adding the readings of the individual wattmeters. Blondel's theorem states that real power can be measured by one less wattmeter element than the number of wires in any polyphase system, provided that one wire can be made common to all the potential circuits. Figure 5-20(a) shows the connection of two wattmeters to measure the power consumption of a balanced three-wire delta-connected three-phase load.

The current coil of wattmeter 1 is connected in line A, and its voltage coil is connected between line A and line C. The current coil of wattmeter 2 is connected in line B, and its voltage coil is connected between line B and line C. The total power, consumed by the balanced three-phase load, equals the algebraic sum of the two wattmeter readings.

The phasor diagram of Fig. 5-20(b) shows the three phase voltages V_{AC}, V_{CB}, and V_{BA}, and the three phase currents I_{AC}, I_{CB}, and I_{BA}. The delta-connected load is assumed to be inductive, and the phase currents lag the phase voltages by an angle θ. The current coil of wattmeter 1 carries the line current $I_{A'A}$, which is the vector sum of the phase currents I_{AC} and I_{AB}. The potential coil of wattmeter 1 is connected across the line voltage V_{AC}. Similarly, the current coil of wattmeter 2 carries the line current $I_{B'B}$, which is the vector sum of the phase currents I_{BA} and I_{BC}, while the voltage across its potential coil is the line voltage V_{BC}. Since the load is balanced, the phase voltages and phase currents are equal in magnitude and we can write

$$V_{AC} = V_{BC} = V \quad \text{and} \quad I_{AC} = I_{CB} = I_{BA} = I$$

The power, represented by the currents and voltages of each wattmeter, is

$$W_1 = V_{AC}I_{A'A} \cos (30° - \theta) = VI \cos (30° - \theta) \qquad \textbf{(5-12)}$$
$$W_2 = V_{BC}I_{B'B} \cos (30° + \theta) = VI \cos (30° + \theta) \qquad \textbf{(5-13)}$$

and
$$\begin{aligned} W_1 + W_2 &= VI \cos (30° - \theta) + VI \cos (30° + \theta) \\ &= (\cos 30° \cos \theta + \sin 30° \sin \theta \\ &\quad + \cos 30° \cos \theta - \sin 30° \sin \theta) VI \\ &= \sqrt{3}\, VI \cos \theta \qquad \textbf{(5-14)} \end{aligned}$$

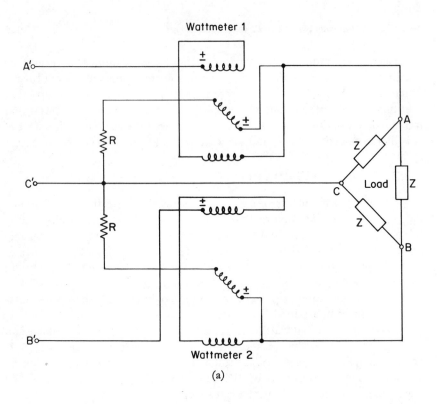

Wattmeter 1

A'o

R

C'o

R

B'o

Wattmeter 2

Z

Z

Z

A

Load

C

B

(a)

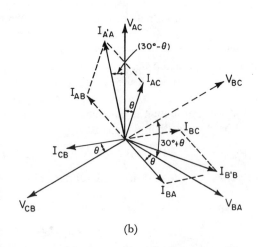

$I_{A'A}$ V_{AC}

$(30°-\theta)$

I_{AC}

I_{AB}

θ

V_{BC}

I_{BC}

$30°+\theta$

I_{CB} θ

θ

$I_{B'B}$

I_{BA}

V_{CB}

V_{BA}

(b)

FIGURE 5-20 (a) Two wattmeters connected to measure the total power in a three-phase three-wire system. (b) Phasor diagram of voltages and currents in the three-phase three-wire system. The angle between phase voltage and phase current is indicated by θ.

Equation (5-14) is the expression for the total power in a three-phase circuit, and the two wattmeters of Fig. 5-20(a) therefore correctly measure this total power. It may be shown that the algebraic sum of the readings of the two wattmeters will give the correct value for power under any condition of unbalance, power factor, or waveform.

If the neutral wire of the three-phase system is also present, as in the case of a four-wire star-connected load—according to Blondel's theorem—three wattmeters would be needed to make the total real power measurement. In Prob. 12 the reader is asked to prove that three wattmeters measure total power in a four-wire system.

5-7.3 Reactive Power Measurement

Reactive power, supplied to an ac circuit, is expressed in a unit called VAR (volt-ampere-reactive), thereby making a distinction between the real power and the quadrature reactive power. Figure 5-21 shows two

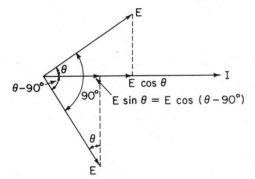

FIGURE 5-21 Vector diagram of voltage and current phasors illustrating a shift of $-90°$ of the voltage phasor.

phasors E and I, representing the voltage and the current, at a phase angle θ. Real power is the product of the in-phase components of voltage and current ($EI \cos \theta$), and reactive power, the product of the quadrature components, equals $EI \sin \theta$ or $EI \cos (\theta - 90°)$. If the voltage is shifted through 90° from its true value, the in-phase component of the shifted voltage will be $E \cos (\theta - 90°)$ and the product of the in-phase components will be $EI \cos (\theta - 90°)$, which is the reactive power.

Any ordinary wattmeter, together with a suitable phase-shifting network, may be used to measure reactive power. In a single-phase circuit a 90°-phase shift can be produced by R, L, and C components, carefully proportioned. The common application of VAR measurement, however, is found in three-phase systems, where the required phase-shifting is done with two auto-transformers connected in an "open-delta" configuration (Fig. 5-22). The current coils of the wattmeters are connected in series with the line, as usual. The potential coils are connected to the auto-transformers in

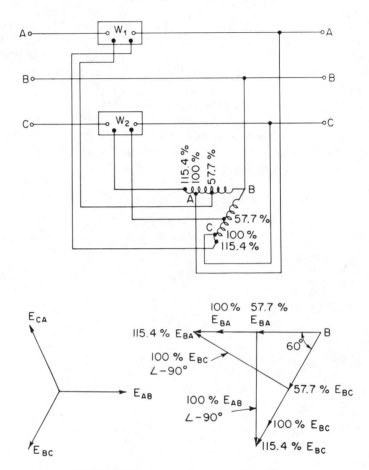

FIGURE 5-22   Reactive power measurement.

the manner indicated. Phase line *B* is connected to the common terminal of the two transformers, and the phase *A* and *C* lines are connected to the 100-per cent taps on the transformers. Both transformers will produce 115.4 per cent of the line voltage across the total winding. The potential coil of watt-meter 1 is connected from the 57.7-per cent tap of transformer 1 to the 115.4-per cent tap on transformer 2, which produces a voltage equal to the line voltage but shifted by 90°. This is shown in the phasor diagram of Fig. 5-22. The voltage coil of wattmeter 2 is connected in a similar way. Since both volt-age coils now receive an emf equal to the line voltage but displaced by 90°, the wattmeters will read the reactive power consumed by the load. The arithmetic sum of the two wattmeter readings represents the total reactive power supplied to the load. In a single instrument package, the combination of wattmeter and phase-shifting transformer is called the *VARmeter*.

5-8 WATTHOUR METER

The watthour meter is not often found in a laboratory situation but it is widely used for the commercial measurement of electrical energy. In fact, it is evident wherever a power company supplies the industrial or domestic consumer with electrical energy. Figure 5-23 shows the elements of a single-phase watthour meter in schematic form.

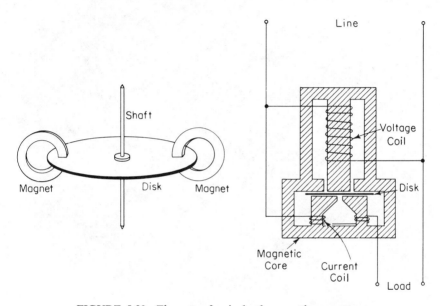

FIGURE 5-23   Elements of a single-phase watthour meter.

The current coil is connected in series with the line, and the voltage coil is connected across the line. Both coils are wound on a metal frame of special design, providing two magnetic circuits. A light aluminum disk is suspended in the air gap of the current-coil field, which causes eddy currents to flow in the disk. The reaction of the eddy currents and the field of the voltage coil creates a torque (motor action) on the disk, causing it to rotate. The developed torque is proportional to the fieldstrength of the voltage coil and the eddy currents in the disk which are in turn a function of the fieldstrength of the current coil. The number of rotations of the disk is therefore proportional to the energy consumed by the load in a certain time interval, and is measured in terms of kilowatthours (kWh). The shaft that supports the aluminum disk is connected by a gear arrangement to the clock mechanism on the front of the meter, providing a decimally calibrated readout of the number of kWh.

Damping of the disk is provided by two small permanent magnets located opposite each other at the rim of the disk. Whenever the disk rotates,

the permanent magnets induce eddy currents in it. These eddy currents reac'
with the magnetic fields of the small permanent magnets, damping the motioɪ
of the disk. A typical single-phase watthour meter is shown in Fig. 5-24.

FIGURE 5-24 Watthour meter for industrial or domestic application (courtesy
Westinghouse Electric Corporation).

Calibration of the watthour meter is performed under conditions of full
rated load and 10 per cent of rated load. At full load, the calibration consists
of adjustment of the position of the small permanent magnets until the meter
reads correctly. At very light loads, the voltage component of the field produces a torque that is not directly proportional to the load. Compensation
for the error is provided by inserting a shading coil or plate over a portion of
the voltage coil, with the meter operating at 10 per cent of rated load. Calibration of the meter at these two positions usually provides satisfactory readings
at all other loads.

The *floating-shaft* watthour meter uses a unique design to suspend the
disk. The rotating shaft has a small magnet at each end. The upper magnet of

the shaft is attracted to a magnet in the upper bearing, and the lower magnet of the shaft is attracted to a magnet in the lower bearing. The movement thus floats without touching either bearing surface, and the only contact with the movement is that of the gear connecting the shaft with the gear train.

Measurements of energy in three-phase systems are performed with polyphase watthour meters. The current coils and voltage coils are connected similar to those of the three-phase wattmeter of Fig. 5-20. Each phase of the watthour meter has its own magnetic circuit and its own disk, but all the disks are mounted on a common shaft. The developed torque on each disk is mechanically summed and the total number of revolutions per minute of the shaft is proportional to the total three-phase energy consumed.

5-9 POWER-FACTOR METERS

The power factor, by definition, is the cosine of the phase angle between voltage and current, and power-factor measurements usually involve the determination of this phase angle. This is demonstrated in the operation of the *crossed-coil power-factor meter.* The instrument is basically an electro-dynamometer movement, where the moving element consists of two coils, mounted on the same shaft but at right angles to each other. The moving coils rotate in the magnetic field provided by the field coil that carries the line current.

The connections for this meter in a single-phase circuit are shown in the circuit diagram of Fig. 5-25. The field coil is connected as usual in series with the line and carries the line current. One coil of the movable element is connected in series with a *resistor* across the lines and receives its current from the applied potential difference. The second coil of the movable element is connected in series with an *inductor* across the lines. Since no control springs

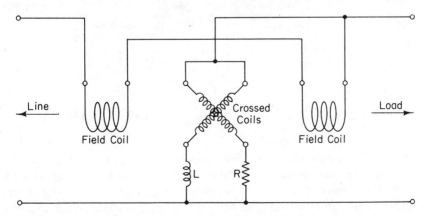

FIGURE 5-25   Connections for a single-phase crossed-coil power-factor meter.

are used, the balance position of the movable element depends on the resulting torque developed by the two crossed coils. When the movable element is in a balanced position, the contribution to the total torque by each element must be equal but of opposite sign. The developed torque in each coil is a function of the current through the coil and therefore depends on the impedance of that coil circuit. The torque is also proportional to the mutual inductance between each part of the crossed coil and the stationary field coil. This mutual inductance depends on the angular position of the crossed-coil elements with respect to the position of the stationary field coil. When the movable element is at balance, it can be shown that its angular displacement is a function of the phase angle between line current (field coil) and line voltage (crossed coils). The indication of the pointer, which is connected to the movable element, is calibrated directly in terms of the phase angle or power factor.

The *polarized-vane power-factor meter* is shown in the construction sketch of Fig. 5-26. This instrument is used primarily in three-phase power

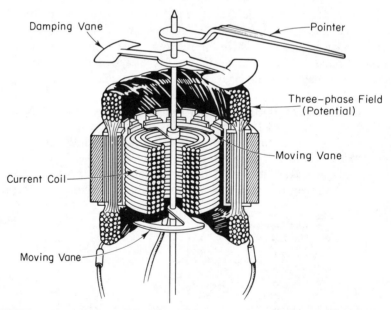

FIGURE 5-26   Polarized-vane power-factor meter (courtesy General Electric Company Limited).

systems, because its operating principle depends on the application of three-phase voltage. The outside coil is the potential coil, which is connected to the three phase lines of the system. The application of three-phase voltage to the potential coil causes it to act like the stator of a three-phase induction motor in setting up a *rotating magnetic flux*. The central coil, or current coil, is connected in series with one of the phase lines, and this *polarizes* the iron vanes. The polarized vanes move in a rotating magnetic field and take up the position that the rotating field has at the instant that the polarizing flux is maximum. This position is an indication of the phase angle and therefore the power factor. The instrument may be used in single-phase systems, provided that a phase-splitting network (similar to that used in single-phase motors) is used to set up the required rotating magnetic field.

Both types of power-factor meter are limited to measurement at comparatively low frequencies and are typically used at the powerline frequency (60 Hz). Phase measurements at higher frequencies often are more accurately and elegantly performed by special electronic instruments or techniques. Methods and instruments for phase measurements at higher frequencies are discussed in Chapters 9, 11, and 12.

5-10 FREQUENCY METERS

Frequency can be determined in a variety of ways, but at the moment we are concerned with indicating instruments; in this category, frequency meters use the effect of frequency upon such factors as mutual inductance, resonance of a tuned circuit, and mechanical resonance.

An example of the use of tuned circuits is found in the *electro-dynamometer-type frequency meter*, shown schematically in Fig. 5-27. In this frequency meter, the field coils form part of two separate resonant circuits.

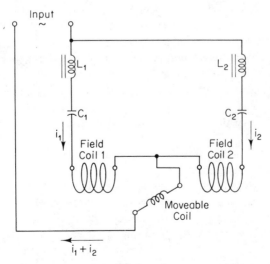

FIGURE 5-27   Circuit arrangement of the electrodynamometer-type frequency meter.

Field coil 1 is in series with inductor L_1 and capacitor C_2, and forms a resonant circuit which is tuned to a frequency slightly below the low end of the instrument scale. Field coil 2 is in series with inductor L_2 and capacitor C_2, and forms a resonant circuit which is tuned to a frequency slightly higher than the high end of the instrument scale. In the case of powerline frequencies, the circuits would be tuned to frequencies of 50 Hz and 70 Hz, respectively, with 60 Hz in the middle of the scale. The two field coils are arranged as shown in the diagram and are returned to the powerline through the winding of the movable coil. The torque on the movable element is proportional to the current through the moving coil. This current consists of the sum of the two field-coil currents. For an applied frequency within the limits of the instrument range, the circuit of field coil 1 operates above the resonant frequency with current i_1 lagging the applied voltage. The circuit of field coil 2 operates below its resonant frequency and is therefore capacitive with current i_2 leading the applied voltage. The torques produced by the two currents on the movable coil are therefore in opposition, and the resulting torque is a function of the frequency of the applied voltage. For each given frequency within the range of the instrument, the resulting torque on the movable element causes the pointer to take up a given position and the pointer deflection is calibrated in terms of the given frequency. The restoring torque is provided by a small iron vane mounted on the moving coil. The range of operation of this instrument is usually limited to the powerline frequencies and it finds its major application in this field where it is used for monitoring the frequency of a power system.

The *tuned-reed frequency meter* operates on the principle of *mechanical* resonance. A series of reeds is fastened to a flexible common base which is mounted on the armature of an electromagnet. The coil of the electromagnet is energized from the ac powerline whose frequency is to be determined. The reeds are tuned to an exact natural frequency by careful selection of their length and mass. The reed, which has a natural frequency equal to the frequency with which the electromagnet is energized, builds up a vibration. The vibration of the reed is visible at the front of the meter where the vibrating tip of the reed is shown through a window. If the frequency to be measured is intermediate between the natural frequencies of two adjacent reeds, both reeds will vibrate and the line frequency will be closest to the reed with the largest vibration. Interpolation between the natural frequencies of the reeds can be made very easily and accurately, since the reed frequencies are exact. This instrument has the advantage of being of very simple construction and very rugged. It maintains its calibration well, provided that the vibration of the reeds is kept within reasonable limits. Although its operation does not depend on the exact value of the voltage, different voltage ranges are usually provided by the addition of a series resistor.

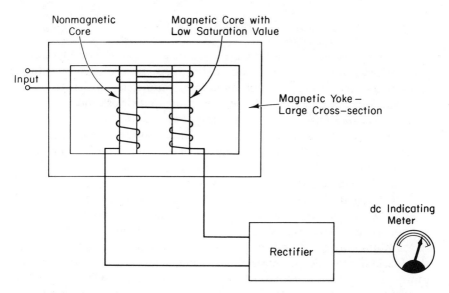

FIGURE 5-28   Schematic representation of the saturable-core frequency meter.

The *saturable-core frequency meter*, which can comfortably handle and measure a wide range of frequencies, is shown schematically in Fig. 5-28. The transformer consists of *two cores* and a *yoke*. One core is of nonmagnetic material; the other core is of magnetic material that saturates at very low values of emf and current. The yoke is made of magnetic material but its cross section is so large that it does not reach saturation. The primary winding of the transformer is wound around both cores simultaneously, as shown in Fig. 5-28. The secondary winding consists of two parts: one half of the winding is wound on the magnetic core and the other half of the winding on the nonmagnetic core. The secondary windings are connected in series in such a way that the voltages induced in the windings oppose each other. When power is supplied to the primary winding, transformer action induces secondary voltages in the secondary windings. Because of the low saturation value of the magnetic core, this core will saturate at very small secondary voltages. As soon as this core is saturated, the rate of increase of induced voltage in that winding will be equal to the rate of increase of the induced voltage in the winding on the nonmagnetic core. Therefore the rate of increase of induced voltages cancels out, since the emfs in the secondary windings oppose each other. The secondary voltage will then not be a function of the primary applied voltage, but will depend only on the frequency of the voltage. The secondary output voltage is rectified and applied to a dc meter, whose deflection is proportional to the frequency. The meter scale is calibrated in terms of frequency.

5-11 INSTRUMENT TRANSFORMERS

Instrument transformers are used to measure ac at generating stations, transformer stations, and at transmission lines, in conjunction with ac measuring instruments (voltmeter, ammeters, wattmeters, VARmeters, etc.). Instrument transformers are classified according to their use and are referred to as *current transformers* (CT) and *potential transformers* (PT).

Instrument transformers perform two important functions: They serve to *extend the range* of the ac measuring instrument, much as the shunt or the multiplier extends the range of a dc meter; they also serve to *isolate* the measuring instrument from the high-voltage powerline.

The *range* of a dc ammeter may be extended by using a shunt that divides the current under measurement between the meter and the shunt. This method is satisfactory for dc circuits, but in ac circuits current division depends not only on the resistances of the meter and the shunt but also on their reactances. Since ac measurements are made over a wide frequency range, it becomes difficult to obtain great accuracy. A CT provides the required range extension through its transformation ratio and in addition produces almost the same reading regardless of the meter constants (reactance and resistance) or, in fact, of the number of instruments (within limits) connected in the circuit.

Isolation of the measuring instrument from the high-voltage powerline is important when we consider that ac power systems frequently operate at voltages of several hundred kilovolts. It would be impractical to bring the high-voltage lines directly to an instrument panel in order to measure voltage or current, not only because of the safety hazards involved but also because of the insulation problems connected with high-voltage lines running closely together in a confined space. When an instrument transformer is used, only the low-voltage wires from the transformer secondary are brought to the instrument panel and only low voltages exist between these wires and ground, thereby minimizing safety hazards and insulation problems.

Many textbooks develop in detail the theory underlying the operation of transformers. Here these instrument transformers are merely described and their use in measurement situations is shown.*

Figure 5-29 shows a *potential transformer*; Fig. 5-30 shows a *current transformer*. The *potential transformer* (PT) is used to transform the high

---

*For fuller treatment of ac machines and circuits, consult textbooks like the following: Michael Liwshitz-Garik and Clyde C. Whipple, *AC Machines*, 2nd ed. (Princeton, N.J.: D. Van Nostrand Company, Inc., 1961), Chaps. 2–5. Russell M. Kerchner and George F. Corcoran, *Alternating Current Circuits*, 4th ed. (New York: John Wiley & Sons, Inc., 1961), pp. 291–317.

FIGURE 5-29   High-voltage potential transformer (courtesy Westinghouse Electric Corporation).

voltage of a powerline to a lower value suitable for direct connection to an ac voltmeter or the potential coil of an ac wattmeter. The usual secondary transformer voltage is 120 V. Primary voltages are standardized to accommodate the usual transmission line voltages which include 2,400 V, 4,160 V, 7,200 V, 13.8 kV, 44 kV, 66 kV, and 220 kV. The PT is rated to deliver a certain power to the secondary load or *burden*. Different load capacities are available to suit individual applications; a general capacity is 200 VA at a frequency of 60 Hz.

The PT must satisfy certain design requirements that include accuracy of the turns ratio, small leakage reactance, small magnetizing current, and minimal voltage drop. Furthermore, since we may be working with very high primary voltages, the insulation between the primary and secondary windings must be able to withstand large potential differences, and the dielectric requirements are very high. In the usual case, the high-voltage coil is of a circular pancake construction, shielded to avoid localized dielectric stresses. The low-voltage coil or coils are wound on a paper form and assembled

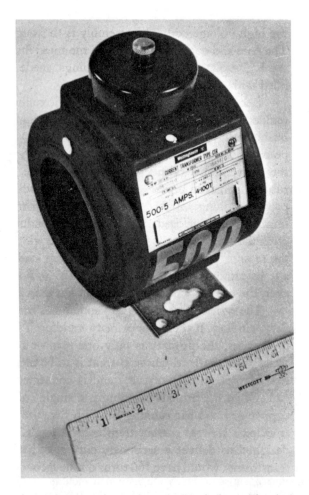

FIGURE 5-30   Current transformer (courtesy Westinghouse Electric Corporation).

inside the high-voltage coil. The assembly is thoroughly dried and oil impreg-
nated. The core and coil assembly is then mounted inside a steel case, which
supports the high-voltage terminals or porcelain bushings. The case is then
filled with an insulating oil.

Recent developments in the synthetic rubber industry have introduced
the molded rubber potential transformer, replacing the insulating oil and
porcelain bushings in some applications. Figure 5-29 shows a rubber-molded
25-kV potential transformer suitable for outdoor use. This unit is less expensive than the conventional oil-filled PT, and since the bushings are made of
molded rubber, porcelain breakage is eliminated. A white polarity dot is

placed on the proper bushing on the front of the transformer. Two stud-type secondary terminals are enclosed in a removable conduit box. The power rating of a potential transformer is based on considerations other than load capacity, for the reasons previously outlined. A typical load rating is 200 VA at 60 Hz for a transformer having a ratio of 2,400/120 V. For most metering purposes, however, the burden will be significantly less than 200 VA.

The *current transformer* (CT) sometimes has a primary and always has a secondary winding. If there is a primary winding, it has a small number of turns. In most cases, the primary is only one turn or a single conductor connected in series with the load whose current is to be measured. The secondary winding has a larger number of turns and is connected to a current meter or a relay coil. Often the primary winding is a single conductor in the form of a heavy copper or brass bar running through the core of the transformer. Such a CT is called a *bar-type* current transformer. The CT secondary winding is usually designed to deliver a secondary current of 5 A. An 800/5-A bar-type current transformer would have 160 turns on the secondary coil.

The primary winding of the current transformer is connected directly in the load circuit. When the secondary winding is open-circuited, the voltage developed across the open terminals may be very high (because of the step-up ratio) and could easily break down the insulation between the secondary windings. The secondary winding of a current transformer should therefore always be short-circuited, or connected to a meter or relay coil. A current transformer should *never* have its secondary open while the primary is carrying current; it should *always* be closed through a current meter, relay coil, wattmeter current coil, or simply a short. Failure to observe this precaution may cause serious damage to either equipment or operating personnel.

The current transformer shown in Fig. 5-30 consists of a core with the secondary winding encased in molded-rubber insulation. The window in the core allows for the insertion of one or more turns of the current-carrying high-voltage conductor. A single conductor constitutes a one-turn primary winding. The nominal ratio of the transformer is given on the nameplate; this is not the turns ratio (since more than one turn can be used as the primary) but only indicates that a primary current of 500 A will cause a secondary current of 5 A when the secondary coil is connected to a 5-A ammeter. Within practical limits, the current in the secondary winding is determined by the primary excitation current and not by the secondary circuit impedance. Since the primary current is determined by the load in the ac system, the secondary current is related to the primary current by approximately the inverse of the turns-ratio. This is true within rather wide limits of the nature of the secondary burden.

Figure 5-31 indicates the use of instrument transformers in a typical measurement application. This diagram illustrates the connection of instru-

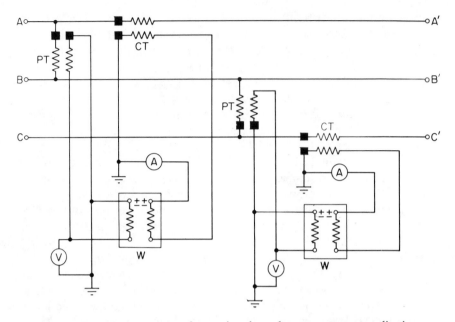

FIGURE 5-31 Instrument transformers in a three-phase measurement application. Polarity markings of the potential and current transformers are indicated by black squares.

ment transformers in a three-wire three-phase circuit, including two watt-meters, two voltmeters, and two ammeters. The potential transformers are connected across phase lines *A* and *B*, and phase lines *C* and *B*; the current transformers are in phase lines *A* and *D*. The secondary windings of the potential transformers are connected to the voltmeter coils and the potential coils of the wattmeters; the current transformer secondaries feed the ammeters and the current coils of the wattmeters.

The polarity markings on the transformers, indicated by a dot at the transformer leads, aid in making the correct polarity connections to the measuring instruments. At any given instant of the ac cycle, the dot-marked terminals have the *same* polarity and the marked wattmeter terminals must be connected to these transformer leads as shown.

REFERENCES

1. Stout, Melville B., *Basic Electrical Measurements*, 2nd ed., chap. 17. Englewood Cliffs, N.J.: Prentice-Hall, Inc., 1960.

2. Bartholomew, Davis, *Electrical Measurements and Instrumentation*, chap. 5. Boston: Allyn and Bacon, Inc., 1963.

PROBLEMS

1. Which of the following movements will measure ac without resorting to the use of rectification:
   (a) radial-vane moving-iron meters,
   (b) electrodynamometers,
   (c) core-magnet moving-coil mechanisms,
   (d) bridge-type thermocouple instruments?

2. (a) What is meant by *transfer instrument*?
   (b) Explain why the electrodynamometer can be used as a transfer instrument.

3. Explain why the ohms-per-volt rating of the ac section of a commercial multimeter is lower than that of its dc section.

4. (a) What is meant by *waveform error* in an ac voltmeter reading?
   (b) Which voltmeters may be affected by waveform errors?

5. (a) What is the principal advantage of the electrostatic voltmeter?
   (b) Explain why this instrument has a "square-law" scale.
   (c) Can this instrument be used as a transfer instrument? Why or why not?

6. Outline a calibration procedure for an electrodynamometer-type ac voltmeter. State what laboratory equipment is required for this calibration and indicate the expected accuracy.

7. The circuit diagram of Fig. 5-5 shows a full-wave rectifier ac voltmeter. The meter movement has an internal resistance of 250 Ω and requires 1 mA for full-scale deflection. The diodes each have a forward resistance of 50 Ω and infinite reverse resistance. Calculate (a) the series resistance R_s required for full-scale meter deflection when 25 V rms is applied to the meter terminals, (b) the ohms-per-volt rating of this ac voltmeter.

8. Calculate the indication of the meter of Prob. 7 when a triangular waveform with a peak value of 20 V is applied to the meter terminals.

9. A resistance of 250 Ω is placed in parallel with the meter movement of the instrument of Prob. 7.
   (a) What is the function of this resistor?
   (b) What effect does this resistance have on the ohms-per-volt rating of the voltmeter?
   (c) Calculate the new value of resistor R_s to give full-scale deflection for an input voltage of 25 V rms.

10. The commercial voltmeter of Fig. 5-7 uses a 1-mA meter movement with an internal resistance of 100 Ω. The shunting resistance across the movement is 200 Ω. Diodes D_1 and D_2 each have a forward resistance of 200 Ω and infinite reverse resistance.
    (a) Explain the function of the shunting resistor across the meter movement.
    (b) Explain the function of diode D_2.

(c) Calculate the values of series resistors R_1, R_2, and R_3 if the required meter ranges are 10 V, 50 V, and 100 V, respectively.

(d) Determine the ohms-per-volt rating of this ac voltmeter.

11. A thermocouple instrument reads 10 A at full-scale deflection. Calculate the current that causes half-scale deflection.

12. Prove that three wattmeters correctly measure the total power in a three-phase four-wire system. Assume that the load is star-connected, balanced, and purely resistive. Draw a complete phasor diagram of all line and phase voltages and currents.

13. How many wattmeters are needed to measure the total power in a three-phase three-wire circuit when the load consists of a Y-connected induction motor? Assume that it is necessary to use current and potential transformers, and draw a complete circuit diagram of the measurement setup.

14. What is the significance of the marking dots on a current or potential transformer?

6

Principles and Applications of Potentiometers

6-1 INTRODUCTION

A potentiometer is an instrument designed to measure an unknown voltage by *comparing* it with a known voltage. The known voltage may be supplied by a standard cell or any other known voltage-reference source. Measurements using the *comparison method* are capable of a very high degree of accuracy because the result obtained does not depend on an actual pointer deflection, as is the case with a moving-coil instrument, but only on the accuracy of the known voltage standard to which the comparison is made.

Since the potentiometer makes use of a *balance* or *null* condition, no power is consumed from the circuit containing the unknown emf when the instrument is balanced; as a result, the voltage determination is independent of the source resistance. Although the potentiometer measures voltage, it can be used to determine current simply by measuring the voltage drop produced by the unknown current through a known resistance.

The potentiometer is used extensively for the calibration of voltmeters and ammeters, and provides the *standard* method for the calibration of these instruments. The potentiometer is therefore an important instrument in the field of electrical measurement and calibration.

6-2 POTENTIOMETER CIRCUITS

6-2.1 Basic Circuit

The principle of operation of all potentiometers is based on the circuit of Fig. 6-1, which shows the schematic of the basic *slide-wire potentiometer*. We shall study the operation of this basic circuit on a qualitative basis and then proceed to more sophisticated potentiometric instruments.

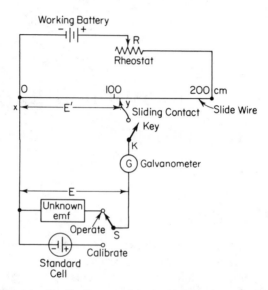

FIGURE 6-1   Circuit diagram of the basic slide-wire potentiometer.

With function switch S in the "operate" position and galvanometer key K open, the working battery supplies current through the rheostat and the slide wire. This working current through the slide wire may be varied by changing the rheostat setting. The method of measuring unknown voltage E depends on finding a position for the sliding contact such that the galvanometer shows *zero deflection* (a *null*) when galvanometer key K is closed. Zero galvanometer current, or a null, means that the unknown voltage E is equal to the voltage drop E' across portion xy of the slide wire. Determination of the value of the unknown voltage now becomes a matter of evaluating the voltage drop E' along the slide wire.

The slide wire is carefully manufactured and has uniform resistance along its entire length. A calibrated scale, usually with centimeter and millimeter divisions, is placed along the slide wire so that the sliding contact can be placed accurately at any desired position along the slide wire. Since the resistance of the slide wire is known accurately, the voltage drop along the entire slide wire, or along any portion of it, can be controlled by adjusting the working current.

As a first step in the measurement procedure, the working current is adjusted or *standardized* by reference to a known voltage source such as the standard cell in Fig. 6-1. The procedure is illustrated in the following discussion:

The slide wire has a total length of 200 cm and a resistance of 200 Ω. The emf of the voltage reference, indicated by the standard cell in Fig. 6-1, is given as 1.019 V. Switch S is thrown to the "calibrate" position and the sliding contact is placed at the 101.9-cm mark on the slide-wire scale. The rheostat is now adjusted to provide a working current such that the galvanometer shows no deflection when key K is depressed. In this condition of balance, or null, the voltage drop along the 101.9-cm portion of the slide wire is equal to the standard cell voltage of 1.019 V. Since the 101.9-cm portion of the slide wire represents a resistance of 101.9 Ω, the working current has in fact been adjusted to 10 mA. The voltage at any point along the slide wire is proportional to the length of the slide wire and is obtained by converting the calibrated length into the corresponding voltage, simply by placing the decimal point in the proper position (e.g., 146.3 cm = 1.463 V). Once calibrated, the working current is never varied.

After the potentiometer has been standardized, any small unknown dc voltage (1.6 V maximum) may be measured. Switch S is thrown to the "operate" position, and the sliding contact is moved along the wire until the galvanometer shows no deflection when key K is closed. At this null condition, the unknown voltage E equals the voltage drop E' across the xy portion of the slide wire, and the slide-wire scale reading is simply converted to its corresponding voltage value.

Example 6-1: The basic slide-wire potentiometer of Fig. 6-1 has a working battery of 3.0 V with negligible internal resistance. The resistance of the slide wire is 400 Ω and its length is 200 cm. A 200-cm scale, placed along the slide wire, has 1-mm scale divisions and interpolation can be made to one-quarter of a division. The instrument is standardized against a voltage-reference source of 1.0180 V with the slider set at the 101.8-cm mark on the scale. Calculate (a) the working current; (b) the resistance of the rheostat; (c) the measurement range; (d) the resolution of the instrument, expressed in mV.

SOLUTION:

(a) When the instrument is standardized, the 101.8-cm mark on the scale corresponds to 1.0180 V (E' in Fig. 6-1). 101.8 cm of the slide wire represents a resistance of $101.8/200 \times 400 \ \Omega = 203.6 \ \Omega$. The working current therefore must be 101.8 V/203.6 Ω = 0.5 mA.

(b) With a working current of 0.5 mA the voltage drop along the entire slide wire is 0.5 mA $\times$ 400 Ω = 2.0 V. The voltage drop across the rheostat then equals $3.0 - 2.0 = 1.0$ V and the rheostat setting is 1.0 V/0.5 mA = 2,000 Ω.

(c) The measurement range is determined by the total voltage across the slide wire which equals 0.5 mA × 400 Ω = 2.0 V.

(d) The resolution of the potentiometer is determined by the voltage represented by one-quarter of a scale division, or 0.25 mm. Since the total length of 200 cm corresponds to 2.0 V, the resolution is

$$\frac{0.25 \text{ mm}}{200 \text{ cm}} \times 2.0 \text{ V} = 0.25 \text{ mV}$$

6-2.2 Single-range Potentiometer

The slide-wire potentiometer is a rather impractical form of construction. Modern laboratory-type potentiometers use calibrated *dial resistors* and a small *circular slide wire* of one or more turns, thereby reducing the size of the instrument. Figure 6-2 shows the schematic diagram of a simple potentiometer where the long slide wire has been replaced by a combination of 15

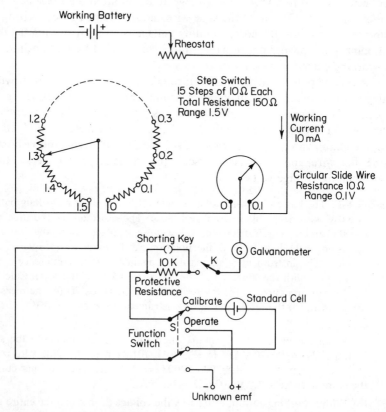

FIGURE 6-2 Circuit diagram of a simple potentiometer illustrating the use of dial resistors and a circular slide wire.

precision resistors and a single-turn circular slide wire. In this case, the resistance of the slide wire is 10 Ω and the dial resistors have a value of 10 Ω each for a total resistance of 150 Ω in the step switch. The slide wire is provided with 200 scale divisions, and interpolation to one-fifth of a division can be estimated comfortably. The working current of this potentiometer is maintained at 10 mA, so that each step of the dial switch corresponds to a voltage step of 0.1 V. Each division on the slide-wire scale corresponds to 0.0005 V and readings can be estimated to approximately 0.0001 V.

The potentiometer is provided with a double-throw function switch which connects either the standard cell or the unknown emf to the circuit. The galvanometer circuit includes a key and a protective series resistance. To operate the galvanometer at its maximum sensitivity the resistance can be shorted out by inserting a *shorting key* in the contact across the resistance.

Example 6-2: The single-range potentiometer of Fig. 6-2 is equipped with a 20-step dial switch where each step represents 0.1 V. The dial resistors are 10 Ω each. The 11-turn slide wire has a resistance of 11 Ω, allowing some overlap between settings of the dial switch. The circular slide-wire scale has 100 divisions and interpolation can be made to one-fifth of a division. The working battery has a terminal voltage of 6.0 V and negligible internal resistance. Calculate (a) the measuring range of the potentiometer; (b) the resolution, in μV; (c) the working current; (d) the setting of the rheostat.

SOLUTION:

(a) The total resistance R_m of the measuring circuit is

$$R_m = R_{\text{dial}} + R_{\text{silde wire}} = (20 \times 10\ \Omega) + 11\ \Omega = 211\ \Omega$$

Since each 10-Ω step of the dial switch represents a voltage of 0.1 V, the total measuring range is $\frac{211}{10} \times 0.1\ V = 2.11\ V$.

(b) The 11-Ω slide wire represents a voltage of 0.11 V. Each turn of the slide wire therefore represents $0.11\ V/11 = 0.01\ V$, or 10 mV. Each division on the slide-wire scale represents $\frac{1}{100} \times 10\ mV = 0.1\ mV$, or $100\ \mu V$. The resolution of the instrument therefore is $\frac{1}{5} \times 100\ \mu V = 20\ \mu V$.

(c) To maintain a voltage of 0.1 V across each 10-Ω dial resistor, the working current must be $0.1\ V/10\ \Omega = 10\ mA$.

(d) Since the voltage across the entire measuring resistance is 2.11 V, the drop across the rheostat must be $6.0\ V - 2.11\ V = 3.89\ V$. The rheostat setting then is $3.89\ V/10\ mA = 389\ \Omega$.

6-2.3 Potentiometric Voltage Measurements

The following steps are required in making a potentiometric measurement:

(a) The combination of dial resistors and slide wire is set to the value of the standard cell voltage. (This value is usually printed on the body of the cell.)

(b) The switch is thrown to the "calibrate" position and the galvanometer key is tapped while the rheostat is adjusted for zero deflection on the galvanometer. The protective resistance is left in the circuit to avoid damage to the galvanometer during the initial stages of adjustment.

(c) As zero deflection is approached, the protective resistance is shorted and final adjustments are made with the rheostat control.

(d) After the standardization has been completed, the switch is thrown to the "operate" position, thus connecting the unknown emf to the circuit. The instrument is balanced by the main dial and the slide wire, leaving the protective resistance again in the circuit.

(e) As balance is approached, the protective resistance is shorted and final adjustments are made to obtain a true balance condition.

(f) The value of the unknown voltage is read directly off the dial settings.

(g) The working current is checked by returning to the "calibrate" position. If the dial settings are exactly the same as in the original calibration procedure, a valid measurement has been made. If the reading does not agree, a second measurement must be made, again returning to a calibration check after completion.

6-2.4 Duo-range Potentiometer

The single-range potentiometer of Sec. 6-2.2 is usually constructed to cover voltage ranges of up to 1.6 V. The circuit can be modified to include a second measuring range of lower value by adding two range resistors and a range switch. Figure 6-3 shows the schematic diagram of a duo-range potentiometer, where R_1 and R_2 are the range resistors and switch S is the range switch. The operation of this potentiometer may more easily be understood and analyzed by redrawing it in simplified form, omitting some of the details of the galvanometer and calibration circuitry. The simplified schematic is shown in Fig. 6-4.

In Fig. 6-4 the total measuring resistance R_m consists of the slide wire in series with the main dial. The main dial has 15 steps of 10 Ω each for a total resistance of 150 Ω. The resistance of the slide wire is 10 Ω. To produce a voltage drop of 1.6 V across the main dial and slide wire, the measuring current I_m must be 10 mA. When the range switch is thrown to the $\times 0.1$ position, the measuring current I_m must be reduced to one-tenth of its original value, or 1 mA, to produce a voltage drop of 0.16 V across measuring resistance R_m.

It is essential in the design of the circuit to be able to change measuring

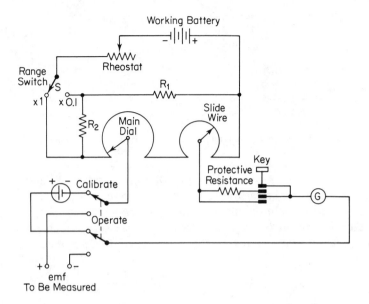

FIGURE 6-3   Schematic diagram of a duo-range potentiometer.

ranges without readjusting the rheostat or changing the voltage of the working battery. Once the instrument has been calibrated on the $\times 1$ range, following the standardization procedure of Sec. 6-2.3, calibration of the $\times 0.1$ range should not be necessary. This requires that voltage E' in Fig. 6-4 remain the same for both positions of the range switch. This condition is satisfied only when the total battery current has the same value for each measuring range.

To analyze the operation of the duo-range potentiometer of Fig. 6-4, we use the elementary circuits of the $\times 1$ range and the $\times 0.1$ range, as shown

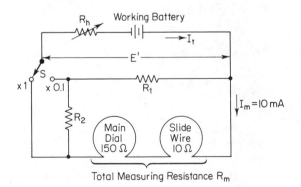

FIGURE 6-4   Simplified schematic diagram of the duo-range potentiometer.

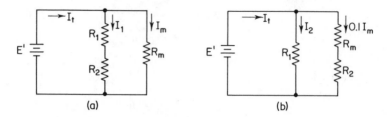

FIGURE 6-5 Elementary circuits of the duo-range potentiometer showing the circuit (a) on the $\times 1$ range and (b) on the $\times 0.1$ range.

in Fig. 6-5. On the $\times 1$ range [Fig. 6-5(a)], range resistors R_1 and R_2 are in parallel with the total measuring resistance R_m. On the $\times 0.1$ range [Fig. 6-5(b)], range resistor R_1 is in parallel with the series combination of R_2 and R_m. A constant battery current is possible only when the total circuit resistance on each range is the same. Equating the resistances of Fig. 6-5(a) and Fig. 6-5(b), we obtain

$$\frac{R_m(R_1 + R_2)}{R_1 + R_2 + R_m} = \frac{R_1(R_2 + R_m)}{R_1 + R_2 + R_m} \tag{6-1}$$

and simplifying,

$$R_2 R_m = R_1 R_2 \quad \text{or} \quad R_1 = R_m \tag{6-2}$$

Equation (6-2) indicates that range resistor R_1 must have the same value as measuring resistance R_m for the battery to supply the same current on both ranges.

Voltage E' must be the same for either position of the range switch in order to change ranges without upsetting the initial calibration. E' can be evaluated by referring to Fig. 6-5. With the range switch in the $\times 1$ position [Fig. 6-5(a)],

$$E' = I_m R_m \tag{6-3}$$

With the range switch in the $\times 0.1$ position [Fig. 6-5(b)],

$$E' = I_2 R_1 \tag{6-4}$$

Combining Eq. (6-3) and Eq. (6-4), we obtain

$$E' = I_m R_m = I_2 R_1 \tag{6-5}$$

Substituting Eq. (6-2) into Eq. (6-5), we obtain

$$I_m = I_2 \tag{6-6}$$

Equation (6-6) indicates that shunt current I_2 on the $\times 0.1$ range must be equal to measuring current I_m on the $\times 1$ range.

The battery current I_t in Fig. 6-5(a) is

$$I_t = I_1 + I_m \tag{6-7}$$

The battery current I_t in Fig. 6-5(b) is

$$I_t = I_2 + 0.1I_m \qquad (6\text{-}8)$$

Combining Eq. (6-7) and Eq. (6-8) and using Eq. (6-6), we obtain

$$I_1 + I_m = I_2 + 0.1I_m$$

or $\qquad\qquad\qquad I_1 = 0.1I_m \qquad\qquad\qquad (6\text{-}9)$

Finally, to establish the resistance of R_2, the only unknown left in the potentiometer circuit, consider again Fig. 6-5(a). The voltage drop across R_m must be equal to the voltage drop across the series combination of R_1 and R_2; therefore

$$I_1(R_1 + R_2) = I_m R_m \qquad (6\text{-}10)$$

Substituting Eq. (6-2) and Eq. (6-9) into Eq. (6-10), we obtain

$$0.1I_m(R_1 + R_2) = I_m R_1$$

or $\qquad\qquad\qquad R_2 = 9R_1 \qquad\qquad\qquad (6\text{-}11)$

For the circuit of Fig. 6-4, where measuring resistance $R_m = 160\,\Omega$, we find that $R_1 = R_m = 160\,\Omega$ and $R_2 = 9R_1 = 9 \times 160 = 1{,}440\,\Omega$. Since we had already assumed a measuring current of 10 mA on the $\times 1$ range, shunt current $I_1 = 0.1 \times 10$ mA $= 1$ mA, and the total battery current $I_t = 11$ mA. On the $\times 0.1$ range, the measuring current is $0.1I_m = 1$ mA and the shunt current $I_2 = I_m = 10$ mA, again giving a total working current of 11 mA. The condition for constant working current on both ranges has therefore been met.

Calibration of the duo-range potentiometer is accomplished in the usual manner on the $\times 1$ range position. Range resistors R_1 and R_2 are both precision resistors and the initial calibration should be valid for the lower range. The potentiometer of Fig. 6-3 can be used to measure voltages up to 0.16 V on the lower range. The dial readings are simply multiplied by the range factor of 0.1. If the slide wire has 100 scale divisions that can be interpolated to one-fifth of a division, the resolution of the potentiometer reading on the $\times 0.1$ range is $1/5 \times 1/100 \times 0.01$ V $= 20\ \mu\text{V}$.

The duo-range potentiometer of Fig. 6-3 was constructed for a voltage ratio of 10/1. A similar arrangement may be used for any other ratio by proper selection of the range resistors R_1 and R_2.

6-2.5 Multiple-range Potentiometer

Precision laboratory potentiometers usually have three voltage ranges: a high range (1.6 V), a middle range (0.16 V), and a low range (0.016 V). A simplified circuit diagram of a three-range instrument is given in Fig. 6-6,

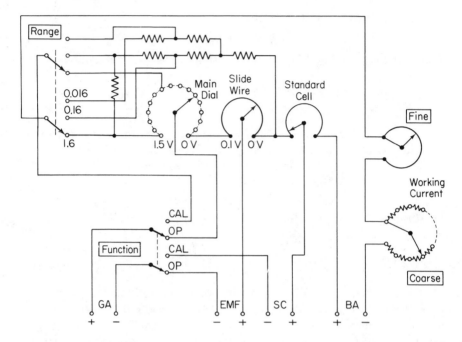

FIGURE 6-6   Three-range laboratory potentiometer.

indicating that the range-switching circuitry is slightly more complex than that of the duo-range potentiometer of Fig. 6-3, although, of course, it performs the same function. To maintain instrument calibration when the operator switches from one range to the next, the total battery current must be held constant on the three measurement ranges. Standardization of the working current is provided by two 10-turn rheostats in series, a "coarse" rheostat and a "fine" rheostat, and the current can be set with a high degree of resolution.

The level of measurement accuracy usually associated with precision potentiometers is such that the circuit requires high-stability components and carefully wired connections; as a result, several circuit elements must be thermally and electrostatically shielded. Additional features, not readily apparent from the circuit diagram but generally incorporated in precision potentiometers, include reversal switches for the detector and the unknown voltage, numerical read out of the measurement settings, and automatic positioning of the decimal point in the read out.

Figure 6-7 shows a portable potentiometer especially designed for calibrating thermocouple-operated instruments and for measuring thermocouple voltages. The simplified circuit diagram of this instrument is given in Fig. 6-8. The reader is encouraged to study this diagram and to relate the various controls shown in the photograph to the actual circuit diagram.

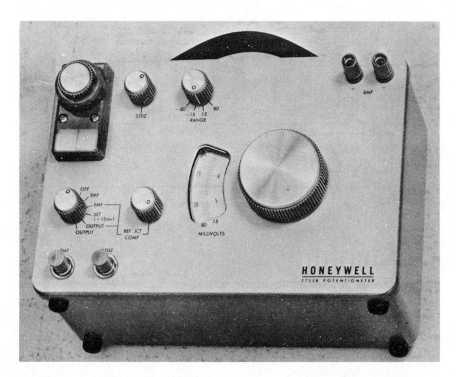

FIGURE 6-7   Portable instrument designed for calibrating thermocouple-operated instruments and measuring thermocouple voltages over the -1 to $+15$ mV and 0 to 80 mV ranges (courtesy Honeywell Test Instruments Division, Denver, Colo.).

6-3 VOLT BOX

The general-purpose potentiometer usually covers a measurement range from 0 V to 1.6 V dc. If higher values of voltage are to be measured, a precision voltage divider, or *volt box*, is used to extend the range of the potentiometer. A typical application, for example, is found in the calibration of dc voltmeters and wattmeters (Sec. 6-6).

Figure 6-9 is a schematic diagram of a volt box with voltage ranges from 3 V to 750 V dc. The voltage to be measured is connected to the "line" terminals and the appropriate voltage range is selected by setting the rotary selector switch. The resistance values are so chosen that the output of the divider, applied to the potentiometer, equals 150 mV at maximum input voltage on each range.

The current drawn from the voltage source under measurement can be made very small by using a high-resistance divider. In practice, however, the choice of resistance values involves a compromise: High resistance is desirable

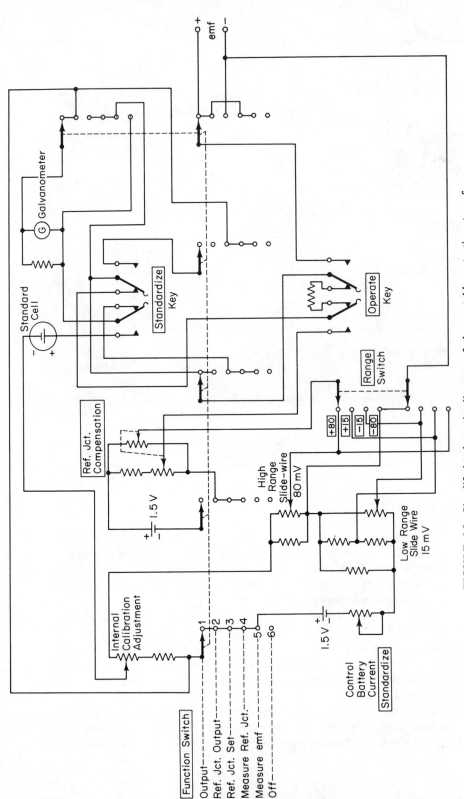

FIGURE 6-8 Simplified schematic diagram of the portable potentiometer of Fig. 6-7 (courtesy Honeywell Test Instruments Division, Denver, Colo.).

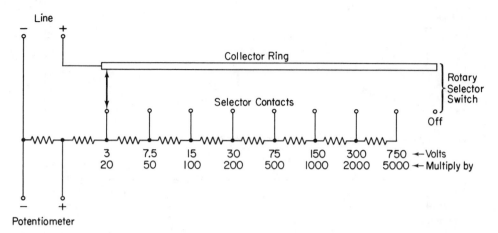

FIGURE 6-9   Schematic diagram of a volt box.

to reduce the current drawn from the voltage source, yet low-value resistors are generally more stable. Also, low-value resistors provide higher galvanometer sensitivity and minimize the effects of high-resistance leakage paths around the binding posts. In Fig. 6-9 the total resistance is relatively high (750 Ω/V), assuring small current drain on the unknown voltage source (maximum 1.33 mA).

Excellent stability and high accuracy can be achieved by using high-quality components such as woven manganin wire resistors and silver-alloy contacts on the rotary switch. For a typical volt box, the limit of error is on the order of ± 0.02 per cent.

6-4   SHUNT BOX

The shunt box is designed for use with potentiometers in the precise measurement of direct currents and in the calibration of dc ammeters and wattmeters (Sec. 6-6).

Figure 6-10 is a schematic diagram of a typical shunt box. The current to be measured is entered through the "line" terminals of the box and develops a voltage drop across the shunt resistor. A rotary switch allows selection of the desired current range from 75 mA to 15 A dc. As in the case of the volt box of Fig. 6-9, the output voltage at the potentiometer terminals of the shunt box equals 150 mV at maximum current on each range. With a total shunt resistance of 2.0 Ω, the maximum power dissipation is only 2.25 W, so that errors due to self-heating of the resistors are kept to a minimum.

In a typical measurement situation the voltage developed across the shunt resistor is measured with a potentiometer. With the selector switch

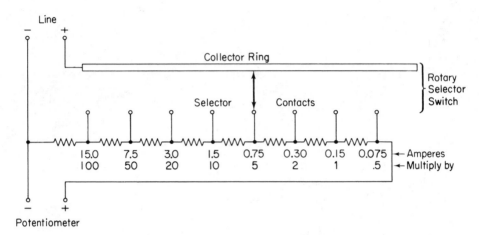

FIGURE 6-10   Schematic diagram of a shunt box.

positioned at the 0.75-A setting, as shown, a current of 600 mA would develop an output voltage of $600/750 \times 150 \text{ mV} = 120 \text{ mV}$. Conversely, a potentiometer reading of 120 mV indicates that the shunt carries a current of $120/150 \times 750 \text{ mA} = 600 \text{ mA}$.

The use of high-quality components such as manganin wire resistors and silver-alloy contacts on the range switch provides excellent electrical stability, resulting in a limit of error on the order of ± 0.02 per cent.

6-5   NULL DETECTORS

Portable potentiometers, such as the one shown in Fig. 6-7, generally contain a built-in null detector that is compatible with the remainder of the potentiometer circuitry. These built-in detectors or galvanometers are usually a fairly rugged version of the well-known d'Arsonval movement and simply indicate an unbalance condition as a pointer deflection on a scale. Precision laboratory potentiometers generally do not have the convenience of a built-in indicator but must be provided with an externally connected galvanometer or null detector.

There are basically three types of null detector:

(a) The *pointer-type galvanometer* with taut-band suspension, generally found in portable instruments. This is a fairly rugged null detector with a sensitivity from 1.0 μA to 0.1 μA per scale division.
(b) The *reflecting galvanometer* with enclosed lamp and scale, typically used in laboratory applications. This galvanometer has high sensitivity, usually in the range of 0.1 μA to 0.01 μA per scale division.

(c) The *electronic null detector* with solid-state circuitry; it has excellent sensitivity at high-input impedance and is extremely rugged but fairly expensive.

Choosing the best type of null detector for a particular application depends on a number of factors. In the case of the pointer or reflecting galvanometer, these factors include galvanometer sensitivity, resistance of the galvanometer coil, period of the galvanometer, and external critical damping resistance of the circuit. Generally speaking, high galvanometer sensitivity is associated with a long period and a large critical damping resistance. A high-sensitivity galvanometer, however, is difficult to set up and tends to be somewhat unstable at its zero-deflection point, so that the choice of galvanometer often involves a trade-off between sensitivity and ease of operation.

Figure 6-11 is a schematic view of a *reflecting galvanometer* with built-in

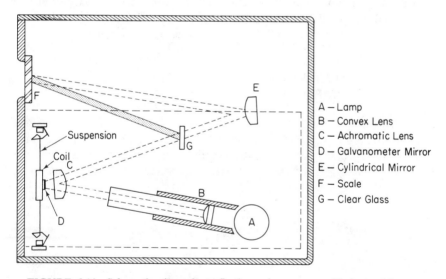

FIGURE 6-11 Schematic view of a reflecting galvanometer with integral lamp and scale.

lamp and integral scale. Lamp A is mounted in a housing near the rear of the instrument. Light shines through a tube which contains a planoconvex lens B. This lens is silvered on its plane face, mounted toward the lamp, except for a narrow rectangular aperture with its long sides vertical. A fine hairline crosses the aperture parallel to the long sides.

The lens projects an image of the lamp filament and hairline through an intermediate achromatic lens C onto the galvanometer mirror D, mounted on the galvanometer coil. The mirror turns proportionally in response to the magnitude of the current through the coil, and reflects the light beam back through lens C to a cylindrical mirror E at the rear of the instrument case. Here the image is amplified and reflected to the reading scale F in the front. The combined effects of the intermediate lens and the cylindrical mirror form the primary hairline image, used as an index to indicate deflections.

As the light beam is reflected by the galvanometer mirror to the cylindrical mirror, it passes through the clear glass G. Almost all the light passes through the glass, but a small amount is reflected to the reading scale to form a secondary image. This secondary image appears as a narrow, bright spot, centered on the primary image. It moves a very small amount for a relatively large deflection of the primary image (1/10 ratio) and is useful in determining the direction of deflection in the event the primary image goes off-scale.

The *electronic null detector* combines the advantages of high-input impedance and high sensitivity. It generally consists of a solid-state dc amplifier with an input attenuator that provides several switch-selected calibrated input ranges. Any deviation from a zero-signal condition is indicated by a pointer deflection on a center-zero output meter.

6-6 CALIBRATION OF VOLTMETERS AND AMMETERS

The potentiometer method is the usual basis for the calibration of voltmeters, ammeters, and wattmeters. Since the potentiometer is a dc measurement device, the instruments to be calibrated must be of the dc or electrodynamometer type. The circuit of Fig. 6-12 shows the measurement setup

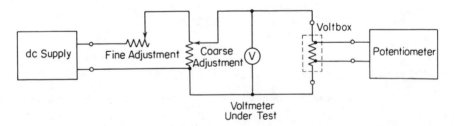

FIGURE 6-12   Calibration of a dc voltmeter by the potentiometer method.

for the calibration of a dc voltmeter. One of the first requirements in this calibration procedure is that a suitable, *stable dc supply* be available, since any variation in the supply voltage causes a corresponding change in the voltmeter calibration voltage.

Figure 6-12 shows that a potential divider network, consisting of two rheostats for coarse and fine control of the calibrating voltage, is connected across the supply source. The voltage across the voltmeter is stepped down to a value suitable for application to a potentiometer by means of a volt box. The voltage applied to the voltmeter is adjusted by the two rheostats until the pointer rests on a major scale division. The potentiometer is used to determine the *true* value of this voltage. If the potentiometer reading does not agree with the voltmeter indication, positive or negative voltmeter *error* is indicated. A selected number of major scale divisions is checked in this way, first with increasing voltage (up-scale) and then with decreasing voltage (down-scale). After these readings have been taken at the selected scale points, a *calibration curve* is plotted. An example of the data needed for the construction of a calibration curve is given in Table 6-1.

Table 6-1

RESULTS OF A CALIBRATION OF A dc VOLTMETER BY THE
POTENTIOMETER METHOD (in volts)

dc *Voltmeter* scale reading	Potentiometer true reading	Correction
0.0	0.00	0.00
1.0	0.95	−0.05
2.0	2.00	0.00
3.0	3.05	+0.05
4.0	4.10	+0.10
5.0	5.10	+0.10
6.0	6.15	+0.15
7.0	7.10	+0.10
8.0	8.15	+0.15
9.0	9.20	+0.20
10.0	10.25	+0.25

The first column of the table shows the major scale divisions at which the calibration readings were made. The second column lists the true values of the calibrating voltages, as measured with the potentiometer. The difference between the two voltages is called the *correction* value, shown in the third column. The correction value is defined as the true voltage reading *minus* the scale reading, and can therefore be positive or negative. The correction, as defined here, must therefore be added to the observed value to obtain the true value.

Figure 6-13 shows the calibration curve constructed from the data given in Table 6-1. The scale readings of column 1 are plotted along the abscissa and the corresponding correction values (column 3) are plotted along the ordinate. Since nothing is known of the variation between the observed scale points, these observations are joined by straight lines which result in the typical calibration curve shown.

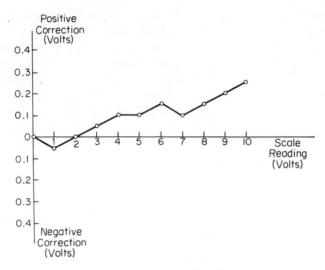

FIGURE 6-13   Typical calibration curve.

When a voltmeter receives its *initial* calibration, for instance, in the case in which the instrument is to be supplied with a new scale, the procedure is as follows: The voltage impressed on the instrument is adjusted by means of the two rheostats and set at one of the major voltage values (e.g., 1.0 V, 2.0 V, etc.). Adjustment is made until the potentiometer indicates that the desired value has indeed been reached, and the division mark is placed on the scale. A number of major voltage points are marked on the scale in this way, and the intermediate values are interpolated.

Since the calibration process is time consuming, the potentiometer method is often used to calibrate a laboratory *standard* voltmeter. A standard voltmeter of this type is a very precise instrument, with a large mirror-backed scale to increase the accuracy of reading. The accuracy of such an instrument is usually better than 0.1 per cent of full-scale reading. Ordinary laboratory meters and panel instruments are then checked by *comparison* with this laboratory standard, instead of against a potentiometer.

Figure 6-14 shows the circuit used for calibrating an ammeter. A shunt box (see Fig. 6-10) is placed in series with the ammeter under calibration. The voltage across the shunt box is measured by a potentiometer, and the current through the shunt, and hence through the ammeter, is calculated. Since the resistance of the shunt is accurately known and the voltage across the shunt is measured by the potentiometer, this method of calibrating an ammeter is very accurate. The actual procedure of calibrating various points on the meter scale is very similar to that for the voltmeter. An error curve (or calibration curve) can be constructed in the manner discussed.

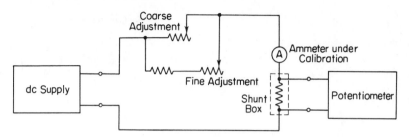

FIGURE 6-14   Calibration of an ammeter by the potentiometer method.

6-7   SELF-BALANCING POTENTIOMETER

The *self-balancing* potentiometer is widely used in industry because it eliminates the constant attention of an operator. In addition to the automatic balancing feature, it draws a curve of the quantity being measured and can be mounted on a switchboard or panel as a monitoring device. In the self-balancing instrument, the unbalance emf, which in a normal potentiometer would produce a galvanometer deflection, is applied to an amplifier via a converter. The output of the amplifier drives a two-phase induction motor that moves the potentiometer slider to balance. The converter, inserted between the potentiometer output and the amplifier input, converts the dc unbalance voltage into an ac unbalance voltage that can easily be amplified to the desired value by an ac amplifier. This scheme avoids use of a dc amplifier with its inherent problems of instability and drift.

The circuit diagram of Fig. 6-15 shows the schematic details of the self-balancing potentiometer which in this case is used for measuring temperature by a thermocouple. The converter consists of a vibrating reed, driven synchronously from the 60-Hz line voltage. The reed operates as a switch that reverses the current through the split winding of the transformer primary for each vibration of the reed. The constant reversal of current for each vibration cycle of the reed converts the unbalance dc voltage of the potentiometer circuit into an alternating voltage at the secondary of the transformer. The ac output of the converter, proportional to the dc input to the converter, is applied to the amplifier. The amplifier output is impressed on the control winding of the two-phase induction motor. The other winding of the motor is supplied by the line voltage. The ac line voltage is shifted 90° in phase with respect to the converter output voltage by the capacitor in the converter driving circuit. Depending on the polarity of the dc unbalance voltage applied to the converter input terminals, the phase of the amplifier output voltage will either lead or lag by 90° the line voltage applied to the induction motor. The

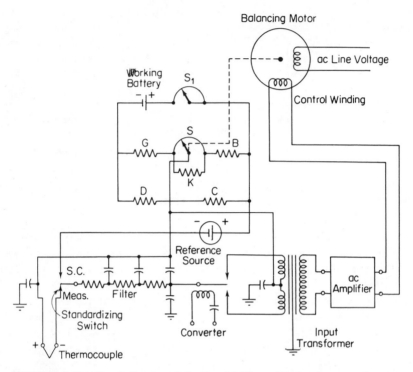

FIGURE 6-15   Circuit diagram of the Speed-0-Max self-balancing potentiometer (courtesy Leeds & Northrup Company).

direction of rotation of the motor is determined by the phase relationship between the two voltages at the two windings, and this in turn is determined by the polarity of the voltage supplied to the converter. Thus if the emf being measured is greater than the balancing voltage produced by the potentiometer, the motor will turn in one direction. If the emf being measured is smaller than the balancing voltage, then the amplifier output will be shifted by 180° and the motor will turn in the opposite direction.

The shaft of the motor is connected mechanically to the slide-wire contact in such a way that the rotation of the motor decreases the unbalance in the potentiometer circuit. When the emf being measured is equal to the potentiometer voltage, the amplifier output voltage is zero and the motor does not rotate. Therefore, under any condition of unbalance, the amplifier output voltage will cause the motor to move the potentiometer to balance.

The motor, which moves the slide-wire contact to maintain potentiometer balance, is mechanically coupled to a pen mechanism, and any movement of the slide-wire contact is followed by a simultaneous movement of the pen on a strip chart. The chart is driven by a separate clockmotor with an adjustable gear train to obtain the desired chart speed.

The emf produced by the thermocouple of Fig. 6-15 is a function of the temperature difference between the hot and the cold junction. (The operation of thermocouples is discussed in Chapter 5.) The variation in temperature of the reference junction is compensated by an electrical *compensating circuit.* The voltage drop across resistor *D*, which is made of a nickel-copper alloy, compensates for the change in temperature of the reference junction. Resistor *G* balances the voltage drop across *D* at the desired base temperature. Resistance *K* and slide wire *S* form the actual measuring circuit, and resistor *B* produces the correct voltage drop for calibration of the circuit with the reference voltage, in this case a Zener-diode reference.

The signal supplied to the input of the potentiometer circuit is passed through a low-pass filter. The filter capacitors have no effect on the dc voltage supplied to the input, but any rapid variations in the input signal and any stray ac signals that may be impressed upon the input signal are smoothed out by the action of the capacitors.

The photographs of Figs. 6-16 and 6-17 show details of the construction of a *self-balancing recording potentiometer.* The balancing- and chart-drive motors, together with two ink-supply vessels, are shown in Fig. 6-16. The

FIGURE 6-16   Inside view of a self-balancing recording potentiometer: Speedomax W/L recorder (courtesy Leeds & Northrup Company).

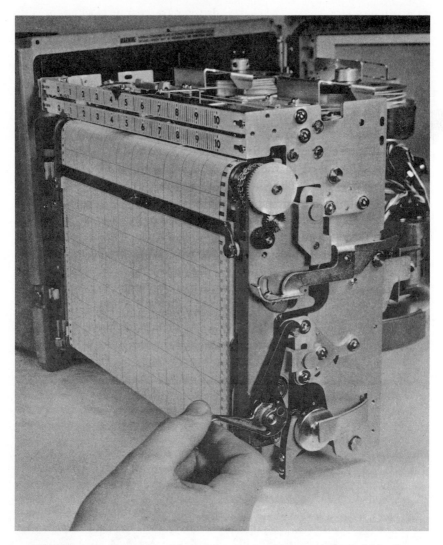

FIGURE 6-17   Chart drive of the Speedomax recorder (courtesy Leeds & Northrup Company).

amplifiers of this dual-recording potentiometer are located in the right-hand corner of the instrument case. The power supply plus the Zener reference source are only partly visible at the left of the amplifiers inside the case. Figure 6-17 shows some of the construction details of the chart-drive mechanism. As can be clearly seen in this photograph, the instrument has two scales and two recording pens.

REFERENCES

1. Bartholomew, Davis, *Electrical Measurements and Instrumentation*, chap. 10. Boston: Allyn and Bacon, Inc., 1963.

2. Frank, Ernest, *Electrical Measurement Analysis*, chap. 9. New York: McGraw-Hill Book Company, Inc., 1959.

3. Stout, Melville B., *Basic Electrical Measurements*, 2nd ed., chap. 7. Englewood Cliffs, N.J.: Prentice-Hall, Inc., 1960.

4. Honeywell Catalog C-15a, *Electronik 15 Potentiometers*. Ft. Washington, Pa: Honeywell, Inc., Industrial Division, 1966.

PROBLEMS

1. The emf of a standard cell is measured with a potentiometer that gives a reading of 1.01892 V. When a 1-MΩ resistor is connected across the standard cell terminals, the potentiometer reading drops to 1.01874 V. Calculate the internal internal resistance R_i of the standard cell.

2. A standard cell has an emf of 1.0190 V and an internal resistance of 250 Ω. A dc voltmeter with a full-scale range of 3 V and internal resistance of 3,000 Ω is connected across the standard cell.
   (a) Calculate the reading of the voltmeter.
   (b) Calculate the current drawn from the standard cell.
   (c) If the standard cell current in (b) exceeds 10 μA, calculate the value of the internal resistance that the voltmeter should have to limit the current to 10 μA.

3. The potentiometer of Fig. 6-1 has a working battery with a terminal voltage of 4.0 V and negligible internal resistance. The 200-cm slide wire has a resistance of 100 Ω and the internal resistance of the galvanometer is 50 Ω. The standard cell has an emf of 1.0191 V and internal resistance of 200 Ω. The rheostat is adjusted so that the potentiometer is standardized with the slider set at the 101.91-cm mark on the slide wire.
   (a) Calculate the working current and the resistance of the rheostat.
   (b) If the connections to the standard cell are accidentally reversed, calculate the current through the standard cell.
   (c) A protective resistance is connected in series with the galvanometer to limit the current through the galvanometer to 10 μA for the conditions of (b). Calculate the resistance of this protective resistor.

4. The voltage between two points in a dc circuit is measured with a slide-wire potentiometer that gives a reading of 1.0 V. A 20-kΩ/V dc voltmeter registers only 0.5 V on its 2.5-V scale when connected to the same two points in the circuit. Calculate the circuit resistance between the two measured points.

5. For the circuit of Fig. 6-3, the main dial consists of 15 steps of 20 Ω each, and the slide-wire resistance is 30 Ω. The standard cell voltage is 1.0190 V. The potentiometer is designed to have a measuring range of 1.65 V dc on the ×1 range. Calculate (a) the value of the measuring current I_m on each range; (b) the current supplied by the battery I_t for each range; (c) the resistances of range resistors R_1 and R_2; (d) the required resistance of the rheostat if the working battery has a voltage of 6.0 V.

6. A potentiometer has 15 steps of 5 Ω each and a slide wire of 5.5 Ω, in series with the working battery of 2.40 V and a rheostat. The maximum range of the instrument is 1.61 V. The galvanometer has a sensitivity of 0.05 μA/mm and an internal resistance of 50 Ω.
   (a) Calculate the resistance setting of the rheostat.
   (b) Determine the resolution of the instrument if the slide wire has 11 turns, 100 divisions per turn, and can be interpolated to one-fifth of a division.
   (c) A 1.10-V source with negligible internal resistance is measured with this potentiometer. Calculate the error (in V) from the true balance necessary to deflect the galvanometer spot 1 mm.

7. The potentiometer of Prob. 6 is first standardized and then balanced correctly against a dc voltage source of 1.50 V and 20-Ω internal resistance. Calculate the deflection of the galvanometer when the slide wire is moved three divisions.

8. The potentiometer of Fig. 6-2 is designed with a dial of 15 steps of 10 Ω, 0.1 V each, and a 10-Ω slide wire. The resistance of the 0–0.1-V dial step, however, is only 9.9 Ω instead of the required 10 Ω. The potentiometer is standardized against a reference voltage of 1.0185 V and is then used to measure an unknown voltage. The instrument reading at balance is 0.6525 V. Calculate (a) the true value of the unknown voltage; (b) the percentage error.

9. Design a voltbox with a resistance of 20 Ω/V and ranges of 3 V, 10 V, 30 V, and 100 V. The voltbox is to be used with a potentiometer having a measuring range of 1.6 V.

10. Design a shunt with ranges of 1 A, 5 A, 10 A, and 20 A. The shunt is to be used with a potentiometer having a measuring range of 1.6 V.

7

dc Bridges and Their Application

7-1 INTRODUCTION

Bridge circuits are extensively used for measuring component values, such as resistance, inductance, or capacitance, and of other circuit parameters directly derived from component values, such as frequency, phase angle, and temperature. Since the bridge circuit merely *compares* the value of an unknown component to that of an accurately known component (a *standard*), its measurement accuracy can be very high indeed. This is so because the readout of this comparison measurement, based on a null indication at bridge balance, is essentially independent of the characteristics of the null detector. The measurement accuracy is therefore directly related to the accuracy of the bridge components, not to that of the null indicator itself.

This chapter introduces some of the basic dc bridges. Starting with the portable test instruments, we present the Wheatstone bridge for the measurement of dc resistance, the Kelvin bridge for low-resistance measurements, and the test set for resistance testing of cables. In the high-precision test and calibration field, we introduce the principle of the guarded Wheatstone bridge and the measurement of very high resistances.

157

7-2 WHEATSTONE BRIDGE

7-2.1 Basic Operation

Figure 7-1(a) is a photograph of a portable, self-contained Wheatstone bridge. Its operation is based on the fundamental diagram of Fig. 7-1(b).

(a) Photograph of the Instrument

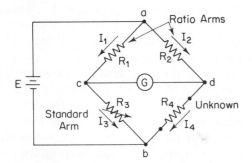

(b) Simplified Schematic of the
Bridge Circuit

FIGURE 7-1 Laboratory-type Wheatstone bridge used for the precision measurement of resistances ranging from fractions of an ohm to several megohms. The ratio control switches the ratio arms in decade steps. The remaining four step switches set the resistance of the standard arm. (courtesy Beckman Instruments, Inc., Cedar Grove Operations).

The bridge has four resistive arms, together with a source of emf (a battery) and a null detector, usually a galvanometer or other sensitive current meter. The current through the galvanometer depends on the potential difference between points c and d. The bridge is said to be *balanced* when the potential difference across the galvanometer is 0 V so that there is no current through the galvanometer. This condition occurs when the voltage from point c to point a equals the voltage from point d to point a; or by referring to the other battery terminal, when the voltage from point c to point b equals the voltage from point d to point b. Hence the bridge is balanced when

$$I_1 R_1 = I_2 R_2 \qquad (7\text{-}1)$$

If the galvanometer current is zero, the following conditions also exist:

$$I_1 = I_3 = \frac{E}{R_1 + R_3} \qquad (7\text{-}2)$$

and

$$I_2 = I_4 = \frac{E}{R_2 + R_4} \qquad (7\text{-}3)$$

Combining Eqs. (7-1), (7-2), and (7-3) and simplifying, we obtain

$$\frac{R_1}{R_1 + R_3} = \frac{R_2}{R_2 + R_4} \qquad (7\text{-}4)$$

from which

$$R_1 R_4 = R_2 R_3 \qquad (7\text{-}5)$$

Equation (7-5) is the well-known expression for balance of the Wheatstone bridge. If three of the resistances have known values, the fourth may be determined from Eq. (7-5). Hence, if R_4 is the unknown resistor, its resistance R_x can be expressed in terms of the remaining resistors as follows:

$$R_x = R_3 \frac{R_2}{R_1} \qquad (7\text{-}6)$$

Resistor R_3 is called the *standard arm* of the bridge, and resistors R_2 and R_1 are called the *ratio arms*.

The measurement of the unknown resistance R_x is independent of the characteristics or the calibration of the null-deflecting galvanometer, provided that the null detector has sufficient sensitivity to indicate the balance position of the bridge with the required degree of precision.

7-2.2   Measurement Errors

The Wheatstone bridge is widely used for precision measurement of resistance from approximately $1\,\Omega$ to the low megohm range. The main source of measurement error is found in the limiting errors of the three known resistors (see Chapter 1, Prob. 11). Other errors may include the following:

(a) Insufficient sensitivity of the null detector. This problem is discussed more fully in Sec. 7-2.3.
(b) Changes in resistance of the bridge arms due to the heating effect of the current through the resistors. Heating effect (I^2R) of the bridge arm currents may change the resistance of the resistor in question. The rise in temperature not only affects the resistance during the actual measurement, but *excessive* currents may cause a permanent change in resistance values. This may not be discovered in time and subsequent measurements could well be erroneous. The power dissipation in the bridge arms must therefore be computed in advance, particularly when low-resistance values are to be measured, and the current must be limited to a safe value.
(c) Thermal emfs in the bridge circuit or the galvanometer circuit can also cause problems when low-value resistors are being measured. To prevent thermal emfs, the more sensitive galvanometers sometimes have copper coils and copper suspension systems to avoid having dissimilar metals in contact with one another and generating thermal emfs.
(d) Errors due to the resistance of leads and contacts exterior to the actual bridge circuit play a role in the measurement of very low-resistance values. These errors may be reduced by using a Kelvin bridge (see Sec. 7-3).

7-2.3 Thévenin Equivalent Circuit

To determine whether or not the galvanometer has the required *sensitivity* to detect an unbalance condition, it is necessary to calculate the galvanometer current. Different galvanometers not only may require different currents per unit deflection (current sensitivity), but they also may have a different internal resistance. It is impossible to say, without prior computation, which galvanometer will make the bridge circuit more sensitive to an unbalance condition. This sensitivity can be calculated by "solving" the bridge circuit for a *small* unbalance. The solution is approached by converting the Wheatstone bridge of Fig. 7-1 to its Thévenin equivalent.

Since we are interested in the current through the galvanometer, the Thévenin equivalent circuit is determined by looking into galvanometer terminals c and d in Fig. 7-1. Two steps must be taken to find the Thévenin equivalent: The first step involves finding the *equivalent voltage* appearing at terminals c and d when the galvanometer is removed from the circuit. The second step involves finding the *equivalent resistance* looking into terminals c and d, with the battery replaced by its internal resistance. For convenience, the circuit of Fig. 7-1 (b) is redrawn in Fig. 7-2 (a).

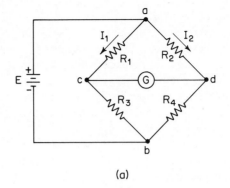

(a)

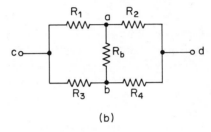

(b)

FIGURE 7-2 Application of Thévenin's theorem to the Wheatstone
bridge. (a) Wheatstone bridge configuration. (b) Thévenin resistance
looking into terminals c and d. (c)
Complete Thévenin circuit, with the
galvanometer connected to terminals
c and d.

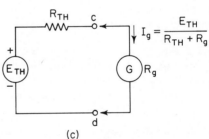

(c)

The Thévenin, or open-circuit, voltage is found by referring to Fig.
7-2(a), and we can write

$$E_{cd} = E_{ac} - E_{ad} = I_1R_1 - I_2R_2$$

where $\qquad I_1 = \dfrac{E}{R_1 + R_3} \quad$ and $\quad I_2 = \dfrac{E}{R_2 + R_4}$

Therefore

$$E_{cd} = E\left(\frac{R_1}{R_1 + R_3} - \frac{R_2}{R_2 + R_4}\right) \qquad (7\text{-}7)$$

This is the voltage of the Thévenin generator.

The resistance of the Thévenin equivalent circuit is found by looking
back into terminals c and d and replacing the battery by its internal resistance.

The circuit of Fig. 7-2(b) represents the Thévenin resistance. Notice that the internal resistance, R_b, of the battery has been included in Fig. 7-2(b). Converting this circuit into a more convenient form requires use of the *delta-wye transformation* theorem. Readers interested in this approach should consult texts on circuit analysis where this theorem is derived and applied.* In most cases, however, the extremely low internal resistance of the battery can be neglected and this simplifies the reduction of Fig. 7-2(a) to its Thévenin equivalent considerably.

Referring to Fig. 7-2(b), we see that a short circuit exists between points a and b when the internal resistance of the battery is assumed to be 0 Ω. The Thévenin resistance, looking into terminals c and d, then becomes

$$R_{TH} = \frac{R_1 R_3}{R_1 + R_3} + \frac{R_2 R_4}{R_2 + R_4} \tag{7-8}$$

The Thévenin equivalent of the Wheatstone bridge circuit therefore reduces to a Thévenin generator with an emf described by Eq. (7-7) and an internal resistance given by Eq. (7-8). This is shown in the circuit of Fig. 7-2(c).

When the null detector is now connected to the output terminals of the Thévenin equivalent circuit, the galvanometer current is found to be

$$I_g = \frac{E_{TH}}{R_{TH} + R_g} \tag{7-9}$$

where I_g is the galvanometer current and R_g its resistance.

Example 7-1: Figure 7-3(a) shows the schematic diagram of a Wheatstone bridge with values of the bridge elements as shown. The battery voltage is 5 V and its internal resistance negligible. The galvanometer has a current sensitivity of 10 mm/μA and an internal resistance of 100 Ω. Calculate the deflection of the galvanometer caused by the 5-Ω unbalance in arm *BC*.

SOLUTION: Bridge balance occurs if arm *BC* has a resistance of 2,000 Ω. The diagram shows arm *BC* as a resistance of 2,005 Ω, representing a small unbalance ($\ll$ 2,000 Ω). The first step in the solution consists of converting the bridge circuit into its Thévenin equivalent circuit. Since we are interested in finding the current in the galvanometer, the Thévenin equivalent is determined with respect to galvanometer terminals *B* and *D*. The potential difference from *B* to *D*, with the galvanometer removed from the circuit, is the Thévenin voltage. Using Eq. (7-7), we obtain

$$E_{TH} = E_{AD} - E_{AB} = 5 \text{ V} \times \left(\frac{100}{100 + 200} - \frac{1,000}{1,000 + 2,005}\right)$$
$$\cong 2.77 \text{ mV}$$

The second step of the solution involves finding the equivalent Thévenin

*Herbert W. Jackson, *Introduction to Electric Circuits*, 3rd ed. (Englewood Cliffs, N.J.: Prentice-Hall, Inc., 1970), pp. 461*ff*.

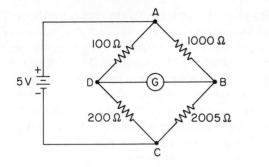

(a) Wheatstone Bridge

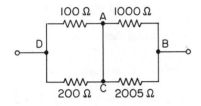

(b) Calculation of the Thévenin Resistance

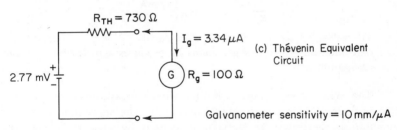

(c) Thévenin Equivalent Circuit

Galvanometer sensitivity = 10 mm/μA

FIGURE 7-3   Calculation of galvanometer deflection caused by a small unbalance in arm *BC*, using the simplified Thévenin approach.

resistance, looking into terminals *B* and *D*, and replacing the battery with its internal resistance. Since the battery resistance is 0 Ω, the circuit is represented by the configuration of Fig. 7-3(b) from which we find

$$R_{\text{TH}} = \frac{100 \times 200}{300} + \frac{1{,}000 \times 2{,}005}{3{,}005} \cong 730 \ \Omega$$

The Thévenin equivalent circuit is given in Fig. 7-3(c). When the galvanometer is now connected to the output terminals of the equivalent circuit, the current through the galvanometer is

$$I_g = \frac{E_{\text{TH}}}{R_{\text{TH}} + R_g} = \frac{2.77 \text{ mV}}{730 \ \Omega + 100 \ \Omega} = 3.34 \ \mu\text{A}$$

The galvanometer deflection is

$$d = 3.34 \ \mu\text{A} \times \frac{10 \text{ mm}}{\mu\text{A}} = 33.4 \text{ mm}$$

At this point the merit of the Thévenin equivalent circuit for the solution of an unbalanced bridge becomes evident. If a different galvanometer is used (with a different current sensitivity and internal resistance), the computation of its deflection is very simple, as is clear from Fig. 7-3(c). Conversely, if the galvanometer sensitivity is given, we can solve for the unbalance voltage needed to give a unit deflection (say 1 mm). This value is of interest when we want to determine the sensitivity of the bridge to unbalance, or in response to the question: "Is the galvanometer selected capable of detecting a certain small unbalance?" The Thévenin method is used to find the galvanometer response, which in most cases is of prime interest.

Example 7-2: The galvanometer of Example 7-1 is replaced by one with an internal resistance of 500 Ω and a current sensitivity of 1 mm/μA. Assuming that a deflection of 1 mm can be observed on the galvanometer scale, determine if this new galvanometer is capable of detecting the 5-Ω unbalance in arm *BC* of Fig. 7-3(a).

SOLUTION: Since the bridge constants have not been changed, the equivalent circuit is again represented by a Thévenin generator of 2.77 mV and a Thévenin resistance of 730 Ω. The new galvanometer is now connected to the output terminals, resulting in a galvanometer current

$$I_g = \frac{E_{TH}}{R_{TH} + R_g} = \frac{2.77 \text{ mA}}{730 \ \Omega + 500 \ \Omega} = 2.25 \ \mu A$$

The galvanometer deflection therefore equals 2.25 μA × 1 mm/μA = 2.25 mm, indicating that this galvanometer produces a deflection that can be easily observed.

The Wheatstone bridge is limited to the measurement of resistances ranging from a few ohms to several megohms. The *upper* limit is set by the reduction in sensitivity to unbalance, caused by high resistance values, because in this case the equivalent Thévenin resistance of Fig. 7-3(c) becomes high, thus reducing the galvanometer current. The *lower* limit is set by the resistance of the connecting leads and the contact resistance at the binding posts. The resistance of the leads could be calculated or measured, and the final result modified, but contact resistance is very hard to compute or measure. For low-resistance measurements, therefore, the Kelvin bridge is generally the preferred instrument.

7-3   KELVIN BRIDGE

7-3.1   Effects of Connecting Leads

The Kelvin bridge is a modification of the Wheatstone bridge and provides greatly increased accuracy in the measurement of *low-value resistances*, generally below 1 Ω. Consider the bridge circuit shown in Fig. 7-4,

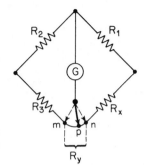

FIGURE 7-4 Wheatstone bridge
circuit, showing resistance R_y of the
lead from point m to point n.

where R_y represents the resistance of the connecting lead from R_3 to R_x. Two
galvanometer connections are possible, to point m or to point n. When the
galvanometer is connected to point m, the resistance R_y of the connecting
lead is added to the unknown R_x, resulting in too high an indication for R_x.
When connection is made to point n, R_y is added to bridge arm R_3 and the
resulting measurement of R_x will be lower than it should be, because now the
actual value of R_3 is higher than its nominal value by resistance R_y. If the
galvanometer is connected to a point p, in between the two points m and n,
in such a way that the ratio of the resistances from n to p and from m to p
equals the ratio of resistors R_1 and R_2, we can write

$$\frac{R_{np}}{R_{mp}} = \frac{R_1}{R_2} \tag{7-10}$$

The balance equation for the bridge yields

$$R_x + R_{np} = \frac{R_1}{R_2}(R_3 + R_{mp}) \tag{7-11}$$

Substituting Eq. (7-10) into Eq. (7-11), we obtain

$$R_x + \left(\frac{R_1}{R_1 + R_2}\right)R_y = \frac{R_1}{R_2}\left[R_3 + \left(\frac{R_2}{R_1 + R_2}\right)R_y\right] \tag{7-12}$$

which reduces to

$$R_x = \frac{R_1}{R_2}R_3 \tag{7-13}$$

Equation (7-13) is the usual balance equation developed for the Wheatstone
bridge and it indicates that the effect of the resistance of the connecting lead
from point m to point n has been eliminated by connecting the galvanometer
to the intermediate position p.

This development forms the basis for construction of the Kelvin double
bridge, commonly known as the *Kelvin bridge*.

7-3.2 Kelvin Double Bridge

The term *double* bridge is used because the circuit contains a second
set of ratio arms, as shown in the schematic diagram of Fig. 7-5. This second

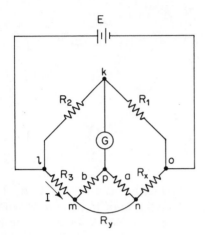

FIGURE 7-5   Basic Kelvin double bridge circuit.

set of arms, labeled a and b in the diagram, connects the galvanometer to a point p at the appropriate potential between m and n, and it eliminates the effect of the yoke resistance R_y. An initially established condition is that the resistance ratio of a and b is the same as the ratio of R_1 and R_2.

The galvanometer indication will be zero when the potential at k equals the potential at p, or when $E_{kl} = E_{lmp}$, where

$$E_{kl} = \frac{R_2}{R_1 + R_2} E = \frac{R_2}{R_1 + R_2} I \left[R_3 + R_x + \frac{(a + b)R_y}{(a + b + R_y)} \right] \quad (7\text{-}14)$$

and

$$E_{lmp} = I \left\{ R_3 + \frac{b}{(a + b)} \left[\frac{(a + b)R_y}{(a + b + R_y)} \right] \right\} \quad (7\text{-}15)$$

We can solve for R_x by equating E_{kl} and E_{lmp} in the following manner:

$$\frac{R_2}{R_1 + R_2} I \left[R_3 + R_x + \frac{(a + b)R_y}{(a + b + R_y)} \right] = I \left[R_3 + \frac{b}{(a + b)} \cdot \frac{(a + b)R_y}{(a + b + R_y)} \right]$$

Or simplifying, we get

$$R_3 + R_x + \frac{(a + b)R_y}{(a + b + R_y)} = \frac{R_1 + R_2}{R_2} \left[R_3 + \frac{bR_y}{(a + b + R_y)} \right]$$

and expanding the right-hand member yields

$$R_3 + R_x + \frac{(a + b)R_y}{(a + b + R_y)} = \frac{R_1 R_3}{R_2} + R_3 + \frac{R_1 + R_2}{R_2} \cdot \frac{bR_y}{(a + b + R_y)}$$

Solving for R_x yields

$$R_x = \frac{R_1 R_3}{R_2} + \frac{R_1}{R_2} \cdot \frac{bR_y}{(a + b + R_y)} + \frac{bR_y}{(a + b + R_y)} - \frac{(a + b)R_y}{(a + b + R_y)}$$

so that

$$R_x = \frac{R_1 R_3}{R_2} + \frac{bR_y}{(a + b + R_y)} \left(\frac{R_1}{R_2} - \frac{a}{b} \right) \quad (7\text{-}16)$$

Using the initially established condition that $a/b = R_1/R_2$, we see that Eq. (7-16) reduces to the well-known relationship

$$R_x = R_3 \frac{R_1}{R_2} \tag{7-17}$$

Equation (7-17) is the usual working equation for the Kelvin bridge. It indicates that the resistance of the yoke has no effect on the measurement, provided that the two sets of ratio arms have equal resistance ratios.

The Kelvin bridge is used for measuring very low resistances, from approximately $1\,\Omega$ to as low as $0.00001\,\Omega$. Figure 7-6 shows the simplified circuit diagram of a commercial Kelvin bridge capable of measuring resistances from $10\,\Omega$ to $0.00001\,\Omega$. In this bridge, resistance R_3 of Eq. (7-17) is represented by the variable *standard* resistor in Fig. 7-6. The ratio arms (R_1 and R_2) can usually be switched in a number of decade steps.

Contact potential drops in the measuring circuit may cause large errors and to reduce this effect the standard resistor consists of nine steps of $0.001\,\Omega$ each plus a calibrated manganin bar of $0.0011\,\Omega$ with a sliding contact. The total resistance of the R_3 arm therefore amounts to $0.0101\,\Omega$ and is variable in steps of $0.001\,\Omega$ plus fractions of $0.0011\,\Omega$ by the sliding contact. When both contacts are switched to select the suitable value of standard resistor, the voltage drop between the ratio-arm connection points is changed, but the total resistance around the battery circuit is unchanged. This arrangement places any contact resistance in series with the relatively high-resistance values of the ratio arms, and the contact resistance has negligible effect.

The ratio R_1/R_2 should be so selected that a relatively large part of the standard resistance is used in the measuring circuit. In this way the value of unknown resistance R_x is determined with the largest possible number of significant figures, and measurement accuracy is improved.

7-4   LOOP TESTS WITH THE PORTABLE TEST SET

7-4.1   Murray-loop Test

The portable Wheatstone bridge is often used to locate faults in multi-core cables, telephone wires, or power transmission lines by means of the so-called Murray-loop and Varley-loop tests. These tests are used in particular to find the location of short circuits or low-resistance faults between a conductor and ground. A commercial Wheatstone bridge, entirely self-contained with batteries and a pointer-type galvanometer, and with the special connections required for these loop tests, is known as a *test set*.

The best known and simplest of the various loop tests is the so-called *Murray-loop* test, principally used to locate ground faults in sheathed cables.

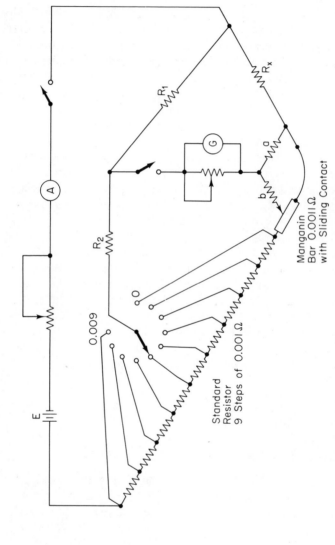

FIGURE 7-6 Simplified circuit of a Kelvin double bridge, used for the measurement of very low resistances.

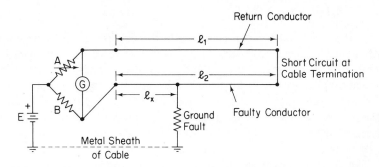

FIGURE 7-7   Locating a ground fault (short circuit) by the Murray-loop test.

The test setup is shown in Fig. 7-7. The faulty conductor with length l_2 is connected at its far end to a healthy conductor with length l_1. The loop formed by these two conductors is connected to the test set in the manner shown in Fig. 7-7, and the bridge is balanced with the adjustable ratio arm A.

At balance,

$$\frac{A}{B} = \frac{R_L - R_x}{R_x} \quad \text{or} \quad R_x = \frac{B}{A+B}R_L \tag{7-18}$$

where R_L is the total resistance of the loop (faulty conductor plus return conductor) and R_x is the resistance of the faulty conductor from the bridge terminal to the location of the ground fault.

Since the wire resistance is proportional to the length and the cross-sectional area of the conductor, we can substitute length for resistance in Eq. (7-18) and write

$$l_x = \frac{B}{A+B}(l_1 + l_2) \tag{7-19}$$

In a multicore cable the return conductor l_1 has the same length and same cross section as the faulty core, so that $l_1 = l_2 = l$ and therefore

$$l_x = 2l\frac{B}{A+B} \tag{7-20}$$

where l is the length of the multicore cable from the bridge terminals to the point of termination. If the return conductor does not have the same characteristics as the faulty core, allowance must be made for the resulting difference in resistance per unit length.

Finally, if the resistance of the ground fault is low (a short circuit, for example), the portable test set can measure the location of the cable fault with reasonable accuracy. If, however, the fault resistance is high, the battery-operated test set is not adequate, and a high-voltage measurement must be made.

7-4.2 Varley-loop Test

One of the most accurate methods of locating grounds, crosses, or short circuits in a multiconductor cable is by the so-called *Varley-loop* test. Essentially a modification of the Murray-loop test of the previous section, this method also uses a Wheatstone bridge, but with two fixed-ratio arms *A* and *B* and a rheostat in the third, or standard, arm. In a typical commercial test set the multiplication ratio of the ratio arms is controlled by a dial switch and generally ranges from 0.001 to 1,000 in decade steps. The rheostat often consists of four decades in series, whose setting can be selected by dial switches.

The three circuit arrangements needed to locate a ground fault are shown in Fig. 7-8(a) to (c). In each measurement the multiplication ratio of

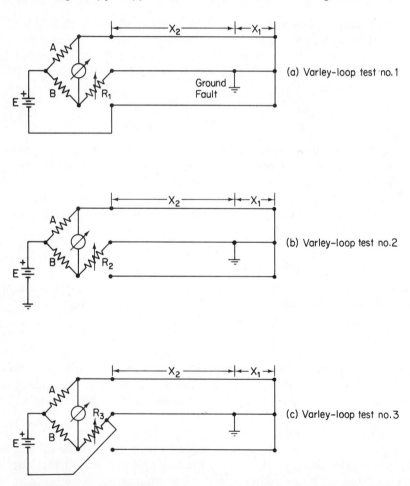

(a) Varley-loop test no. 1

(b) Varley-loop test no. 2

(c) Varley-loop test no. 3

FIGURE 7-8   Varley-loop tests, used to locate grounds, crosses, or short circuits in multiconductor cable.

arms A and B is maintained, and the bridge is balanced for zero galvanometer deflection by the rheostat in the standard arm. The desired results are calculated by conventional circuit analysis and we find that X_1 and X_2, representing the resistances of the cable sections on either side of the fault as indicated in Fig. 7-8, are given by

$$X_1 = \frac{A}{A+B}(R_2 - R_1) \qquad (7\text{-}21)$$

and
$$X_2 = \frac{A}{A+B}(R_3 + R_2) \qquad (7\text{-}22)$$

Since resistance is proportional to length and cross-sectional area, the distance to the fault can be readily determined, with one result serving as a check on the other. This method will locate the trouble spot to within 500 ft in a 50-mi section of cable.

When a circuit consists of conductors of different gages in various sections, the resistance of each section must be considered. For example, when aerial cable is connected to underground cable of another gage, the difference in resistance of the two sections must not only take into account the different sizes of the conductors but also the temperature difference between the aerial cable and the underground cable.

A simpler though less accurate Varley-loop test can be carried out by using only the measurement setup of Fig. 7-8(b), provided that ratio arms A and B are equal and the multiplication ratio one. The usual equation for bridge balance yields

$$\frac{A}{B} = \frac{X_2 + 2X_1}{R_2 + X_2} \qquad (7\text{-}23)$$

Since the ratio arms are equal, $\frac{A}{B} = 1$, and the expression reduces to

$$X_1 = \frac{R_2}{2} \qquad (7\text{-}24)$$

which, in turn, leads to the location of the fault.

7-5   GUARDED WHEATSTONE BRIDGE

7-5.1   Guard Circuits

The measurement of extremely high resistances, such as the insulation resistance of a cable or the leakage resistance of a capacitor (often on the order of several thousands of megohms), is beyond the capability of the ordinary dc Wheatstone bridge. One of the major problems in high-resistance measurements is the leakage that occurs over and around the component or specimen being measured, or over the binding posts by which the component is attached to the instrument, or within the instrument itself. These leakage currents are undesired because they can enter the measuring circuit and affect the measurement accuracy to a considerable extent. Leakage currents, whether inside the instrument itself or associated with the test specimen and its mounting, are particularly noticeable in high-resistance measurements where high voltages are often necessary to obtain sufficient deflection sensitivity. Also, leakage effects are generally variable from day to day, depending on the humidity of the atmosphere.

The effects of leakage paths on the measurement are usually removed by some form of guard circuit. The principle of a simple guard circuit in the R_x arm of a Wheatstone bridge is explained with the aid of Fig. 7-9. Without

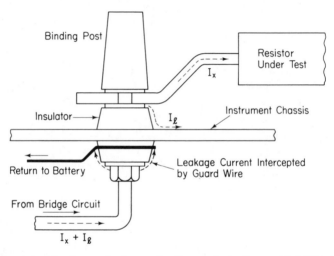

FIGURE 7-9 Simple guard wire on the R_x terminal of a guarded Wheatstone bridge eliminates surface leakage.

a guard circuit, leakage current I_l along the insulated surface of the binding post adds to current I_x through the component under measurement to produce a total circuit current that can be considerably larger than the actual device current. A guard wire, completely surrounding the surface of the insulated post, intercepts this leakage current and returns it to the battery. The guard must be carefully placed so that the leakage current always meets some portion of the guard wire and is prevented from entering the bridge circuit.

In the schematic diagram of Fig. 7-10 the guard around the R_x binding post, indicated by a small circle around the terminal, does not touch any part of the bridge circuitry and is connected directly to the battery terminal. The principle of the guard wire on the binding post can be applied to any internal part of the bridge circuit where leakage affects the measurement; we then speak of a guarded Wheatstone bridge.

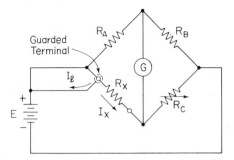

FIGURE 7-10 Guarded terminal returns leakage current to the battery.

7-5.2 Three-terminal Resistance

To avoid the effects of leakage currents external to the bridge circuitry, the junction of ratio arms R_A and R_B is usually brought out as a separate guard terminal on the front panel of the instrument. This guard terminal can be used to connect a so-called *three-terminal resistance*, as shown in Fig. 7-11. The high resistance is mounted on two insulating posts that are fastened to a metal plate. The two main terminals of the resistor are connected to the R_x terminals of the bridge in the usual manner. The third terminal of the resistor is the common point of resistances R_1 and R_2, which represent the leakage paths from the main terminals along the insulating posts to the metal plate, or guard. The guard is connected to the guard terminal on the front panel of the bridge, as indicated in the schematic of Fig. 7-11. This connection puts R_1 in parallel with ratio arm R_A, but since R_1 is very much larger than R_A, its shunting effect is negligible. Similarly, leakage resistance R_2 is in parallel with the galvanometer, but the resistance of R_2 is so much higher than that of the galvanometer that the only effect is a slight reduction in galvanometer sensitivity. The effects of external leakage paths are therefore removed by using the guard circuit on the three-terminal resistance.

If the guard circuit were not used, leakage resistances R_1 and R_2 would be directly across R_x and the measured value of R_x would be considerably in error. Assuming, for example, that the unknown is 100 MΩ and that the leakage resistance from each terminal to the guard is also 100 MΩ, resistance R_x would be measured as 67 MΩ, an error of approximately 33 per cent.

7-5.3 Megohm Bridge

A commercial high-voltage megohm bridge is shown in Fig. 7-12, where the various controls can easily be identified. The large dial in the center of the instrument is the variable ratio arm R_B of Fig. 7-11. The resistance multiplier dial to the right of the large ratio dial corresponds to standard resistor R_C in the circuit diagram and provides for multiplication of the ratio

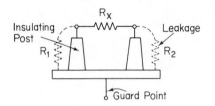

(a)   Three–terminal resistance

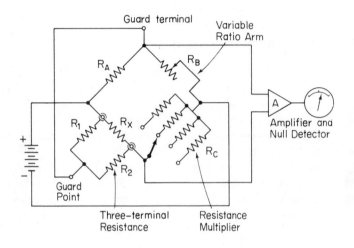

(b) Guarded bridge circuit

FIGURE 7-11   Three-terminal resistance, connected to a guarded high-voltage megohm bridge.

in a number of decade steps. The dc supply voltage is adjustable over several increments from 10 V to 1,000 V, while provision is made to connect an external generator. The null detector is basically a dc amplifier with output meter and the necessary sensitivity to detect small unbalance voltages. The junction of ratio arms R_A and R_B is brought out as a front panel guard terminal, to be used when measuring a three-terminal resistance.

The high-voltage megohm bridge is only one of the instruments used for high-resistance measurements. Other methods may include the use of the well-known *megger* to measure the insulation resistance of electrical machinery, the *direct deflection* method for testing insulation samples, and the *loss-of-charge* method for checking the leakage resistance of capacitors.*

---

*Cf. Melville B. Stout, *Basic Electrical Measurement*, 2nd ed. (Englewood Cliffs, N.J.: Prentice-Hall, Inc., 1960), pp. 126–33.

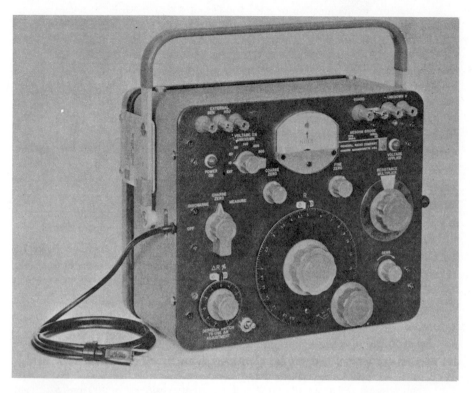

FIGURE 7-12 Commercial megohm bridge, used for the measurement of resistances in the terra-ohm range (courtesy General Radio Company).

REFERENCES

1. Buckingham, H., and E. M. Price, *Principles of Electrical Measurements*, 2nd ed., chap. 9. London: The English Universities Press Limited, 1966.

2. Frank, Ernest, *Electrical Measurement Analysis*, chap. 10. New York: McGraw-Hill Book Company, Inc., 1959.

3. Stout, Melville B., *Basic Electrical Measurements*, 2nd ed., chap. 4. Englewood Cliffs, N.J.: Prentice-Hall, Inc., 1960.

PROBLEMS

1. The four arms of a Wheatstone bridge have resistances of 100 Ω, 1,000 Ω, 500 Ω, and 50.5 Ω, taken in sequence around the bridge. A galvanometer with internal resistance of 75 Ω is connected from the junction of the 100-Ω and 50.5-Ω

resistors to the junction of the 1,000-Ω and 500-Ω resistors. A 4-V battery is connected to the other two corners of the bridge. Use Thévenin's theorem to find (a) the equivalent circuit of this bridge with respect to the galvanometer terminals; (b) the current through the galvanometer.

2. The ratio arms of the Wheatstone bridge of Fig. 7-1 are $R_1 = 1,000\ \Omega$ and $R_2 = 100\ \Omega$, the standard resistance $R_3 = 400\ \Omega$, the unknown $R_x = 41\ \Omega$. A 1.5-V battery with negligible internal resistance is connected from a to b. The galvanometer has an internal resistance of 50 Ω and a current sensitivity of 2 mm/μA.
   (a) Find the equivalent circuit of the bridge with respect to the galvanometer terminals.
   (b) Calculate the deflection of the gálvanometer caused by the unbalance in the circuit.

3. Repeat Prob. 2 with galvanometer and battery interchanged and determine which configuration is more sensitive to unbalance.

4. The standard resistance arm R_3 in Prob. 2 can be adjusted from 0 Ω to 1,000 Ω in steps of 0.1 Ω. The galvanometer deflection can be read within 0.5 mm. When the unknown resistance $R_x = 50\ \Omega$, calculate the resolution of the reading, in ohms and in per cent of the unknown R_x.

5. The three known resistance arms of a Wheatstone bridge have limiting errors of $\pm 0.1\%$. Determine the limiting error of the unknown resistance when measured with this instrument.

6. For the bridge circuit of Fig. 7-1, $R_1 = 1,000\ \Omega$, $R_2 = 4,000\ \Omega$, $R_3 = 100\ \Omega$, and $R_4 = 400\ \Omega$, indicating that the bridge is balanced. The galvanometer has an internal resistance of 100 Ω and a current sensitivity of 100 mm/μA. The battery voltage is 3.0 V. Calculate the galvanometer deflection for an unbalance of 1 Ω in resistance arm R_4. (*Hint:* Determine the Thévenin voltage and resistance in terms of the small unbalance x in R_4. Reduce the expressions for Thévenin voltage and resistance, dropping the increment x from the denominator after reduction.)

7. A Wheatstone bridge has ratio arms of 1,000 Ω and 100 Ω and is used to measure an unknown resistance of 25 Ω. The battery has negligible internal resistance and an emf of 1 V. Two galvanometers are available. Galvanometer A has an internal resistance of 25 Ω and a current sensitivity of 20 mm/μA. Galvanometer B has an internal resistance of 200 Ω and a current sensitivity of 100 mm/μA. Calculate (a) the sensitivity of each galvanometer to unbalance in the R_x arm, expressed in mm/Ω; (b) the ratio of the galvanometer sensitivities to unbalance.

8. A Wheatstone bridge is used to measure high resistances (in the megohm range). The bridge has ratio arms of 10 kΩ and 10 Ω. The standard arm is variable and may be adjusted from 1 Ω to 10 kΩ. A battery of 10 V and negligible internal resistance is connected from the junction of the ratio arms to the opposite corner.
   (a) Calculate the maximum resistance that can be measured by this arrangement.

(b) If the galvanometer has a sensitivity of 200 mm/μA and a resistance of 50 Ω, and the maximum resistance of part (a) is connected to the R_x terminals, calculate the unbalance necessary to produce a galvanometer deflection of 1 mm.

(c) If the galvanometer is replaced by one with a current sensitivity of 1,000 mm/μA and internal resistance of 1,000 Ω, calculate the unbalance in R_x needed to produce a galvanometer deflection of 1 mm.

9. Each of the ratio arms of a laboratory-type Wheatstone bridge has a guaranteed accuracy of $\pm 0.05\%$, while the standard arm has a guaranteed accuracy of $\pm 0.1\%$. The ratio arms are both set at 1,000 Ω (a 1/1 ratio), and the bridge is balanced with the standard arm adjusted to 3,154 Ω. Determine the upper and lower limits of the unknown resistance, based on the guaranteed accuracies of the known bridge arms.

10. The ratio arms of the Kelvin bridge of Fig. 7-5 are 100 Ω each. The galvanometer has an internal resistance of 500 Ω and a current sensitivity of 200 mm/μA. The unknown resistance $R_x = 0.1002\ \Omega$, and the standard resistance is set at 0.1000 Ω. A dc current of 10 A is passed through the standard and the unknown from a 2.2-V battery in series with a rheostat. The resistance of the yoke may be neglected. Calculate (a) the deflection of the galvanometer, in millimeters; (b) the resistance unbalance required to produce a galvanometer deflection of 1 mm. (*Hint:* In the calculation of the Thévenin voltage and resistance, assess the effects of the ratio arms and the rheostat, and neglect the appropriate terms.)

11. The ratio arms of a Kelvin bridge are 1,000 Ω each. The galvanometer has an internal resistance of 100 Ω and a current sensitivity of 500 mm/μA. A dc current of 10 A is passed through the standard arm and the unknown from a 2.2-V battery in series with a rheostat. The standard resistance is set at 0.1000 Ω and the galvanometer deflection is 30 mm. Neglecting the resistance of the yoke, determine the value of the unknown.

8

ac Bridges and Their Application

8-1 GENERAL FORM OF THE AC BRIDGE

8-1.1 Conditions for Bridge Balance

The ac bridge is a natural outgrowth of the dc bridge and in its basic form consists of four bridge arms, a source of excitation, and a null detector. The power source supplies an ac voltage to the bridge at the desired frequency. For measurements at low frequencies, the powerline may serve as the source of excitation; at higher frequencies, an oscillator generally supplies the excitation voltage. The null detector must respond to ac unbalance currents and in its cheapest (but very effective) form consists of a pair of headphones. In other applications, the null detector may consist of an ac amplifier with an output meter, or an electron ray tube (tuning eye) indicator.

The general form of an ac bridge is shown in Fig. 8-1. The four bridge arms Z_1, Z_2, Z_3, and Z_4 are indicated as unspecified impedances and the detector is represented by headphones. As in the case of the Wheatstone bridge for dc measurements, the balance condition in this ac bridge is reached when the detector response is zero, or indicates a null. Balance adjustment to obtain a null response is made by varying one or more of the bridge arms.

The general equation for bridge balance is obtained by using *complex notation* for the impedances of the bridge circuit. (Boldface type is used to indicate quantities in complex notation.) These quantities may be impedances

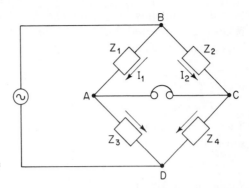

FIGURE 8-1   General form of the
ac bridge.

or admittances as well as voltages or currents. The condition for bridge balance requires that the potential difference from A to C in Fig. 8-1 be zero.
This will be the case when the voltage drop from B to A equals the voltage
drop from B to C, in both *magnitude* and *phase*. In complex notation we
can write

$$\mathbf{E_{BA}} = \mathbf{E_{BC}} \quad \text{or} \quad \mathbf{I_1 Z_1} = \mathbf{I_2 Z_2} \tag{8-1}$$

For zero detector current (the balance condition), the currents are

$$\mathbf{I_1} = \frac{\mathbf{E}}{\mathbf{Z_1 + Z_3}} \tag{8-2}$$

and

$$\mathbf{I_2} = \frac{\mathbf{E}}{\mathbf{Z_2 + Z_4}} \tag{8-3}$$

Substitution of Eqs. (8-2) and (8-3) into Eq. (8-1) yields

$$\mathbf{Z_1 Z_4} = \mathbf{Z_2 Z_3} \tag{8-4a}$$

or when using admittances instead of impedances,

$$\mathbf{Y_1 Y_4} = \mathbf{Y_2 Y_3} \tag{8-4b}$$

Equation (8-4a) is the most convenient form in most cases and is the *general
equation for balance* of the ac bridge. Equation (8-4b) can be used to advantage when dealing with parallel components in bridge arms.

   Equation (8-4a) states that the product of impedances of one pair of
opposite arms must equal the product of impedances of the other pair of
opposite arms, with the impedances expressed in complex notation. If the
impedance is written in the form $\mathbf{Z} = Z \angle \theta$, where Z represents the magnitude
and θ the phase angle of the complex impedance, Eq. (8-4a) can be rewritten
in the form

$$(Z_1 \angle \theta_1)(Z_4 \angle \theta_4) = (Z_2 \angle \theta_2)(Z_3 \angle \theta_3) \tag{8-5}$$

Since in multiplication of complex numbers the magnitudes are *multiplied*
and the phase angles *added*, Eq. (8-5) can also be written as

$$Z_1 Z_4 \angle (\theta_1 + \theta_4) = Z_2 Z_3 \angle (\theta_2 + \theta_3) \tag{8-6}$$

Equation (8-6) shows that *two conditions* must be met *simultaneously* when balancing an ac bridge. The first condition is that the magnitudes of the impedances satisfy the relationship

$$Z_1 Z_4 = Z_2 Z_3 \qquad (8\text{-}7)$$

or, in words:

The products of the magnitudes of the opposite arms must be equal.

The second condition requires that the phase angles of the impedances satisfy the relationship

$$\angle\theta_1 + \angle\theta_4 = \angle\theta_2 + \angle\theta_3 \qquad (8\text{-}8)$$

Again, in words:

The sum of the phase angles of the opposite arms must be equal.

8-1.2   Application of the Balance Equations

The two balance conditions expressed in Eqs. (8-7) and (8-8) can be applied when the impedances of the bridge arms are given in polar form, with both magnitude and phase angle. In the usual case, however, the component values of the bridge arms are given, and the problem is solved by writing the balance equation in complex notation. The following examples illustrate the procedure.

Example 8-1: The impedances of the basic ac bridge of Fig. 8-1 are given as follows:

$$Z_1 = 100\ \Omega\ \angle 80° \text{ (inductive impedance)}$$

$$Z_2 = 250\ \Omega \text{ (pure resistance)}$$

$$Z_3 = 400\ \Omega\ \angle 30° \text{ (inductive impedance)}$$

$$Z_4 = \text{unknown}$$

Determine the constants of the unknown arm.

SOLUTION: The first condition for bridge balance requires that

$$Z_1 Z_4 = Z_2 Z_3 \qquad (8\text{-}7)$$

Substituting the magnitudes of the known components and solving for Z_4, we obtain

$$Z_4 = \frac{Z_2 Z_3}{Z_1} = \frac{250 \times 400}{100} = 1{,}000\ \Omega$$

The second condition for bridge balance requires that the sums of the phase angles of opposite arms be equal, or

$$\theta_1 + \theta_4 = \theta_2 + \theta_3 \qquad (8\text{-}8)$$

Substituting the known phase angles and solving for θ_4, we obtain

$$\theta_4 = \theta_2 + \theta_3 - \theta_1 = 0 + 30 - 80 = -50°$$

Hence the unknown impedance Z_4 can be written in polar form as

$$Z_4 = 1,000 \ \Omega \ \angle -50°$$

indicating that we are dealing with a capacitive element, possibly consisting of a series combination of a resistor and a capacitor.

The problem becomes slightly more complex when the component values of the bridge arms are specified and the impedances are to be expressed in complex notation. In this case, the inductive or capacitive reactances can only be calculated when the frequency of the excitation voltage is known, as Example 8-2 shows.

Example 8-2: The ac bridge of Fig. 8-1 is in balance with the following constants: arm AB, $R = 450 \ \Omega$; arm BC, $R = 300 \ \Omega$ in series with $C = 0.265 \ \mu F$; arm CD, unknown; arm DA, $R = 200 \ \Omega$ in series with $L = 15.9$ mH. The oscillator frequency is 1 kHz. Find the constants of arm CD.

SOLUTION: The general equation for bridge balance states that

$$Z_1 Z_2 = Z_2 Z_3 \qquad \text{(8-4a)}$$

The impedances of the bridge arms, expressed in complex notation, are:

$$Z_1 = R = 450 \ \Omega$$
$$Z_2 = R - j/\omega C = (300 - j600) \ \Omega$$
$$Z_3 = R + j\omega L = (200 + j100) \ \Omega$$
$$Z_4 = \text{unknown}$$

Substituting the known values in Eq. (8-4a) and solving for the unknown yields

$$Z_4 = \frac{450 \times (200 + j100)}{(300 - j600)} = +j150 \ \Omega$$

This result indicates that Z_4 is a pure inductance with an inductive reactance of 150 Ω at a frequency of 1 kHz. Since the inductive reactance $X_L = 2\pi f L$, we solve for L and obtain $L = 23.9$ mH.

8-2 COMPARISON BRIDGES

8-2.1 Capacitance Comparison Bridge

In its basic form the ac bridge can be used for the measurement of an unknown inductance or capacitance by *comparing* it with a known inductance or capacitance. A basic capacitance comparison bridge is shown in Fig. 8-2. The ratio arms are both resistive and are represented by R_1 and R_2. The standard arm consists of capacitor C_s in series with resistor R_s, where C_s is a high-quality standard capacitor and R_s a variable resistor. C_x represents the unknown capacitance and R_x is the leakage resistance of the capacitor.

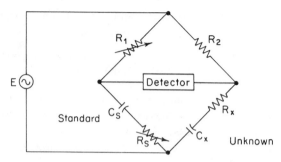

FIGURE 8-2   Capacitance comparison bridge.

To write the balance equation, we first express the impedances of the four bridge arms in complex notation and we find that

$$\mathbf{Z}_1 = R_1; \quad \mathbf{Z}_2 = R_2; \quad \mathbf{Z}_3 = R_s - \frac{j}{\omega C_s}; \quad \mathbf{Z}_4 = R_x - \frac{j}{\omega C_x}$$

Substituting these impedances in Eq. (8-4a), the general equation for bridge balance, we obtain

$$R_1\left(R_x - \frac{j}{\omega C_x}\right) = R_2\left(R_s - \frac{j}{\omega C_s}\right) \tag{8-9}$$

which can be expanded to

$$R_1 R_x - R_1\frac{j}{\omega C_x} = R_2 R_s - R_2\frac{j}{\omega C_s} \tag{8-10}$$

Two complex numbers are equal when both their real terms and their imaginary terms are equal. Equating the real terms of Eq. (8-10), we obtain

$$R_1 R_x = R_2 R_s \quad \text{or} \quad R_x = R_s\frac{R_2}{R_1} \tag{8-11}$$

Equating the imaginary terms of Eq. (8-10), we obtain

$$\frac{jR_1}{\omega C_x} = \frac{jR_2}{\omega C_s} \quad \text{or} \quad C_x = C_s\frac{R_1}{R_2} \tag{8-12}$$

Equations (8-11) and (8-12) describe the two balance conditions that must be met simultaneously and they also show the two unknowns C_x and R_x expressed in terms of the known bridge components.

To satisfy both balance conditions, the bridge must contain two variable elements in its configuration. Any two of the available four elements could be chosen, although in practice capacitor C_s is a high-precision standard capacitor of fixed value and is not available for adjustment. Inspection of the balance equations shows that R_s does not appear in the expression for C_x. Hence to eliminate any interaction between the two balance controls, R_s

is an obvious choice as a variable element. We further accept that R_1 is the second variable element, as indicated in Fig. 8-2.

Since we are measuring an unknown capacitor whose resistive effects could be very small, the first adjustment should be made for the capacitive term, and R_1 is therefore adjusted for minimum sound in the headphones. In most cases, the sound will not altogether disappear, because the second balance condition has not yet been met. Hence R_s is adjusted for balance of the resistive term and the sound is made to decrease further. It is found that alternate adjustment of the two resistors is necessary to produce zero output in the headphones and to achieve the true balance condition. The need for alternate adjustment becomes clear when we realize that any change in R_1 not only affects the capacitive balance equation but also the resistive balance equation, since R_1 appears in both expressions.

The process of alternate manipulation of R_1 and R_s is typical of the general balancing procedure for ac bridges and is said to cause *convergence* of the balance point. It should also be noted that the frequency of the voltage source does not enter either of the balance equations and the bridge is therefore said to be *independent* of the frequency of the applied voltage.

8-2.2 Inductance Comparison Bridge

The general configuration of the inductance comparison bridge is similar to that of the capacitance comparison bridge. The unknown inductance is determined by comparison with a known standard inductor, as shown in the diagram of Fig. 8-3. The derivation of the balance equations essentially

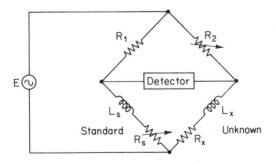

FIGURE 8-3   Inductance comparison bridge.

follows the same steps as for the capacitance comparison bridge and will not be presented in detail.

It can be shown that the inductive balance equation yields

$$L_x = L_s \frac{R_2}{R_1} \qquad (8\text{-}13)$$

and the resistive balance equation gives

$$R_x = R_s \frac{R_2}{R_1} \qquad\qquad (8\text{-}14)$$

In this bridge, R_2 is chosen as the inductive balance control, and R_s is the resistive balance control.

The measurement range of the standard comparison bridge of Fig. 8-3 can be extended by modifying the circuit slightly. This is shown in Fig. 8-4, where variable resistance r can be connected by means of switch S to

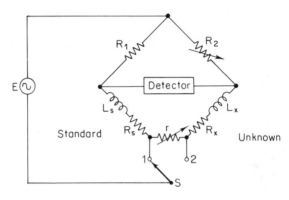

FIGURE 8-4   Inductance comparison bridge with extended measurement range.

either the standard arm (position 1) or the unknown arm (position 2). With the switch in position 1, the solution for R_x is

$$R_x = (R_s + r)\frac{R_2}{R_1} \qquad\qquad (8\text{-}15)$$

With the switch in position 2, the solution for R_x is

$$R_x = R_s \frac{R_2}{R_1} - r \qquad\qquad (8\text{-}16)$$

Since the resistive component of an inductor is usually much larger than that of a capacitor, the resistive adjustment becomes rather important and should be carried out first. The addition of resistor r gives the option of extending the adjustment range for the resistive balance equation.

8-3   MAXWELL BRIDGE

The Maxwell bridge, whose schematic diagram is shown in Fig. 8-5, measures an unknown *inductance* in terms of a known *capacitance*. One of the ratio arms has a resistance and a capacitance in *parallel*, and it may now prove somewhat easier to write the balance equations using the *admittance* of arm 1 instead of its impedance.

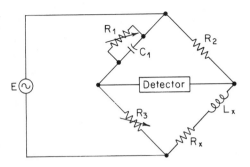

FIGURE 8-5   Maxwell bridge for inductance measurements.

Rearranging the general equation for bridge balance, as expressed in Eq. (8-4a), we obtain

$$\mathbf{Z}_x = \mathbf{Z}_2 \mathbf{Z}_3 \mathbf{Y}_1 \qquad (8\text{-}17)$$

where $\mathbf{Y}_1$ is the admittance of arm 1. Reference to Fig. 8-5 shows that

$$\mathbf{Z}_2 = R_2; \qquad \mathbf{Z}_3 = R_3; \qquad \text{and} \quad \mathbf{Y}_1 = \frac{1}{R_1} + j\omega C_1$$

Substitution of these values in Eq. (8-17) gives

$$\mathbf{Z}_x = R_x + j\omega L_x = R_2 R_3 \left(\frac{1}{R} + j\omega C_1 \right) \qquad (8\text{-}18)$$

Separation of the real and imaginary terms yields

$$R_x = \frac{R_2 R_3}{R_1} \qquad (8\text{-}19)$$

and

$$L_x = R_2 R_3 C_1 \qquad (8\text{-}20)$$

where the resistances are expressed in ohms, inductance in henrys, and capacitance in farads.

The Maxwell bridge is limited to the measurement of *medium-Q coils* $(1 < Q < 10)$. This can be shown by considering the second balance condition which states that the sum of the phase angles of one pair of opposite arms must be equal to the sum of the phase angles of the other pair. Since the phase angles of the resistive elements in arm 2 and arm 3 add up to 0°, the sum of the angles of arm 1 and arm 4 must also add up to 0°. The phase angle of a *high-Q* coil will be very nearly 90° (positive), which requires that the phase angle of the capacitive arm must also be very nearly 90° (negative). This in turn means that the resistance of R_1 must be very large indeed, which can be very impractical. High-Q coils are therefore generally measured on the Hay bridge, presented in Sec. 8-4.

The Maxwell bridge is also unsuited for the measurement of coils with a very low Q-value $(Q < 1)$ because of balance convergence problems. Very low Q-values occur in inductive resistors, for example, or in an RF coil if measured at low frequency. As can be seen from the equations for R_x and L_x, adjustment for inductive balance by R_3 upsets the resistive balance by R_1 and gives the effect known as *sliding balance*. Sliding balance describes the interaction between controls, so that when we balance with R_1 and then with R_3, then go back to R_1, we find a new balance point. The balance point appears to move or *slide* toward its final point after many adjustments. Interaction does not occur when R_1 and C_1 are used for the balance adjustments, but a variable capacitor is not always suitable.

The usual procedure for balancing the Maxwell bridge is by first adjusting R_3 for inductive balance and then adjusting R_1 for resistive balance. Returning to the R_3 adjustment, we find that the resistive balance is being disturbed and moves to a new value. This process is repeated and gives *slow* convergence to final balance. For medium-Q coils, the resistance effect is not pronounced, and balance is reached after a few adjustments.

8-4 HAY BRIDGE

The Hay bridge of Fig. 8-6 differs from the Maxwell bridge by having resistor R_1 in series with standard capacitor C_1 instead of in parallel. It is immediately apparent that for large phase angles, R_1 should have a very low value. The Hay circuit is therefore more convenient for measuring high-Q coils.

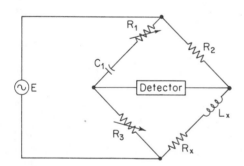

FIGURE 8-6 Hay bridge for inductance measurements.

The balance equations are again derived by substituting the values of the impedances of the bridge arms into the general equation for bridge balance. For the circuit of Fig. 8-6 we find that

$$\mathbf{Z}_1 = R_1 - \frac{j}{\omega C_1}; \qquad \mathbf{Z}_2 = R_2; \qquad \mathbf{Z}_3 = R_3; \qquad \mathbf{Z}_x = R_x + j\omega L_x$$

Substituting these values in Eq. (8-4a), we get

$$\left(R_1 - \frac{j}{\omega C_1}\right)(R_x + j\omega L_x) = R_2 R_3 \qquad \textbf{(8-21)}$$

which expands to

$$R_1R_x + \frac{L_x}{C_1} - \frac{jR_x}{\omega C_1} + j\omega L_x R_1 = R_2 R_3$$

Separating the real and imaginary terms, we obtain

$$R_1 R_x + \frac{L_x}{C_1} = R_2 R_3 \tag{8-22}$$

and

$$\frac{R_x}{\omega C_1} = \omega L_x R_1 \tag{8-23}$$

Both Eq. (8-22) and Eq. (8-23) contain L_x and R_x, and we must solve these equations simultaneously. This yields

$$R_x = \frac{\omega^2 C_1^2 R_1 R_2 R_3}{1 + \omega^2 C_1^2 R_1^2} \tag{8-24}$$

$$L_x = \frac{R_2 R_3 C_1}{1 + \omega^2 C_1^2 R_1^2} \tag{8-25}$$

These expressions for the unknown inductance and resistance both contain the angular velocity ω and it therefore appears that the frequency of the voltage source must be known accurately. That this is not true when a high-Q coil is being measured follows from the following considerations: Remembering that the sum of the opposite sets of phase angles must be equal, we find that the inductive phase angle must be equal to the capacitive phase angle, since the resistive angles are zero. Figure 8-7 shows that the tangent of the

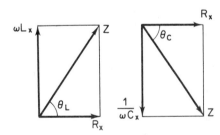

FIGURE 8-7   Impedance triangles illustrate inductive and capacitive phase angles.

inductive phase angle equals

$$\tan \theta_L = \frac{X_L}{R} = \frac{\omega L_x}{R_x} = Q \tag{8-26}$$

and that of the capacitive phase angle is

$$\tan \theta_C = \frac{X_C}{R} = \frac{1}{\omega C_1 R_1} \tag{8-27}$$

When the two phase angles are equal, their tangents are also equal and we can write

$$\tan \theta_L = \tan \theta_C \quad \text{or} \quad Q = \frac{1}{\omega C_1 R_1} \tag{8-28}$$

Returning now to the term $(1 + \omega^2 C_1^2 R_1^2)$ which appears in Eqs. (8-24) and (8-25), we find that, after substituting Eq. (8-28) in the expression for L_x, Eq. (8-25) reduces to

$$L_x = \frac{R_2 R_3 C_1}{1 + (1/Q)^2} \tag{8-29}$$

For a value of Q greater than ten, the term $(1/Q)^2$ will be smaller than $\frac{1}{100}$ and can be neglected. Equation (8-25) therefore reduces to the expression derived for the Maxwell bridge,

$$L_x = R_2 R_3 C_1$$

The Hay bridge is suited for the measurement of high-Q inductors, especially for those inductors having a Q greater than ten. For Q-values smaller than ten, the term $(1/Q)^2$ becomes important and cannot be neglected. In this case, the Maxwell bridge is more suitable.

8-5 SCHERING BRIDGE

The *Schering bridge*, one of the most important ac bridges, is used extensively for the measurement of *capacitors*. It offers some decided advantages over the capacitance comparison bridge, discussed in Sec. 8-2.1. Although the Schering bridge is used for capacitance measurements in a general sense, it is particularly useful for measuring insulating properties, i.e., for phase angles very nearly 90°.

The basic circuit arrangement is shown in Fig. 8-8, and inspection of

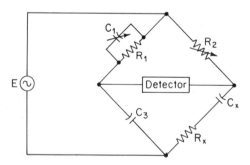

FIGURE 8-8 Schering bridge for the measurement of capacitance.

the circuit shows a strong resemblance to the comparison bridge. Notice that arm 1 now contains a parallel combination of a resistor and a capacitor, and the standard arm contains only a capacitor. The standard capacitor is usually a high-quality mica capacitor for general measurement work or an air capacitor for insulation measurements. A good-quality mica capacitor has very low losses (no resistance) and therefore a phase angle of approximately 90°. An air capacitor, when designed carefully, has a very stable value and a very small electric field; the insulating material to be tested can easily be kept out of any strong fields.

The balance conditions require that the sum of the phase angles of arms 1 and 4 equals the sum of the phase angles of arms 2 and 3. Since the standard capacitor is in arm 3, the sum of the phase angles of arm 2 and arm 3 will be $0° + 90° = 90°$. In order to obtain the 90°-phase angle needed for balance, the sum of the angles of arm 1 and arm 4 must equal 90°. Since in general measurement work the unknown will have a phase angle smaller than 90°, it is necessary to give arm 1 a small capacitive angle by connecting capacitor C_1 in parallel with resistor R_1. A small capacitive angle is very easy to obtain, requiring a small capacitor across resistor R_1.

The balance equations are derived in the usual manner, and by substituting the corresponding impedance and admittance values in the general equation, we obtain

$$\mathbf{Z}_x = \mathbf{Z}_2 \mathbf{Z}_3 \mathbf{Y}_1$$

or

$$R_x - \frac{j}{\omega C_x} = R_2 \left(\frac{-j}{\omega C_3} \right) \left(\frac{1}{R_1} + j\omega C_1 \right)$$

and expanding

$$R_x - \frac{j}{\omega C_x} = \frac{R_2 C_1}{C_3} - \frac{jR_2}{\omega C_3 R_1} \qquad (8\text{-}30)$$

Equating the real terms and the imaginary terms, we find that

$$R_x = R_2 \frac{C_1}{C_3} \qquad (8\text{-}31)$$

$$C_x = C_3 \frac{R_1}{R_2} \qquad (8\text{-}32)$$

As can be seen from the circuit diagram of Fig. 8-8, the two variables chosen for the balance adjustment are capacitor C_1 and resistor R_2. There seems to be nothing unusual about the balance equations or the choice of variable components, but consider for a moment how the quality of a capacitor is defined.

The *power factor* (PF) of a series RC combination is defined as the cosine of the phase angle of the circuit. Therefore the PF of the unknown equals $PF = R_x/Z_x$. For phase angles very close to 90°, the reactance is almost equal to the impedance and we can approximate the power factor to

$$PF \simeq \frac{R_x}{X_x} = \omega C_x R_x \qquad (8\text{-}33)$$

The *dissipation factor* of a series RC circuit is defined as the cotangent of the phase angle and therefore, by definition, the dissipation factor

$$D = \frac{R_x}{X_x} = \omega C_x R_x \qquad (8\text{-}34)$$

Incidentally, since the quality of a coil is defined by $Q = X_L/R_L$, we find that the dissipation factor, D, is the reciprocal of the quality factor, Q, and therefore $D = 1/Q$. The dissipation factor tells us something about the quality of a capacitor; i.e., how close the phase angle of the capacitor is to the ideal value of 90°. By substituting the value of C_x in Eq. (8-32) and of R_x in Eq. (8-31) into the expression for the dissipation factor, we obtain

$$D = \omega R_1 C_1 \tag{8-35}$$

If resistor R_1 in the Schering bridge of Fig. 8-8 has a fixed value, the dial of capacitor C_1 may be calibrated directly in dissipation factor D. This is the usual practice in a Schering bridge. Notice that the term ω appears in the expression for the dissipation factor [Eq. (8-35)]. This means, of course, that the calibration of the C_1 dial holds for only one particular frequency at which the dial is calibrated. A different frequency can be used, provided that a correction is made by multiplying the C_1 dial reading by the ratio of the two frequencies.

8-6   UNBALANCE CONDITIONS

It sometimes happens that an ac bridge cannot be balanced at all simply because one of the stated balance conditions (Sec. 8-1) cannot be met. Consider, for example, the circuit of Fig. 8-9, where $\mathbf{Z}_1$ and $\mathbf{Z}_4$ are induc-

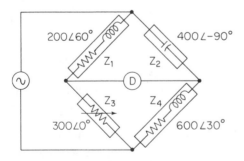

FIGURE 8-9   An ac bridge that cannot be balanced.

tive elements (positive phase angles), $\mathbf{Z}_2$ is a pure capacitance ($-90°$ phase angle), and $\mathbf{Z}_3$ is a variable resistance (zero phase angle). The resistance of R_3 needed to obtain bridge balance can be determined by applying the first balance condition (magnitudes) and we find that

$$R_3 = \frac{\mathbf{Z}_1 \mathbf{Z}_4}{\mathbf{Z}_2} = \frac{200 \times 600}{400} = 300 \ \Omega$$

Hence adjusting R_3 to a value of 300 Ω will satisfy the first condition.

Considering the second balance condition (phase angles) yields the following situation:

$$\theta_1 + \theta_4 = +60° + 30° = +90°$$
$$\theta_2 + \theta_3 = -90° + \ \ 0° = -90°$$

Obviously, $\theta_1 + \theta_4 \neq \theta_2 + \theta_3$, and the second condition is not satisfied. In this case, bridge balance cannot be obtained.

An interesting illustration of a bridge balancing problem is given in Example 8-3, where minor adjustments to one or more of the bridge arms results in a situation where balance can be obtained.

Example 8-3: Consider the circuit of Fig. 8-10(a) and determine whether or not the bridge is in complete balance. If not, show two ways in which it can be made to balance and specify numerical values for any additional components. Assume that bridge arm 4 is the unknown that cannot be modified.

SOLUTION: Inspection of the circuit shows that the first balance condition (magnitudes) can easily be met by slightly increasing the resistance of R_3. The second balance condition requires that $\theta_1 + \theta_4 = \theta_2 + \theta_3$

where
$$\theta_1 = -90° \text{ (pure capacitance)}$$
$$\theta_2 = \theta_3 = 0° \text{ (pure resistance)}$$
$$\theta_4 < +90° \text{ (inductive impedance)}$$

Obviously, balance is not possible with the configuration of Fig. 8-10(a) because the sum of θ_1 and θ_4 will be slightly negative while $\theta_2 + \theta_3$ will be exactly 0°. Balance can be restored by modifying the circuit in such a way that the phase angle condition is satisfied. There are basically two methods to accomplish this: The first option is to modify $\mathbf{Z}_1$ so that its phase angle is decreased to less than 90° (equal to θ_4) by placing a resistor in parallel with the capacitor. This modification results in a Maxwell bridge configuration, as shown in Fig. 8-10(b). The resistance of R_1 can be determined by the standard approach of Sec. 8-3, using the admittance of arm 1, and we can write

$$\mathbf{Y}_1 = \frac{\mathbf{Z}_4}{\mathbf{Z}_2 \mathbf{Z}_3}$$

where
$$\mathbf{Y}_1 = \frac{1}{R_1} + \frac{j}{1,000}$$

Substituting the known values and solving for R_1, we obtain

$$\frac{1}{R_1} + \frac{j}{1,000} = \frac{100 + j500}{500 \times 1,000}$$

and
$$R_1 = 5,000 \ \Omega$$

It should be noted that the addition of R_1 upsets the first balance condition of the circuit (the magnitude of $\mathbf{Z}_1$ has changed) and variable resistor R_3 should be adjusted to compensate for this effect.

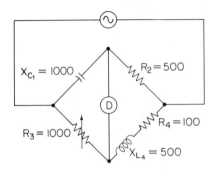

(a) Unbalanced
    condition

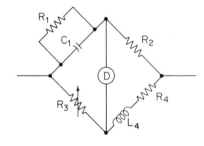

(b) Bridge balance is
    restored by adding
    a resistor to arm 1.
    (Maxwell configuration).

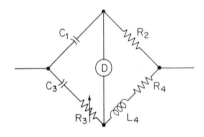

(c) Alternative method
    of restoring bridge
    balance, by adding
    a capacitor to arm 3.

FIGURE 8-10   A bridge balancing problem.

The second option is to modify the phase angle of arm 2 or arm 3 by adding a series capacitor, as shown in Fig. 8-10(c). Again writing the general balance equation, using impedances this time, we obtain

$$\mathbf{Z}_3 = \frac{\mathbf{Z}_1 \mathbf{Z}_4}{\mathbf{Z}_2}$$

Substituting the component values and solving for X_C yields

$$1,000 - jX_C = \frac{-j1,000(100 + j500)}{500}$$

or $X_C = 200 \ \Omega$

In this case, also, the magnitude of $\mathbf{Z}_3$ has increased so that the first balance condition has changed. A small readjustment of R_3 is necessary to restore balance.

8-7 WIEN BRIDGE

The Wien bridge is presented here not only for its use as an ac bridge to measure *frequency*, but also for its application in various other useful circuits. We find, for example, a Wien bridge in the harmonic distortion analyzer, where it is used as a *notch filter*, discriminating against one specific frequency. The Wien bridge also finds application in audio- and HF oscillators as the *frequency-determining* element. In this chapter, however, the Wien bridge is discussed in its basic form, designed to measure frequency; in other chapters it is shown as an element of different types of instrument.

The Wien bridge has a series RC combination in one arm and a parallel RC combination in the adjoining arm (see Fig. 8-11). The impedance of arm

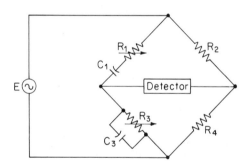

FIGURE 8-11 Frequency measurement with the Wien bridge.

1 is $\mathbf{Z}_1 = R_1 - j/\omega C_1$. The admittance of arm 3 is $\mathbf{Y}_3 = 1/R_3 + j\omega C_3$. Using the basic equation for bridge balance and substituting the appropriate values, we obtain

$$R_2 = \left(R_1 - \frac{j}{\omega C_1}\right) R_4 \left(\frac{1}{R_3} + j\omega C_3\right) \tag{8-36}$$

Expanding this expression, we get

$$R_2 = \frac{R_1 R_4}{R_3} + (j\omega C_3 R_1 R_4) - \frac{jR_4}{\omega C_1 R_3} + \frac{R_4 C_3}{C_1} \tag{8-37}$$

Equating the *real* terms, we obtain

$$R_2 = \frac{R_1 R_4}{R_3} + \frac{R_4 C_3}{C_1} \tag{8-38}$$

which reduces to

$$\frac{R_2}{R_4} = \frac{R_1}{R_3} + \frac{C_3}{C_1} \tag{8-39}$$

Equating the *imaginary* terms, we obtain

$$\omega C_3 R_1 R_4 = \frac{R_4}{\omega C_1 R_3} \qquad (8\text{-}40)$$

where $\omega = 2\pi f$,

and solving for f, we get

$$f = \frac{1}{2\pi \sqrt{C_1 C_3 R_1 R_3}} \qquad (8\text{-}41)$$

Notice that the two conditions for bridge balance now result in an expression determining the required resistance ratio, R_2/R_4, and another expression determining the frequency of the applied voltage. In other words, if we satisfy Eq. (8-39) and also excite the bridge with a frequency described by Eq. (8-41), the bridge will be in balance.

In most Wien bridge circuits, the components are chosen such that $R_1 = R_3$ and $C_1 = C_3$. This reduces Eq. (8-39) to $R_2/R_4 = 2$ and Eq. (8-41) to

$$f = \frac{1}{2\pi RC} \qquad (8\text{-}42)$$

which is the general expression for the frequency of the Wien bridge. In a practical bridge, capacitors C_1 and C_3 are fixed capacitors, and resistors R_1 and R_3 are variable resistors controlled by a common shaft. Provided now that $R_2 = 2R_4$, the bridge may be used as a frequency-determining device balanced by a single control. This control may be calibrated directly in terms of frequency.

Because of its frequency sensitivity, the Wien bridge may be difficult to balance (unless the waveform of the applied voltage is purely sinusoidal). Since the bridge is *not* balanced for any harmonics present in the applied voltage, these harmonics will sometimes produce an output voltage masking the true balance point.

8-8   WAGNER GROUND CONNECTION

The discussion so far has assumed that the four bridge arms consist of simple lumped impedances which do not interact in any way. In practice, however, *stray capacitances* exist between the various bridge elements and ground, and also between the bridge arms themselves. These stray capacitances shunt the bridge arms and cause measurement errors, particularly at the higher frequencies or when small capacitors or large inductors are measured. One way to control stray capacitances is by shielding the arms and connecting the shields to ground. This does not eliminate the capacitances but at least makes them constant in value, and they can therefore be compensated.

One of the most widely used methods for eliminating some of the effects of stray capacitance in a bridge circuit is the *Wagner ground connection*. This circuit eliminates the troublesome capacitance which exists between the detector terminals and ground. Figure 8-12 shows the circuit of a capacitance

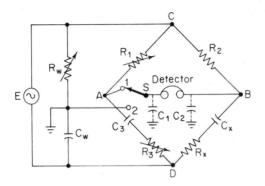

FIGURE 8-12   The Wagner ground connection eliminates the effect of stray capacitances across the detector.

bridge, where C_1 and C_2 represent these stray capacitances. The oscillator is removed from its usual ground connection and bridged by a series combination of resistor R_w and capacitor C_w. The junction of R_w and C_w is grounded and is called the Wagner ground connection. The procedure for initial adjustment of the bridge is as follows: The detector is connected to point 1, and R_1 is adjusted for null or minimum sound in the headphones. The switch is then thrown to position 2, which connects the detector to the Wagner ground point. Resistor R_w is now adjusted for minimum sound. When the switch is thrown to position 1 again, some unbalance will probably be shown. Resistors R_1 and R_3 are then adjusted for minimum detector response, and the switch is again thrown to position 2. A few adjustments of R_w and R_1 (and R_3) may be necessary before a null is reached on both switch positions. When null is finally obtained, points 1 and 2 are at the same potential, and this is ground potential. Stray capacitances C_1 and C_2 are then effectively shorted out and have no effect on normal bridge balance. There are also capacitances from points C and D to ground, but the addition of the Wagner ground point eliminates them from the detector circuit, since current through these capacitances will enter through the Wagner ground connection.

The capacitances across the bridge arms are not eliminated by the Wagner ground connection and they will still affect the accuracy of the measurement. The idea of the Wagner ground can also be applied to other bridges, as long as care is taken that the grounding arms duplicate the impedance of one pair of bridge arms across which they are connected. Since the addition of the Wagner ground connection does not affect the balance conditions, the procedure for measurement remains unaltered.

8-9 UNIVERSAL IMPEDANCE BRIDGE

One of the most useful and versatile laboratory bridges is the universal impedance bridge. Several of the bridge configurations discussed so far are combined in a single instrument capable of measuring both dc and ac resistance, the inductance and storage factor Q of an inductor, and the capacitance and dissipation factor D of a capacitor. A representative example of a universal impedance bridge is given in Fig. 8-13 which clearly shows the arrangement of the various front panel controls.

The universal bridge consists of four basic bridge circuits, together with suitable switching, ac and dc detectors, ac and dc generators, and impedance standards. The Wheatstone bridge circuit is used for both ac and dc resistance measurements. Capacitance is measured in terms of a standard capacitor and precision resistors in a four-arm network, with means for determining the losses in the unknown capacitor. The Maxwell configuration is used for low-Q inductor measurements and the Hay bridge for measurements of inductors with a Q above ten. For dc resistance measurements, a suspension galvanometer with a current sensitivity of 0.5 μA per division is used. A selective

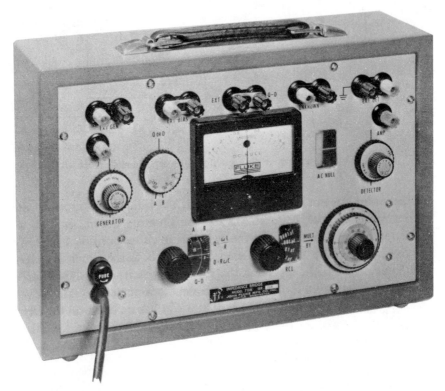

FIGURE 8-13   Universal impedance bridge (courtesy John Fluke Manufacturing Company).

amplifier operating an electron ray tube is used as a null indicator for all ac measurements. Terminals are provided for the connection of external ac and dc null detectors. High-impedance headphones may also be connected and used as an ac detector. The dc generator is a simple dc power supply. The ac generator consists of an oscillator using plug-in RC networks for frequency selection, with a frequency of 10 kHz as the standard frequency.

Figure 8-14 shows the various bridge configurations used in this impedance bridge. It will be found that most general-purpose bridges incorporate the same ideas as the instrument described here.

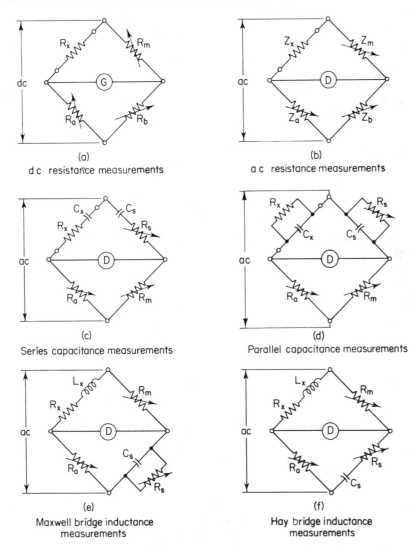

FIGURE 8-14   Bridge configurations of the universal impedance bridge of Fig. 8-13.

REFERENCES

1. Stout, Melville B., *Basic Electrical Measurements*, 2nd ed., chaps. 9, 10. Englewood Cliffs, N.J.: Prentice-Hall, Inc., 1960.

2. Frank, Ernest, *Electrical Measurement Analysis*, chaps. 9, 13. New York: McGraw-Hill Book Company, Inc., 1959.

PROBLEMS

1. A balanced ac bridge has the following constants: arm AB, $R = 2,000\ \Omega$ in parallel with $C = 0.047\ \mu F$; BC, $R = 1,000\ \Omega$ in series with $C = 0.47\ \mu F$; CD, unknown; DA, $C = 0.5\ \mu F$. The frequency of the oscillator is 1,000 Hz. Find the constants of arm CD.

2. A bridge is balanced at 1,000 Hz and has the following constants: AB, $0.2\ \mu F$ pure capacitance; BC, $500\ \Omega$ pure resistance; CD, unknown; DA, $R = 300\ \Omega$ in parallel with $C = 0.1\ \mu F$. Find the R and C or L constants of arm CD, considered as a series circuit.

3. A 1,000-Hz bridge has the following constants: arm AB, $R = 1,000\ \Omega$ in parallel with $C = 0.5\ \mu F$; BC, $R = 1,000\ \Omega$ in series with $C = 0.5\ \mu F$; CD, $L = 30$ mH in series with $R = 200\ \Omega$. Find the constants of arm DA to balance the bridge. Express the result as a pure R in series with a pure C or L, and also as a pure R in parallel with a pure C or L.

4. An ac bridge has in arm AB a pure capacitance of $0.2\ \mu F$; in arm BC, a pure resistance of $500\ \Omega$; in arm CD, a series combination of $R = 50\ \Omega$ and $L = 0.1$ H. Arm DA consists of a capacitor $C = 0.4\ \mu F$ in series with a variable resistor R_s. $\omega = 5,000$ rad/s.
   (a) Find the value of R_s to obtain bridge balance.
   (b) Can complete balance be attained by adjustment of R_s? If not, specify the position and the value of an adjustable resistance to complete the balance.

5. A balanced ac bridge has the following constants: AB, $R = 500\ \Omega$; BC, $R = 1,000\ \Omega$; CD, unknown; DA, $C = 0\ 2\ \mu F$. A voltage of 10 V at 1,000 Hz is impressed on the bridge at points A and C.
   (a) Find the constants of the unknown.
   (b) The 1,000-Ω resistor is changed to $1,002\ \Omega$. Find the voltage across the high-impedance detector.
   (c) Repeat (b), with the detector and the generator interchanged.

6. An unbalanced ac bridge has the following constants: arm AB, $R = 2,000\ \Omega$ in parallel with $C = 0.2\ \mu F$; BC, $R = 1,500\ \Omega$; CD, $L = 0.8$ H in series with $R = 500\ \Omega$; DA, $R = 2,000\ \Omega$. The oscillator has an output of 20 V and is connected to A and C. The frequency is 1,000 Hz.
   (a) What should the values of the constants in arm CD be for bridge balance?
   (b) For the original bridge constants given in this problem, find the voltage across the high-impedance detector.

7. A bridge is balanced at 1,000 Hz and has pure resistance ratio arms, AB 1,500 Ω, BC 1,000 Ω. The unknown is connected from C to D. Arm DA has a standard capacitor of 0.1 μF and negligible internal resistance, to which is added a series resistance of 10 Ω to give balance. The generator has an output of 15 V and is connected from B to D. The detector is a high-impedance voltmeter.
   (a) Find the constants of arm CD.
   (b) Find the detector voltage for an increase of 10 Ω in arm BC.

8. In the ac bridge of Fig. 8-3, $R_1 = 521$ Ω, $R_2 = 1,200$ Ω, $C_s = 0.045$ μF, and $R_s = 12.1$ Ω. The frequency of the oscillator is 10 kHz.
   (a) Determine the values of R_x and C_x.
   (b) It is found that R_1 has 2 μH series inductance and 550 pF shunt capacitance, R_2 has 5 μH series inductance and 1,050 pF shunt capacitance, C_s has 1.5 MΩ shunt resistance, and R_s is unchanged. Determine the error in measuring R_x and C_x as in part (a).

9. An ac bridge has the following constants: arm AB, $R = 1,000$ Ω in parallel with $C = 0.159$ μF; BC, $R = 1,000$ Ω; CD, $R = 500$ Ω; DA, $C = 0.636$ μF in series with an unknown resistance. Find the frequency for which this bridge is in balance and determine the value of the resistance in arm DA to produce this balance.

10. An ac bridge has the following constants: arm AB, $R = 800$ Ω in parallel with $C = 0.4$ μF; BC, $R = 500$ Ω in series with $C = 1.0$ μF; CD, $R = 1,200$ Ω; DA, pure R of unknown value.
   (a) Find the frequency for which the bridge is in balance.
   (b) Find the resistance required in arm DA to produce balance.

11. An ac bridge has pure resistances in three of its arms: R_1 in arm AB, R_2 in arm BC, and R_3 in arm DA. Arm CD consists of a coil with series components of R and L, shunted by a variable capacitance, C. Derive the balance equations for this bridge, measuring the constants of the coil in terms of the other components. Express the Q of the coil in terms of bridge-arm constants at balance.

12. An ac bridge, marked ABCD around the corners, has the following constants: AB, pure capacitance 0.01 μF; BC, pure resistance 2,500 Ω; CD, unknown; DA, capacitance 0.02 μF in series with a resistance of 7,500 Ω. The bridge is balanced at a frequency such that $\omega = 50,000$ rad/s.
   (a) Find the unknown for the bridge constants stated.
   (b) There is a stray capacitance of 100 pF across arm DA, in addition to the constants above. Find the true value of the unknown.

9

Oscilloscopes

9-1 INTRODUCTION

The *cathode ray oscilloscope* (hereafter CRO) is an extremely useful and versatile laboratory instrument used for measurement and analysis of waveforms and other phenomena in electronic circuits. CROs are basically very fast X-Y plotters that display an input signal versus another signal or versus time. The "stylus" of this plotter is a luminous spot that moves over the display area in response to input voltages.

In the usual CRO application the X axis, or horizontal input, is an internally generated linear ramp voltage, or time base, which moves the luminous spot periodically from left to right over the display area or screen. The voltage under examination is applied to the Y axis, or vertical input, of the CRO, moving the spot up and down in accordance with the instantaneous value of the input voltage. The spot then traces a pattern on the screen which shows the input voltage variation as a function of time. When the input voltage is repetitive at a fast enough rate, the display appears as a stationary pattern on the screen. The CRO therefore provides a means of observing time-varying voltages.

In addition to voltages, the CRO can present visual representations of many dynamic phenomena by means of transducers that convert current, pressure, strain, temperature, acceleration, and many other physical quantities into voltages.

200

CROs are used to investigate waveforms, transient happenings, and other time-varying quantities from the very low-frequency range to the very high frequencies. Recordings of these happenings can be made by a special camera attached to the CRO for quantitative interpretation.

The principles upon which the CRO operates are discussed in detail in the following sections.

9-2 BASIC CRO OPERATION

The major subsystems of a general-purpose CRO are shown in the simplified block diagram of Fig. 9-1. They are:

(a) Cathode ray tube, or CRT
(b) Vertical amplifier
(c) Delay line
(d) Time base generator
(e) Horizontal amplifier
(f) Trigger circuit
(g) Power supply

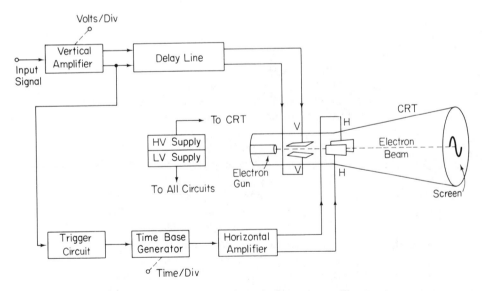

FIGURE 9-1   Block diagram of a general-purpose oscilloscope.

The *cathode ray tube*, or CRT, is the heart of the oscilloscope, with the rest of the CRO consisting of circuitry to operate the CRT. Basically, the CRT produces a sharply focused beam of electrons, accelerated to a very high velocity. This focused and accelerated beam of electrons travels from

its source (the *electron gun*) to the front of the CRT, where it strikes the fluorescent material deposited on the inside face of the CRT (the *screen*) with
sufficient energy to cause the screen to light up in a small spot.

While traveling from its source to the screen, the electron beam passes
between a set of vertical *deflection plates* and a set of horizontal deflection
plates. Voltages applied to the vertical deflection plates can move the beam
in the vertical plane, and the CRT spot moves up and down. Voltages applied
to the horizontal deflection plates can move the beam in the horizontal
plane, and the CRT spot moves from side to side. These movements are
independent of one another so that the CRT spot can be positioned anywhere
on the screen by the simultaneous application of appropriate vertical and
horizontal voltage inputs.

The signal waveform to be viewed on the CRT screen is applied to the
vertical amplifier input. The gain of this amplifier is set by a calibrated input
attenuator, usually marked VOLTS/DIV. The push-pull output of the
amplifier is fed to the vertical deflection plates of the CRT, via a so-called
delay line, with sufficient power to drive the CRT spot in the vertical direction.

The *time base generator*, or sweep generator, develops a sawtooth waveform that is used as the horizontal deflection voltage of the CRT. The positive-going part of the sawtooth waveform is linear, and its rate of rise is set
by a front panel control marked TIME/DIV. The sawtooth voltage is fed to
the horizontal amplifier. This amplifier includes a phase inverter and produces
two simultaneous output waveforms: a positive-going sawtooth (run-up) and
a negative-going sawtooth (run-down). The positive-going sawtooth is applied
to the right-hand horizontal deflection plate of the CRT, and the negative-
going sawtooth to the left-hand horizontal deflection plate. These voltages
cause the electron beam to the "swept" across the CRT screen, from left to
right, in time units that are controlled by the TIME/DIV control.

The simultaneous application of deflection voltages to both sets of
plates causes the CRT spot to trace an image on the screen. This is shown in
Fig. 9-2, where a sawtooth, or *sweep*, voltage is applied to the horizontal
plates and a sinewave signal is applied to the vertical plates. Since the horizontal sweep voltage increases linearly with time, the CRT spot moves across
the screen at a constant velocity, from left to right. At the end of the sweep,
when the sawtooth voltage suddenly drops from its maximum value to zero,
the CRT spot returns rapidly to its starting position on the left-hand side of
the screen and remains there until a new sweep is initiated. When, simultaneously with the horizontal sweep voltage, an input signal is applied to the
vertical deflection plates, the electron beam will be under the influence of
two forces: one in the horizontal plane, moving the CRT spot across the
screen at a linear rate, and one in the vertical plane, moving the CRT spot
up and down in accordance with the magnitude and polarity of the input

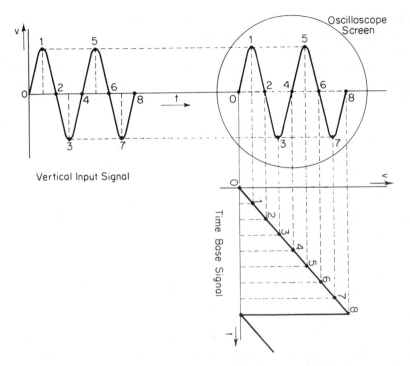

FIGURE 9-2 The CRO spot traces an image on the screen when horizontal and vertical deflection voltages are applied.

signal. The resultant motion of the electron beam therefore produces a CRT display of the vertical input signal as a function of time.

If the input signal is of a recurrent nature, a stable CRT display can be maintained by starting each horizontal sweep at the same point on the signal waveform. To achieve this, a sample of the input waveform is fed to a *trigger* circuit which produces a trigger pulse at some selected point on the input waveform. This trigger pulse is used to start the time base generator, which in turn starts the horizontal sweep of the CRT spot from the left-hand side of the screen.

In the usual case, the leading edge of the input waveform is used to activate the trigger generator which produces the trigger pulse and starts the sweep. This action takes place over a definite time interval (0.15 μs), so that the sweep is not initiated until after the leading edge of the input signal has passed. This then prevents the leading edge of the waveform to be displayed on the screen. The purpose of the *delay line* is to retard the arrival of the input waveform at the vertical deflection plates until the trigger and time base circuits have had a chance to start the sweep of the beam. The delay line introduces a total delay of approximately 0.25 μs in the vertical deflection channel, so that the leading edge of the waveform can be viewed even though it was used to trigger the sweep.

The *power supply* consists of a high-voltage section to operate the CRT, and a low-voltage section to supply the electronic circuitry of the oscilloscope. These supplies are of conventional design and need no further elaboration.

9-3 CATHODE RAY TUBE (CRT)

9-3.1 CRT Operation

The internal structure of a cathode ray tube, or CRT, is shown in the schematic view of Fig. 9-3. The main components of this general-purpose CRT are:

(a) Electron gun assembly
(b) Deflection plate assembly
(c) Fluorescent screen
(d) Glass envelope and base of the tube

In summary, the *electron gun* assembly produces a narrow and sharply focused beam of electrons that leaves the gun with a very high velocity and travels toward the *fluorescent screen*. On striking the screen, the kinetic energy of the high-velocity electrons is converted into light emission, and the beam produces a small light spot on the CRT screen. On its way toward the screen, the electron beam passes between two pairs of electrostatic *deflection plates*, indicated in Fig. 9-3 as the deflection plate assembly. If voltages are applied to the deflection plates, the electron beam can be deflected in both the vertical and the horizontal direction, so that the light spot traces a pattern on the screen in accordance with these voltage inputs.

A conventional electron gun, used in a general-purpose CRT, is shown in Fig. 9-4. The name "electron gun" is derived from the analogy between the motion of an electron emitted from the CRT gun structure and the travel path of a bullet fired from a gun. In fact, the study of the motion of charged particles (electrons) in an electric field is often called *electron ballistics*.

In the schematic CRT view of Fig. 9-3, electrons are emitted from an indirectly heated thermionic *cathode*. The cathode is completely surrounded by a *control grid*, which consists of a nickel cylinder with a small centrally located hole, coaxial with the tube axis. The electrons that manage to pass through this small hole in the grid together make up the so-called *beam current*. The magnitude of the beam current can be adjusted by a front panel control marked INTENSITY, which varies the negative voltage (bias) of the control grid with respect to the cathode. An increase in control grid bias

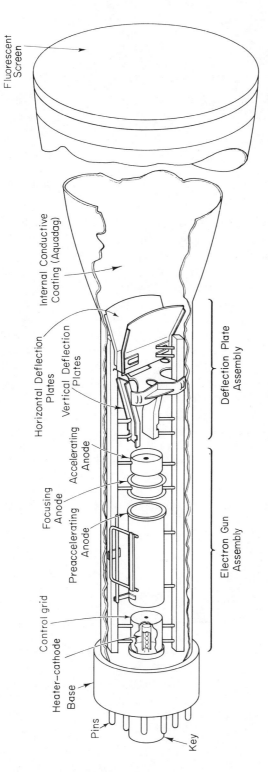

FIGURE 9-3 Internal structure of a cathode ray tube.

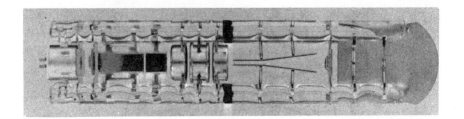

FIGURE 9-4   The electron gun and deflection plate assembly of a general-purpose
CRT (courtesy Tektronix, Inc.).

reduces the beam current, and hence the intensity of the CRT image, while
a decrease in grid bias increases the beam current. This action is identical to
that of the control grid in a conventional vacuum triode.

The electrons, emitted by the cathode and passing through the small
hole in the control grid, are accelerated by the high positive potential applied
to two *accelerating anodes*. These two anodes are separated by a *focusing
anode*, which provides a method for focusing the electrons into a narrow and
sharply defined beam. The two accelerating anodes and the focusing anode
are also cylindrical in form, with small openings in the center of each cylinder,
coaxial with the CRT axis. The holes in these electrodes allow the accelerated
and focused electron beam to travel past the vertical and horizontal deflection plates to the fluorescent screen.

9-3.2 Electrostatic Focusing

Electrostatic focusing is used in all CROs. In order to examine the
operation of the electrostatic focusing method, it is useful to consider first
the behavior of an individual particle in an electric field. Consider the diagram of Fig. 9-5, which shows a hypothetical electron situated at rest in an

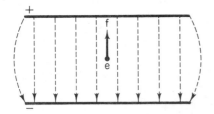

FIGURE 9-5   Force f on an electron in a uniform electric field.

electric field. The definition of electric field intensity states that the force on
a unit positive charge at any point in an electric field is the electric field
intensity at that point. By definition therefore

$$\epsilon = \frac{f}{q} \, (\text{V/m}) \qquad\qquad (9\text{-}1)$$

where ϵ = electric field intensity, in V/m

  f = force on the charge, in N

  q = charge, in C

An electron is a *negatively* charged particle and its charge is

$$e = 1.602 \times 10^{-19} \text{ C} \tag{9-2}$$

The force on the electron in an electric field is, from Eq. (9-1),

$$f_e = -e\epsilon \quad \text{N} \tag{9-3}$$

where the minus sign indicates that the force acts in a direction *opposite* to the direction of the electric field.

This is valid only when the electric field in which the charged particle is situated is of *uniform* intensity. That this is not always so can be seen from Fig. 9-6, where the electric field between two parallel plates of finite dimensions is shown. In Fig. 9-6 the field intensity is directed from the positive to

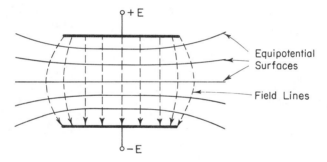

FIGURE 9-6   Electric field and equipotential surfaces for two parallel plates.

the negative plate. The lateral repulsion of the electric field lines causes a spreading of the space between the lines, resulting in a curvature of the field at the ends of the plates. The density of the field lines therefore will be less at the ends of the plates than in the center region between the plates. When points of equal potential are connected on each of the field lines, we obtain *equipotential surfaces*, shown as the solid lines in Fig. 9-6. Since the force on an electron acts in a direction opposite to the direction of the field, we can also state that the force on an electron is in the direction *normal* to the equipotential surfaces.

When two cylinders are placed end to end and a potential difference is applied to them, the resulting electric field between the two cylinders is not of uniform density. Lateral repulsion will cause spreading of the lines, resulting in a field as shown in Fig. 9-7. The equipotential surfaces are shown as solid lines. Because of the varying density of the electric field in the area between the cylinders, the equipotential surfaces are *curved*.

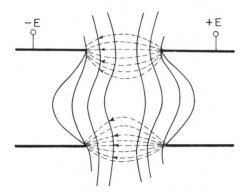

Consider now the regions on both sides of an equipotential surface, S, as shown in Fig. 9-8. The potential to the left of the surface S is V^- and to the right of S is V^+. An electron, which is moving in a direction AB at an angle with the normal to the equipotential surface and entering the area to the left of S with a velocity v_1, experiences a force at the surface, S. This

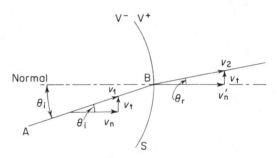

FIGURE 9-8   Refraction of an electron ray at an equipotential surface.

force acts in a direction normal to the equipotential surface. Because of this force, the velocity of the electron increases to a new value, v_2, after it has passed S. The tangential component, v_t, of the velocity on both sides of S remains the same. Only the normal component of the velocity, v_n, is increased by the force at the equipotential surface to a new value, v_n'. From Fig. 9-8, then, it follows that

$$v_t = v_1 \sin \theta_i = v_2 \sin \theta_r \qquad (9\text{-}4)$$

where θ_i is the angle of incidence and θ_r the angle of refraction of the electron ray. Rearranging Eq. (9-4) we obtain

$$\frac{\sin \theta_i}{\sin \theta_r} = \frac{v_2}{v_1} \qquad (9\text{-}5)$$

Equation (9-5) is identical to the expression relating the refraction of a light beam in geometrical optics. The refraction, or bending, of an electron beam at an equipotential surface therefore follows the same laws as the bending of a light beam at a refracting surface, such as an optical lens. For this reason, an electrostatic focusing system in a CRT is sometimes called an *electron lens*.

Consider now the three elements of the electrostatic focusing system shown in the functional diagram of Fig. 9-9. The first electrode of this elec-

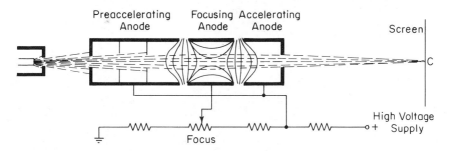

FIGURE 9-9   Electrostatic focusing system of a CRT.

tron lens is the preaccelerating anode, a metal cylinder containing several baffles to collimate the electron beam that enters through the small opening on the left-hand side. The second electrode is the focusing anode, and the third electrode is the accelerating anode. The three electrodes are cylindrical in form and coaxial with the CRT axis.

The preaccelerating anode and the accelerating anode are connected together to a high positive potential (say, $+1,500$ V) supplied by the CRT high-voltage supply. The focusing anode, located between the two accelerating anodes, is connected to a lower positive potential (say, $+500$ V). The potential difference between the focusing anode and the accelerating anodes creates electric fields between the cylindrical elements. Since the field lines are nonuniformly spaced, as was shown in Fig. 9-7, the equipotential surfaces are curved to form a *double concave lens* system. This is indicated in Fig. 9-9 by the field lines in the areas between the electrodes.

Electrons are emitted by the cathode as a slightly divergent beam. Those electrons that enter the electric field between the preaccelerating anode and the focusing anode at angles other than normal to the equipotential surface will be refracted toward the normal. The electron beam therefore tends to become parallel to the CRT axis, as shown. This approximately parallel beam then enters the second concave lens and is refracted once again to become slightly convergent and focused on the screen at the center of the CRT axis.

The focal length of this double concave lens system can be increased or decreased by varying the voltage on the focusing anode, so that the focal point of the beam is moved along the CRT axis. The potentiometer that provides adjustment of the voltage on the focusing anode is a CRO front panel control marked FOCUS.

9-3.3 Electrostatic Deflection

In discussing the electrostatic deflection method of an electron beam in an oscilloscope we return to the statement made in Sec. 9-3.2 regarding the force on the electron in a uniform electric field. For convenience, the diagram of Fig. 9-5 is reproduced in Fig. 9-10. By definition of the electric

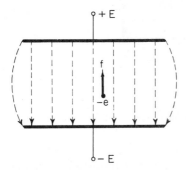

FIGURE 9-10 Force f on an electron in a uniform electric field.

field intensity, ϵ, the force on the electron is $f_e = -e\epsilon$ Newton. The action of the force on the electron will accelerate it in the direction of the positive electrode, along the lines of the field flux. Newton's second law of motion allows us to calculate this acceleration since

$$f = ma \tag{9-6}$$

Substituting Eq. (9-3) into Eq. (9-6), we obtain

$$a = \frac{f}{m} = \frac{-e\epsilon}{m} \text{ m/s}^2 \tag{9-7}$$

where a = acceleration of the electron, in m/s²

f = force on the electron, in N

m = mass of the electron, in kg

When the motion of an electron in an electric field is discussed it is usually specified with respect to the customary Cartesian axes, as shown in Fig. 9-11. In discussing the concepts which follow, we shall use subscript

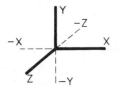

FIGURE 9-11 Cartesian coordinate system.

notation for the *vector* components of velocity, field intensity, and acceleration. For example, the velocity component along the X axis will be written
v_x (m/s); the component of the force along the Y axis is written f_y (N), etc.
The motion of an electron in a given electric field cannot be determined
unless the initial values of velocity and displacement are known. The term
initial represents the value of velocity or displacement at the time of observation, or time $t = 0$. The subscript 0 will be used to indicate these initial values.
For example, the initial velocity component along the X axis is written as
v_{0x}.

Consider now an electric field of constant intensity with the lines of
force pointing in the negative Y direction, shown in Fig. 9-12. An electron

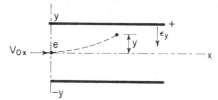

FIGURE 9-12   Path of a moving
electron in a uniform electric field.

entering this field in the positive X direction with an initial velocity v_{0x} will
experience a force. Since the field acts only along the Y axis, there will be no
force along either the X axis or the Z axis, and the acceleration of the electron
along these axes must be zero. Zero acceleration means constant velocity,
and since the electron enters the field in the positive X direction with an initial
velocity v_{0x}, it will continue to travel along the X axis at that velocity. Since
the velocity along the Z axis was zero at time $t = 0$, there will be no movement of the electron along the Z axis.

Newton's second law of motion, applied to the force on the electron
acting in the Y direction, yields

$$f = ma_y \quad \text{or} \quad a_y = \frac{f}{m} = \frac{-e\epsilon_y}{m} = \text{constant} \qquad \textbf{(9-8)}$$

Equation (9-8) indicates that the electron moves with a *constant acceleration*
in the Y direction of the uniform electric field. To find the displacement of
the electron due to this accelerating force, we use the well-known expressions
for velocity and displacement:

$$v = v_0 + at \quad \text{(m/s)} \qquad \text{(velocity)} \qquad \textbf{(9-9)}$$

$$x = x_0 + v_0 t + \tfrac{1}{2}at^2 \quad \text{(m)} \quad \text{(displacement)} \qquad \textbf{(9-10)}$$

Subject to the initial condition of zero velocity in the Y direction, $(v_{0y} = 0)$,
Eq. (9-9) yields

$$v_y = a_y t \quad \text{(m/s)}$$

which, after substitution of Eq. (9-8), results in

$$v_y = \frac{-e\epsilon_y t}{m} \quad \text{(m/s)} \tag{9-11}$$

The displacement of the electron in the Y direction follows from Eq. (9-10), which yields, applying the initial conditions of zero displacement ($y_0 = 0$) and zero velocity ($v_{0y} = 0$),

$$y = \tfrac{1}{2}a_y t^2 \quad \text{(m)}$$

which, after substitution of Eq. (9-8), results in

$$y = \frac{-e\epsilon_y t^2}{2m} \quad \text{(m)} \tag{9-12}$$

The X distance, traveled by the electron in the time interval t, depends on the initial velocity v_{0x} and we can write, again using Eq. (9-10),

$$x = x_0 + v_{0x}t + \tfrac{1}{2}a_x t^2 \quad \text{(m)}$$

which, after applying the initial conditions for the X direction ($x_0 = 0$ and $a_x = 0$), becomes

$$x = v_{0x}t \quad \text{or} \quad t = \frac{x}{v_{0x}} \quad \text{(s)} \tag{9-13}$$

Substituting Eq. (9-13) into Eq. (9-12), we obtain an expression of the vertical deflection as a function of the horizontal distance traveled by the electron:

$$y = \left[\frac{-e\epsilon_y}{2v_{0x}^2 m}\right]x^2 \quad \text{(m)} \tag{9-14}$$

Equation 9-14 shows that the path of an electron, traveling through an electric field of constant intensity and entering the field at right angles to the lines of flux, is *parabolic* in the X-Y plane.

In Fig. 9-13 two parallel plates, called *deflection plates*, are placed a

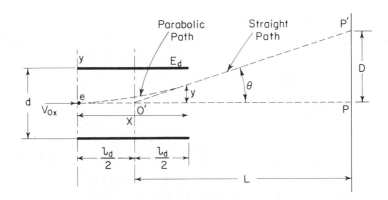

FIGURE 9-13   Deflection of the cathode ray beam.

distance d apart and are connected to a source of potential difference E_d, so that an electric field ϵ exists between the plates. The intensity of this electric field is given by

$$\epsilon = \frac{E_d}{d} \quad \text{(V/m)} \tag{9-15}$$

An electron entering the field with an initial velocity v_{0x} is deflected toward the positive plate following the parabolic path of Eq. (9-14), as indicated in Fig. 9-13. When the electron leaves the region of the deflection plates, the deflecting force no longer exists, and the electron travels in a straight line toward point P', a point on the fluorescent screen. The slope of the parabola at a distance $x = l_d$, where the electron leaves the influence of the electric field, is defined as

$$\tan \theta = \frac{dy}{dx} \tag{9-16}$$

where y is given by Eq. (9-14). Differentiating Eq. (9-16) with respect to x and substituting $x = l_d$ yields

$$\tan \theta = \frac{dy}{dx} = -\frac{e\epsilon_y l_d}{mv_{0x}^2} \tag{9-17}$$

The straight line of travel of the electron is tangent to the parabola at $x = l_d$, and this tangent intersects the X axis at point O'. The location of this *apparent* origin $0'$ is given by Eq. (9-14) and Eq. (9-17) since

$$x - 0' = \frac{y}{\tan \theta} = \frac{e\epsilon_y l_d^2 / 2mv_{0x}^2}{e\epsilon_y l_d / mv_{0x}^2} = \frac{l_d}{2} \quad \text{(m)} \tag{9-18}$$

The apparent origin $0'$ is therefore at the center of the deflection plates and a distance L from the fluorescent screen.

The deflection on the screen is given by

$$D = L \tan \theta \quad \text{(m)} \tag{9-19}$$

Substituting Eq. (9-17) for $\tan \theta$, we obtain

$$D = L \frac{e\epsilon_y l_d^2}{mv_{0x}^2} \quad \text{(m)} \tag{9-20}$$

The potential energy of the electron entering the area between the deflection plates with an initial velocity v_{0x} is

$$\tfrac{1}{2} m v_{0x}^2 = eE_a \tag{9-21}$$

where E_a is the accelerating voltage in the electron gun. Rearranging (9-21), we obtain

$$v_{0x}^2 = \frac{2eE_a}{m} \tag{9-22}$$

Substituting Eq. (9-15) for the field intensity ϵ_y, and Eq. (9-22) for the velocity of the electron in the X direction v_{0x} into Eq. (9-20), we obtain

$$D = L \frac{e\epsilon_y l_d^2}{mv_{0x}^2} = \frac{Ll_d E_d}{2dE_a} \quad \text{(m)} \tag{9-23}$$

where $D =$ deflection on the fluorescent screen (meters)

$L =$ distance from center of deflection plates to screen (meters)

$l_d =$ effective length of the deflection plates (meters)

$d =$ distance between the deflection plates (meters)

$E_d =$ deflection voltage (volts)

$E_a =$ accelerating voltage (volts)

Equation (9-23) indicates that for a given accelerating voltage E_a and for the particular dimensions of the CRT, the deflection of the electron beam on the screen is directly proportional to the deflection voltage E_d. This direct proportionality indicates that the CRT may be used as a *linear voltage-indicating device*. This discussion assumed that E_d was a fixed dc voltage. However, the deflection voltage usually is a varying quantity and the image on the screen follows the variations of the deflection voltage in a linear manner, according to Eq. (9-23).

The *deflection sensitivity S* of a CRT is defined as the deflection on the screen (in meters) per volt of deflection voltage. By definition, therefore

$$S = \frac{D}{E_d} = \frac{Ll_d}{2dE_a} \quad (\text{m/V}) \qquad (9\text{-}24)$$

where $S =$ deflection sensitivity (m/V)

The *deflection factor G* of a CRT, by definition, is the reciprocal of the sensitivity S and is expressed as

$$G = \frac{1}{S} = \frac{2dE_a}{Ll_d} \quad (\text{V/m}) \qquad (9\text{-}25)$$

with all terms defined as for Eqs. (9-23) and (9-24). The expressions for deflection sensitivity S and deflection factor G indicate that the sensitivity of a CRT is independent of the deflection voltage but varies linearly with the accelerating potential. High accelerating voltages therefore produce an electron beam that requires a high deflection potential for a given excursion on the screen. A highly accelerated beam possesses more kinetic energy and therefore produces a brighter image on the CRT screen, but this beam is also more difficult to deflect and we sometimes speak of a *hard* beam. Typical values of deflection factors range from 10 V/cm to 100 V/cm, corresponding to sensitivities of 1.0 mm/V to 0.1 mm/V, respectively.

9-3.4 Screens for CRTs

When the electron beam strikes the screen of the CRT, a spot of light is produced. The screen material on the inner surface of the CRT that produces this effect is the *phosphor*. The phosphor absorbs the kinetic energy of the bombarding electrons and reemits energy at a lower frequency in the visual spectrum. The property of some crystalline materials, such as phosphor or zinc oxide, to emit light when stimulated by radiation is called *fluorescence*. Fluorescent materials have a second characteristic, called *phosphorescence*, which refers to the property of the material to continue light emission even after the source of exitation (in this case the electron beam) is cut off. The length of time during which phosphorescence, or afterglow, occurs is called the *persistence* of the phosphor. Persistence is usually measured in terms of the time required for the CRT image to decay to a certain percentage (usually 10 per cent) of the original light output.

The intensity of the light emitted from the CRT screen, called *luminance*, depends on several factors. First, the light intensity is controlled by the number of bombarding electrons striking the screen per second. If this so-called *beam current* is increased, or the same amount of beam current is concentrated in a smaller area by reducing the spot size, the luminance will increase. Second, luminance depends on the energy with which the bombarding electrons strike the screen, and this, in turn, is determined by the accelerating potential. An increase in accelerating potential will yield an increase in luminance. Third, luminance is a function of the time the beam strikes a given area of the phosphor; therefore sweep speed will affect the luminance. And finally, luminance is a function of the physical characteristics of the phosphor itself. Almost all manufacturers provide their customers with a choice of phosphor materials. Table 9-1 summarizes the characteristics of some of the commonly used phosphors.

Table 9-1

PHOSPHOR DATA CHART

Phosphor type	Fluorescence	Phosphorescence	Relative luminance(*)	Decay to 0.1 % (ms)	Comments
P1	yellow-green	yellow-green	50%	95	General-purpose; replaced by P31 in most applications
P2	blue-green	yellow-green	55%	120	Good compromise for high- and low-speed applications
P4	white	white	50%	20	Television displays
P7	blue	yellow-green	35%	1,500	Long decay; observation of low-speed phenomena
P11	purple-blue	purple-blue	15%	20	Photographic applications
P31	yellow-green	yellow-green	100%	32	General-purpose; brightest available phosphor

*Luminance is the photometric equivalent of brightness and is based on measurements made with a sensor having spectral sensitivity approximating the human eye. P31 is the reference phosphor.

As Table 9-1 shows, a number of factors must be considered in selecting a phosphor for a given application. For example, P11 phosphor, with its short persistence, is excellent for waveform photography but not at all suitable for visual observation of low-speed phenomena. P31 phosphor, with its high luminance and medium persistence, is the best compromise for general-purpose viewing and is therefore found in the majority of standard laboratory-type CROs.

It is possible to inflict serious damage to the CRT screen by incorrect handling of the front panel controls. When a phosphor is excited by an electron beam with excessive current density, permanent damage of the phosphor may occur through *burning*, and the light output will be reduced. Two factors control the occurrence of burning: beam density and duration of excitation. Beam density is controlled with the INTENSITY, FOCUS, and ASTIGMA-TISM controls on the CRO front panel. The length of time that the beam excites a certain area on the phosphor can be adjusted with the sweep or TIME/DIV control. Burning, and possibly complete *destruction* of the phosphor, can be avoided by keeping the beam intensity down and the exposure time short.

The bombarding electrons striking the phosphor release secondary-emission electrons, thus keeping the screen in a state of electrical equilibrium. These secondary-emission low-velocity electrons are collected by a conductive coating, known as *aquadag*, on the inside surface of the glass tube, which is electrically connected to the second anode. In some tubes, particularly CRTs with magnetic focusing (such as TV picture tubes), the accelerating anode is dispensed with entirely and the conductive coating is used as the final accelerating anode.

9-3.5 Graticules

The waveform display on the face of the CRT can be visually measured against a set of horizontal and vertical scale marks, called a *graticule*, as shown in Fig. 9-14.

These scale marks can be placed external to the face of the CRT tube, in which case we speak of an *external* graticule, or they can be applied to the inside face of the CRT, in which case we speak of an *internal* graticule.

The external graticule consists of a clear or tinted plastic plate with scribed division marks on it. It is mounted on the outside of the CRT face. The external graticule has the advantage of being easily replaced by one having a special pattern, such as degree marks for color TV vector analysis. Also, the position of the external graticule can easily be adjusted to align with the trace of the CRT. The major disadvantage of the external graticule is *parallax*, because the scale marks are not in the same plane as the waveform image written on the phosphor. As a result, the alignment of trace and graticule will vary with viewing position.

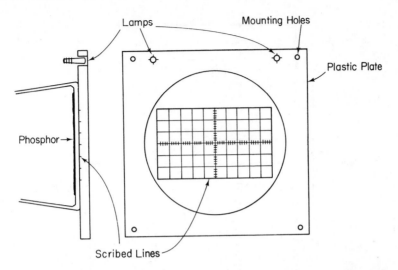

FIGURE 9-14   External graticule (6 × 10 cm) and its illumination. The fluorescent screen is not in the same plane as the graticule lines inscribed into the back of the plastic graticule plate, so that the alignment of CRT trace and graticule will vary with the viewing position (parallax).

A graticule on the inside of the CRT face is called an internal graticule. With the internal graticule there is no parallax, because the CRT image and the graticule are in the same plane. CRTs with an internal graticule, however, are more costly to manufacture and the graticule cannot be changed without changing the CRT. In addition, CRTs with internal graticules must have some method of trace alignment, which adds to the overall cost of the CRO.

9-3.6   CRT Connections

Electrical connection to the various elements inside the glass envelope of the CRT is made through the base of the tube. Figure 9-15 shows typical CRT connections for a general-purpose oscilloscope.

The various supply voltages for the electron gun assembly are developed by two power supplies connected in series: a high-voltage supply for the accelerating voltage, and a low-voltage supply for the auxiliary circuitry. A divider network is connected across the two supplies to provide the necessary operating voltages to the system.

The intensity of the electron beam is adjusted by varying the cathode-to-grid voltage of the triode section. In Fig. 9-15 this adjustment is made by the 500-kΩ potentiometer, which is available as a front panel control marked INTENSITY. The 2-MΩ potentiometer in the divider network is also a front panel control, marked FOCUS. It adjusts the negative voltage on the focus ring of the lens section between −500 V and −900 V. The lens effect becomes stronger (shorter focal length) as the voltage on the focus ring is made more

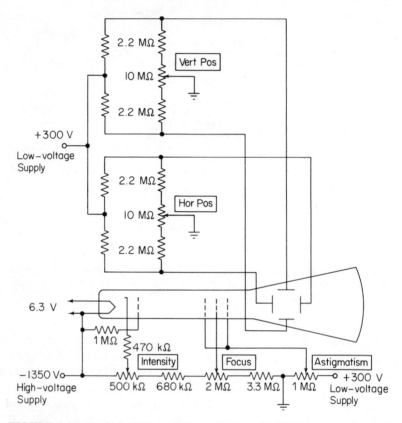

FIGURE 9-15 Typical CRT connections, showing the CRO front panel controls that determine the intensity, focus, and position of the CRT spot on the screen.

negative with respect to the two outside anodes. The ASTIGMATISM control on the CRT front panel sets the voltage on the accelerating anode with respect to the vertical deflection plates that follow the lens section. This forms a cylindrical lens that corrects for any defocusing that might be present, and adjustment is made for the roundest spot on the CRT screen.

The beam can be positioned anywhere on the screen by manipulating two separate front panel controls marked VERTICAL POSITION and HORIZONTAL POSITION. With the VERT POS control in its midposition, the vertical deflection plates are connected to identical dc voltages, and there is no electric field between them. Hence the electron beam is not deflected and simply travels down the center of the CRT. A slight adjustment of the VERT POS control introduces an unbalance in the dc voltages applied to the plates, and the resulting potential difference establishes an electric field between them. This field affects the deflection of the beam as it passes between the plates and brings the CRT spot to a new position on the screen. In a similar manner, the HOR POS control can move the CRT spot in any horizontal direction on the screen. It then follows that simultaneous adjustment of both position controls can bring the spot to any desired location on the screen.

9-4   VERTICAL DEFLECTION SYSTEM

9-4.1   Basic Elements

The vertical deflection system must meet fairly stringent performance requirements, which can be summarized by stating that the system must faithfully reproduce the input voltage waveform, within specified limits of bandwidth, risetime, and amplitude. The vertical deflection system also provides a buffer (isolation) between the signal source and the CRT deflection plates. In some cases, the vertical system provides for various modes of operation, such as dc or ac coupling, multiple trace operation, multiple display modes, differential input capability, and so on. These special features are generally available on the more sophisticated laboratory-type CROs that use so-called *plug-in* units.

The vertical deflection system normally comprises the elements shown in the block diagram of Fig. 9-16. They are:

(a) CRO probe
(b) Input selector
(c) Input attenuator
(d) Vertical amplifier

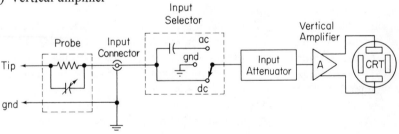

FIGURE 9-16   Functional block diagram of the vertical deflection system.

The CRO probe performs the important function of connecting the vertical amplifier to the circuit under measurement, without loading or otherwise disturbing the circuit. Different kinds of probes are available for various measurement applications. They are described in some detail in Sec. 9-7. The general-purpose probe in Fig. 9-16 is called a *passive probe*. It consists of a series resistor (signal attenuation) and a variable shunt capacitor (probe compensation), both contained in the probe body, plus a probe tip and a ground connector. The probe body is connected to the vertical input terminal through a cable with a BNC connector or, in the case of a low-cost low-frequency CRO, with banana plugs or some other simple connector.

9-4.2 Input Selector

The input selector in Fig. 9-16 is shown as a three-position switch: *ac-gnd-dc*. Placing the input selector to the *ac* position capacitively couples the signal voltage to the attenuator. The capacitor blocks the dc component of the input waveform and allows only the ac component to enter the amplifier. This is a useful feature that allows the measurement of ac signal voltages that are superimposed on dc bias or supply voltages. Placing the input selector to the *dc* position connects the signal voltage directly to the attenuator so that both ac and dc components are applied to the amplifier. This measurement mode is especially useful for determining total instantaneous voltage values.

The *ground* connection on the input selector, available on some CROs as an intermediate position between *ac* and *dc*, is a safety feature that removes any stored charge in the input attenuator by momentarily grounding the attenuator input when switching from the *dc* to the *ac* mode.

9-4.3 Input Attenuator

The input attenuator consists of a number of RC voltage dividers, controlled on the CRO front panel by the VOLTS/DIV selector. This selector is calibrated in terms of the deflection factor (V/div), usually in the 1–2–5 sequence. A typical range of attenuator settings is 0.1, 0.2, 0.5, 1, 2, 5, 10, 20, and 50 volts per division, with maximum attenuation at the 50-V/div setting.

To insure linear CRO operation over the specified frequency range (a typical bandwidth is dc-to-25 MHz), the attenuation of the input signal must be independent of the frequency, and this requires a so-called *compensated* attenuator. Figure 9-17 shows such an attenuator, together with the input stage of the vertical amplifier, whose input impedance is represented by resistor R_i in parallel with capacitor C_i.

With the attenuator switch in the upper position, input signal v_i is connected directly to the input of the vertical amplifier, without attenuation. In our example this would correspond to the 0.1-V/div setting, or maximum sensitivity of the vertical deflection system. With the switch in the lower position, as indicated in Fig. 9-17, attenuator network R_a–C_a is connected into the circuit and voltage division takes place. It follows that output voltage v_o is proportional to the ratio of amplifier input impedance and total circuit

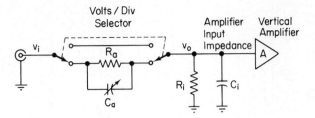

FIGURE 9-17   Input attenuator (VOLTS/DIV selector).

impedance. In the compensated attenuator this impedance ratio is kept constant, and independent of the frequency of the signal voltage, by adjusting C_a in such a way that time constant $R_a C_a$ equals time constant $R_i C_i$. That this is so is shown in Fig. 9-18, where R_a, C_a, R_i, and C_i are presented in the familiar bridge configuration.

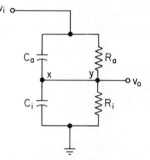

FIGURE 9-18   Input attenuator presented as a bridge circuit. At bridge balance, branch *xy* can be removed from the circuit,

and $$v_o = \frac{R_i}{R_a + R_i} v_i$$

The bridge is in balance when $R_a X_{C_i} = R_i X_{C_a}$ or when $R_a C_a = R_i C_i$. At balance, there is no current in connecting branch *xy*, and the *xy* connection can be omitted from the circuit. Hence the output voltage at bridge balance is determined by the resistive voltage divider and equals

$$v_o = \frac{R_i}{R_a + R_i} v_i \qquad (9\text{-}26)$$

independent of the frequency of the signal input.

A practical method of balancing the bridge and compensating the attenuator consists of applying a square-wave test signal (calibrator) to the attenuator input and adjusting C_a until the output voltage observed on the CRT screen is an exact replica of the input signal. The effects of attenuator compensation are shown in Fig. 9-19. With a calibrator signal applied to the input, and C_a correctly adjusted, the output voltage is a square wave, as shown in Fig. 9-19(a). *Overcompensation* occurs when C_a is too large; this shows as overshoot in the pulse waveform, while a high-frequency sine wave appears larger than life, as indicated in Fig. 9-19(b). *Undercompensation*

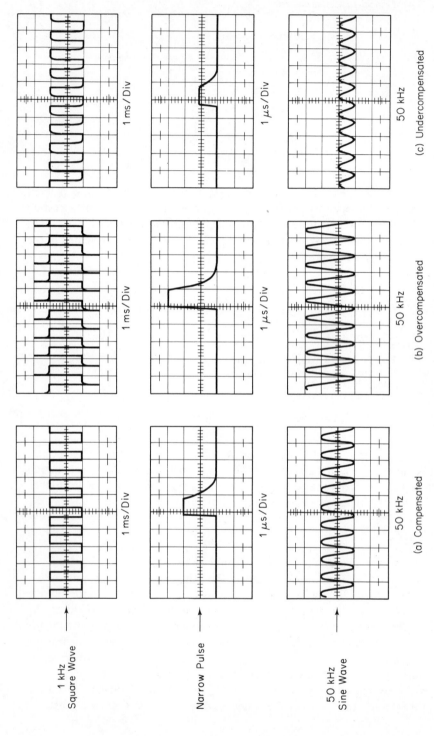

FIGURE 9-19 Effects of attenuator compensation.

occurs when C_a is too small; this results in rounded corners on the calibrator waveform, while high-frequency signals suffer excessive attenuation, as shown in Fig. 9-19(c).

The effects of an incorrectly compensated attenuator (overshoot or rounding) can be explained with the aid of Fig. 9-20. In Fig. 9-20(a), a step

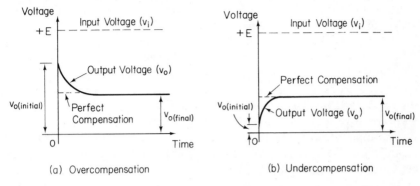

(a) Overcompensation   (b) Undercompensation

FIGURE 9-20   Response of an attenuator to a step voltage input. Perfect compensation occurs when the time constant of the attenuator $\tau_a = R_a C_a$ equals the time constant of the amplifier input $\tau_i = R_i C_i$ (see also Fig. 9-18). Overcompensation occurs when $\tau_a > \tau_i$. Undercompensation occurs when $\tau_a < \tau_i$.

voltage is applied to the attenuator so that the input changes abruptly from 0 V to $+E$ V at time $t = 0$. An infinite current exists at instant $t = 0$ for an infinitesimal time, and a charge $q = \int_{0-}^{0+} i \, dt$ is delivered to each capacitor. According to Kirchhoff's voltage law, the applied voltage at $t = 0$ is

$$E = \frac{q}{C_a} + \frac{q}{C_i} = \frac{C_i + C_a}{C_i C_a} q \quad \text{(V)} \tag{9-27}$$

The initial output voltage at $t = 0$, when the instantaneous change in input voltage is infinitely large (from 0 to $+E$), is determined by the capacitive voltage divider and equals

$$v_{o(\text{initial})} = \frac{q}{C_i} = \frac{C_a}{C_i + C_a} E \quad \text{(V)} \tag{9-28}$$

The output voltage changes exponentially from its initial value to a final steady-state value, with a time constant $\tau = R_{\text{TH}} C_{\text{TH}}$, where R_{TH} and C_{TH} are the Thévenin parameters of the attenuator looking back into the output terminals with the input short-circuited. The final output voltage at $t = \infty$ is determined by the resistors only, because the capacitors act as open circuits under the steady-state condition of dc voltage. Therefore

$$v_{o(\text{final})} = \frac{R_i}{R_i + R_a} E \quad \text{(V)} \tag{9-29}$$

When the attenuator is correctly compensated,

$$v_{o(\text{initial})} = v_{o(\text{final})}$$

or using Eqs. (9-28) and (9-29), we get

$$R_a C_a = R_i C_i \qquad\qquad (9\text{-}30)$$

which, of course, is the condition for bridge balance.

The vertical attenuator can be moved through a number of VOLTS/ DIV settings, and each time another R_a–C_a attenuator network is selected. All these networks use the same principle: simple RC voltage dividers that maintain a set ratio to one another and that are frequency-compensated by the small variable capacitor C_a. In the laboratory-type CROs the resistive and capacitive components of the attenuator are so selected that the CRO vertical input always presents the same impedance to the circuit under test, regardless of the VOLTS/DIV setting. Typical values of these input parameters are 1 MΩ shunted by 33 pF.

9-4.4 Vertical Amplifier

The vertical amplifier consists of several stages, with fixed overall sensitivity or gain, usually expressed in terms of the deflection factor (V/div). The advantage of fixed gain is that the amplifier can more easily be designed to meet and maintain the requirements of stability and bandwidth. The vertical amplifier is kept within its signal handling capability by proper selection of the input attenuator. With the attenuator in its most sensitive position, the overall gain of the amplifier corresponds to the lowest reading of the VOLTS/DIV selector.

The vertical amplifier generally consists of two major circuit blocks: a *preamplifier* and a *main vertical amplifier*. In laboratory-type CROs the preamplifier is often provided as a plug-in unit, which can quickly and easily be connected into the *main frame* of the CRO. The main amplifier then forms an integral part of the main frame. Different types of vertical plug-in units, designed for specific measurement applications, can greatly extend the capabilities of the CRO at reasonable cost.

Figure 9-21 shows a functional block diagram of the vertical amplifier. The first element of the preamplifier is the input stage, often consisting of a FET source follower, whose high-input impedance virtually isolates the amplifier from the attenuator. This FET input stage is sometimes followed by a BJT emitter follower, which acts as an impedance transformer to match the medium impedance of the FET output to the low-impedance input of the phase inverter that follows. The phase inverter, or paraphase amplifier, provides two antiphase output signals which are required to operate the push-pull output amplifier. This final stage of the preamplifier provides the

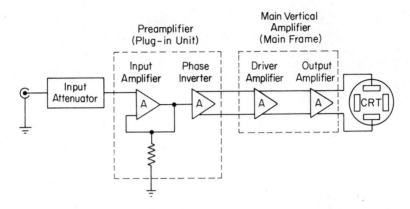

FIGURE 9-21   Block diagram of the vertical amplifier.

necessary drive for the main vertical amplifier. A push-pull output signal of 100 mV/div is representative for a standard preamplifier of the plug-in type.

The main vertical amplifier shown in the block diagram of Fig. 9-21 consists of a driver amplifier and a push-pull output stage that delivers equal signal voltages of opposite polarity to the vertical deflection plates of the CRT. Push-pull circuits are almost always used in vertical amplifiers as well as in horizontal amplifiers because they improve the deflection linearity of the CRT. The main amplifier also includes circuitry required to center the beam, plus additional amplifiers to drive the vertical delay line.

9-5   DELAY LINE

9-5.1   Function of the Delay Line

All electronic circuitry in the CRO (attenuators, amplifiers, pulse shapers, generators, and indeed the circuit wiring itself) causes a certain amount of time delay in the transmission of signal voltages to the deflection plates. Almost all of this delay is created in circuits that switch, shape, or generate. Comparing the vertical and horizontal deflection circuits in the CRO block diagram of Fig. 9-22, we observe that the horizontal signal (time base, or sweep voltage) is initiated, or *triggered*, by a portion of the output signal applied to the vertical CRT plates. Signal processing in the horizontal channel consists of generating and shaping a trigger pulse (trigger pickoff) that starts the sweep generator, whose output is fed to the horizontal amplifier and then to the horizontal deflection plates. This whole process takes time: on the order of 80 ns or so.

To allow the operator to observe the leading edge of the signal waveform, the signal drive for the vertical CRT plates must therefore be *delayed* by

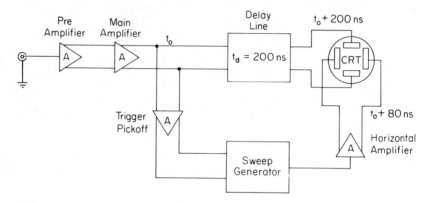

FIGURE 9-22   Delay of the vertical signal allows the horizontal sweep to start prior to vertical deflection.

at least the same amount of time. This is the function of the vertical delay line. We observe that in Fig. 9-22 a 200-ns delay line has been added to the vertical channel, so that the signal voltage to the CRT plates is delayed by 200 ns, and the horizontal sweep is started prior to the vertical deflection. Although the delay line can appear almost anywhere along the vertical signal path, the trigger pickoff *must* precede the delay line.

There are basically two kinds of delay line: the lumped-parameter delay line and the distributed-parameter delay line.

9-5.2 Lumped-parameter Delay Line

The lumped-parameter delay line consists of a number of cascaded symmetrical LC networks, such as the so-called *T-section* of Fig. 9-23.

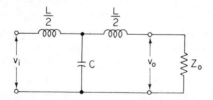

FIGURE 9-23   T-filter section.

If the T-section is terminated in its *characteristic impedance* Z_o, then, by definition, the impedance looking back into the input terminals is also Z_o. This condition of termination gives the T-section the characteristics of a low-pass filter whose attenuation and phase shift are a function of frequency, and whose passband is defined by the frequency range over which the attenuation is zero. The upper limit of the passband is called the *cutoff frequency* of the filter, given by

$$f_c = \frac{1}{\pi\sqrt{LC}}$$
(9-31)

If the spectrum of input signal v_i consists of frequencies much less than the cutoff frequency, output signal v_o will be a faithful reproduction of v_i, but delayed by a time

$$t_s \approx \frac{1}{\pi f_c} = \sqrt{LC}$$
(9-32)

where t_s is the time delay for a single T-section. A number of T-sections, cascaded into a so-called *lumped-parameter* delay line, increases the total delay time to

$$t_d = n t_s$$
(9-33)

where n is the number of cascaded T-sections.

Because of the sharp cutoff frequency of the lumped-parameter delay line, amplitude and phase distortion become a problem when the frequency of the input signal increases. The application of a step-voltage input, for example, which contains high-frequency components (odd harmonics), causes an output voltage that suffers from transient response distortion in the form of overshoot and ringing, as shown in Fig. 9-24. This kind of

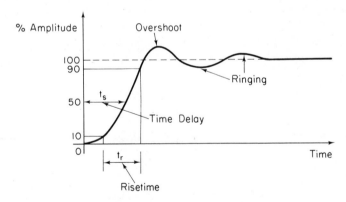

FIGURE 9-24   Step-voltage response of a T-section filter terminated in its characteristic impedance $Z_o = \sqrt{L/C}$.

response can be improved to more closely resemble the original step-voltage input by modifying the design of the filter sections into, for example, m-derived sections. The m-derived section is a popular circuit that uses mutual coupling between the two inductors of the T-section.

It is important to *match* the delay line as closely as possible to its characteristic impedance Z_o, at both input and output ends. This requirement often leads to complex termination circuitry in an effort to optimize the balance between amplitude and phase distortion and to obtain better transient response.

A practical delay line circuit in a CRO is driven by a push-pull amplifier and then consists of a symmetrical arrangement of cascaded filter sections, as in Fig. 9-25. Optimum response of the delay line requires precise proportioning of the L and C components in each section; the variable capacitors must be carefully adjusted to be effective.

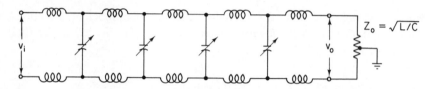

FIGURE 9-25   Push-pull transmission line with single termination.

9-5.3 Distributed-parameter Delay Line

The distributed-parameter delay line consists of a specially manufactured coaxial cable with a high value of inductance per unit length. For this type of delay line, the straight center conductor of the normal coaxial cable is replaced with a continuous coil of wire, wound in the form of a helix on a flexible inner core. To reduce eddy currents, the outer conductor is usually made of braided insulated wire, electrically connected at the ends of the cable. Construction details are shown schematically in Fig. 9-26.

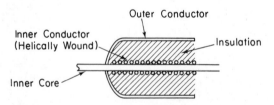

FIGURE 9-26   Helical high-impedance delay line.

The inductance of the delay line is produced by the inner coil, and it equals that of a solenoid with n turns per meter. The inductance can be increased by winding the helical inner conductor on a ferromagnetic core, which has the effect of increasing the delay time t_d and the characteristic impedance Z_o. The capacitance of the delay line is that of two coaxial cylinders separated by a polyethylene dielectric. The capacitance can be increased by using a thinner dielectric spacing between the inner and outer conductors.

Typical parameters for a helical, high-impedance delay line are $Z_o = 1,000\ \Omega$ and $t_d = 180$ ns/m. The coaxial delay line is advantageous because

it does not require the careful adjustment of a lumped-param
occupies much less space.

9-6   HORIZONTAL DEFLECTION SYSTEM

9-6.1   Basic Sweep Generator

The CRO usually displays the vertical input waveform as a function of
time. This requires a horizontal deflection voltage that moves, or sweeps, the
CRT spot across the screen, from left to right, with a constant velocity, and
then rapidly returns the spot to its starting position on the left-hand side of
the screen, ready for the next sweep. This sweep voltage, or time base, is produced within the horizontal deflection system of the CRO by the sweep
generator.

The ideal sweep voltage rises at a linear rate from some minimum
value to some maximum value, and then returns abruptly to its original level,
as shown in the sawtooth waveform of Fig. 9-27.

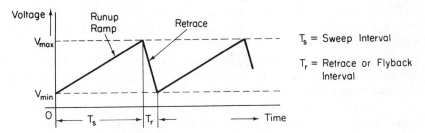

FIGURE 9-27   Linear sawtooth waveform.

The linearly rising portion of the sawtooth is called the *ramp voltage.*
During time interval T_s, when the ramp voltage rises from V_{min} to V_{max}, the CRT
spot is swept across the screen from left to right. In the retrace, or flyback,
interval T_r, the sweep voltage drops rapidly to its minimum value, and the
CRT spot flies back to its starting point on the screen. In almost all CROs
the electron beam is cut off during the retrace time interval, so that the CRT
spot cannot produce a retrace image on the screen.

All sweep generators are refinements of the basic RC charging circuit
of Fig. 9-28(a). In this circuit, switch S is initially closed, and voltage e_c across
the capacitor is zero. When the switch is opened, capacitor voltage e_c rises
exponentially from zero toward the positive supply voltage E, as shown in
Fig. 9-28(b). The instantaneous voltage across the capacitor is given by the
equation

$$e_c = E(1 - \epsilon^{-t/RC}) \qquad\qquad (9\text{-}34)$$

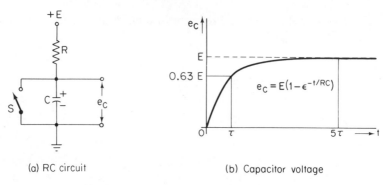

(a) RC circuit   (b) Capacitor voltage

FIGURE 9-28   Basic *RC* charging circuit.

where *t* is the total time of the charging process and *RC* is the time constant of the charging circuit.

The rise in capacitor voltage is extremely nonlinear: e_C reaches 63 per cent of its final value in one time constant and it reaches its full value E in five time constants. It is obvious that e_C cannot be used as a *linear* sweep voltage. If, however, the charging process is terminated early by closing switch *S*, so that only the initial, steeply rising portion of the voltage waveform is used as a sweep voltage, reasonable linearity can be obtained. For example, when *S* is closed at $t = 0.2\tau$, capacitor voltage $e_C = 0.1E$, and the slope error (deviation from linearity) is less than 10 per cent. In some applications this amount of nonlinearity is acceptable, and the simple *RC* circuit is therefore used in some low-cost, low-frequency CROs.

In a practical *RC* sweep circuit the function of switch *S* in Fig. 9-28(a) is performed by an electronic switching device, such as a unijunction transistor, a silicon controlled switch, a thyristor, a gas thyratron, etc. Figure 9-29(a) shows the well-known relaxation oscillator, in which the unijunction transistor (UJT) acts as the switch. When the power is first applied, capacitor *C* charges exponentially through resistor *R*, and UJT emitter voltage V_E rises toward the supply voltage V_{BB}. When V_E reaches the peak voltage V_P of the UJT, the emitter-to-base-1 diode becomes forward biased and the UJT triggers on. This provides a low-resistance discharge path between E and B1, so that the capacitor discharges rapidly through the UJT. Emitter voltage V_E therefore decreases abruptly until it can no longer sustain the minimum bias required for UJT conduction. At this point the low-resistance E-B1 path is broken, and the capacitor recharges. This cycle of charging and recharging repeats in a continuous or free-running process and produces a sawtooth waveform as in Fig. 9-29(b).

To improve sweep linearity, an actual UJT relaxation circuit would probably use two separate voltage supplies: a low-voltage supply for the UJT and a high-voltage supply for the *RC* circuit.

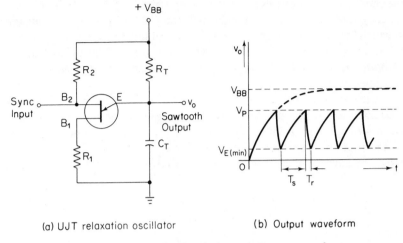

(a) UJT relaxation oscillator   (b) Output waveform

FIGURE 9-29 Practical sawtooth generator.

The frequency of oscillation can be varied by changing the values of R and/or C (change in time constant). In a practical CRO sweep circuit, resistor R is used for continuous control of the frequency (VARIABLE control) and capacitor C is changed in steps to provide a number of frequency ranges (TIME/DIV selector switch). Since both R and C can change the sweep frequency or time base, they are often called the *timing resistor* and *timing capacitor*, respectively.

9-6.2 Synchronization of the Sweep

The sawtooth generator of Fig. 9-29 is said to be free running because there is no external control that starts the generation of each new sweep. The new sweep simply starts as soon as the capacitor has discharged sufficiently to turn the UJT off. It is possible to use a free-running sweep to produce a stable CRT display, provided that the frequency of the vertical input signal is a whole multiple of the sweep frequency ($f_v = nf_s$). This situation is shown in Fig. 9-30, where two cycles of the signal waveform occur in the

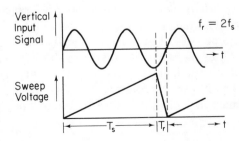

FIGURE 9-30 Vertical input signal and a free-running sweep.

same time interval as one cycle of the sweep voltage ($f_v = 2f_s$). If this exact
frequency relationship is not maintained, the CRT display will be unstable
and will drift across the screen. To produce a stable display, the sweep generator must run synchronously, or "in step," with the vertical signal source, so
that both vertical and horizontal signals arrive at some reference point in
their cycles at the same time.

In the relaxation oscillator of Fig. 9-29(a) synchronization of the sweep
can be obtained by applying a so-called *sync signal* to the sync input terminal
in such a manner as to lower the peak voltage of the UJT and thereby end
the run-up ramp prematurely. This situation is clarified in Fig. 9-31, where a
train of negative sync pulses is superimposed on the peak voltage of the
UJT.

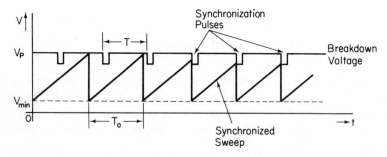

FIGURE 9-31   Principle of sweep synchronization.

The first several pulses have no effect on the frequency of the sawtooth
waveform, and the sweep generator continues to run unsynchronized at its
own natural frequency. Eventually, the capacitor charging process is stopped
prematurely by a sync pulse that occurs at the exact moment that the run-up
ramp voltage equals the (momentarily) reduced peak voltage of the UJT.
At that instant the capacitor discharges rapidly through the UJT, and the
run-up ramp is terminated. When the capacitor voltage has dropped to the
minimum voltage required to maintain UJT conduction, the transistor
switches off and the capacitor charges again to produce the next ramp
voltage.

It is clear that the process of synchronization can only take place
because the sync pulse terminates the sweep prematurely. This means that
the period (T) of the sync signal must be shorter than the natural period
(T_o) of the sawtooth waveform. It also means that when the sweep is synchronized, it assumes the frequency of the sync signal, which is slightly lower
than its own natural frequency. In addition, the amplitude of the sync signal
must be sufficiently large to bridge the gap between the actual capacitor

voltage and the quiescent peak voltage of the UJT. Small amplitude sync pulses simply will not synchronize the sweep.

Instead of using a negative sync pulse as in Fig. 9-31, sweep synchronization can also be obtained with a sinusoidal sync signal of sufficient amplitude. The negative sync pulse prematurely stops the charging process of the capacitor, thereby shortening the natural period of the sweep signal. The sine wave sync signal can either shorten or lengthen the natural period of the sawtooth. This is shown in Fig. 9-32, where two sweep voltages of different frequency

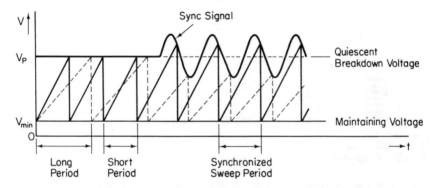

FIGURE 9-32 A sinusoidal sync signal can be used to synchronize sweep voltages whose natural period is longer or shorter than the period of the sync signal.

are synchronized to the same sine wave sync signal. One sawtooth waveform (drawn as a solid line), whose natural period is shorter than the period of the sync signal, is lengthened until it is "in step" with the sine wave. The other sawtooth (drawn as a broken line), whose natural period is longer than the period of the sync signal, is shortened until it is in sync with the sine wave. In both cases, though, the synchronized sweep assumes the frequency of the synchronization signal.

The sync signal for the sweep generator can be obtained from various sources and is selected by a CRO front panel control called the SYNC SELECTOR. In Fig. 9-33 this selector is shown as a three-position switch marked INT-EXT-LINE. On the INT, or internal, position a sample of the vertical amplifier signal, provided by a voltage divider, is used to develop the sync pulse. This therefore relates the start of the sweep to the vertical input signal under observation. On the EXT, or external, position of the SYNC selector, the sweep generator can be synchronized to an externally applied signal connected to a front panel jack marked EXT. In the LINE position a sample of the powerline voltage is applied to the sweep generator, so that the signal under observation is synchronized to the frequency of the powerline.

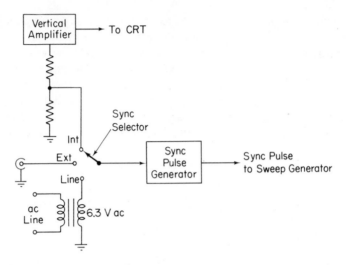

FIGURE 9-33   A sync selector circuit.

9-6.3   Triggered Sweep

Laboratory-type CROs are usually provided with a time base system that uses a so-called *triggered* sweep. With the triggered sweep, the sawtooth generator does not develop a ramp voltage unless told to do so by a trigger pulse. A triggered sweep extends the versatility of the CRO in the sense that it allows CRT displays of vertical input signals of very short duration (for example, a narrow pulse), stretched out over an appreciable area of the screen, simply because the sweep is initiated by a trigger pulse derived from the waveform under observation.

Figure 9-34(a) shows a modification of the basic relaxation oscillator of Fig. 9-29(a) as a practical example of a triggered sweep circuit. The two resistors R_3 and R_4 form a voltage divider across supply voltage V_{BB}. Their resistances are selected such that voltage V_D at the cathode of diode D is below the peak voltage V_P for UJT conduction. When the circuit is first switched on and the UJT is in the nonconducting state, timing capacitor C_T charges exponentially through timing resistor R_T toward V_{BB}, until a point is reached where the diode becomes forward-biased and conducts. The capacitor then never reaches the peak voltage required for UJT conduction but is clamped at V_D and cannot discharge. If a negative trigger pulse of sufficient amplitude is now applied to base-2 of the UJT, peak voltage V_P is momentarily lowered and the UJT fires. As a result, C_T discharges rapidly through the UJT until the maintaining voltage of the UJT is reached. At this point the UJT switches off, and C_T charges toward supply voltage V_{BB} until it is clamped again at V_D where it awaits the arrival of the next trigger pulse.

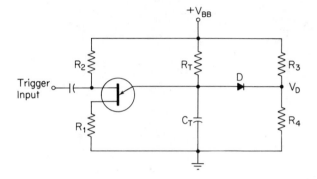

(a) Triggered sweep circuit

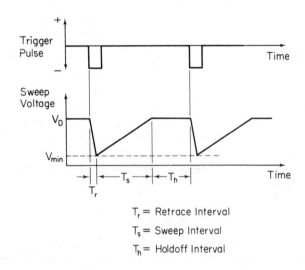

T_r = Retrace Interval

T_s = Sweep Interval

T_h = Holdoff Interval

(b) Trigger pulse and sweep waveforms

FIGURE 9-34   Triggered sweep generator.

The output waveform of the triggered sweep generator is shown in Fig. 9-34(b). Note that the trigger pulse initiates the retrace before the sweep can be generated, so that the initial part of the waveform under observation will be lost in the short retrace time interval unless sufficient signal delay is provided by the vertical delay line.

The block diagram of Fig. 9-35 shows a typical trigger circuit for a CRO with triggered sweep. The trigger circuit accepts input signals of a variety of shapes and amplitudes, and from various sources, and converts them into pulses of uniform amplitude for reliable sweep operation. The

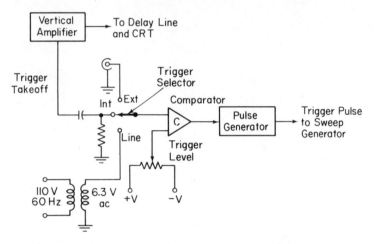

FIGURE 9-35   Block diagram of a sweep trigger circuit.

trigger selector is shown as a three-position switch marked INT-EXT-LINE and provides the operator with a choice of trigger input signals in the same manner as the SYNC selector of Fig. 9-33. This trigger input signal is applied to a voltage comparator whose reference level is set by the TRIGGER LEVEL control on the CRO front panel. The comparator circuit reacts to variations of trigger input signal which exceed the preset value selected by the trigger level control. The pulse generator (Schmitt trigger) that follows the comparator produces a negative trigger pulse each time the comparator output crosses its quiescent level, which in turn will trigger the sweep generator to start a new sweep.

9-6.4 Improvement of Sweep Linearity

Laboratory oscilloscopes are designed to make accurate measurements with respect to time and therefore require a sweep with excellent linearity. Several methods may be used to improve sweep linearity. Among the more important ones are the following:

(a) Constant current charging, whereby the timing capacitor is charged linearly from a constant current source
(b) The Miller sweep circuit, whereby a step input is converted into a linear ramp by using an operational integrator
(c) The phantastron circuit, which is a variation of the Miller circuit
(d) The bootstrap circuit, whereby constant charging current is maintained by keeping the voltage across the charging resistor, and hence the charging current through it, constant
(e) Compensating networks, which are used to improve the linearity of the Miller and bootstrap circuits

Detailed analysis of these circuits falls outside the scope of this text. Students who wish to pursue this matter may consult the many excellent textbooks* in the field of pulse and switching circuits.

9-6.5   Horizontal Amplifier

In a conventional CRO the performance requirements (gain/bandwidth) of the horizontal amplifier are of a lower order than those for the vertical amplifier. While the vertical amplifier must be able to handle small-amplitude, fast-risetime signals, the horizontal amplifier only has to process a sweep signal with its fairly high amplitude and relatively slow risetime. The gain of the horizontal amplifier, however, is larger than that of the vertical amplifier, because the horizontal deflection sensitivity of the CRT is smaller than the vertical deflection sensitivity.

Figure 9-36 shows the block diagram of a basic horizontal amplifier

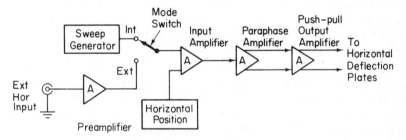

FIGURE 9-36   Block diagram of a basic horizontal amplifier.

that is generally used in a simple, low-frequency CRO. This amplifier consists of three stages: an input amplifier, a paraphase amplifier, and a push-pull output stage. In the usual application, the input amplifier receives its signal from the sweep generator, which typically produces a 10-V time-base ramp. Together with the sweep voltage, the input stage also receives a dc offset voltage that allows horizontal positioning of the CRT spot on the screen. The single-ended output is connected to a negative-feedback paraphase amplifier, which produces two balanced output signals to feed the final stage. The push-pull output amplifier provides two ramp voltages, a positive-going ramp and a negative-going ramp, amplified to the required level, for simultaneous application to the horizontal deflection plates of the CRT.

There are a number of useful applications in which the CRO is placed in the so-called *X-Y mode* of operation instead of the usual Y-t mode. In the X-Y mode the vertical input signal is applied to the CRO in the usual way,

---

*Millman and Taub, *Pulse, Digital, and Switching Waveforms* (New York: McGraw-Hill Book Company, Inc., 1965), pp. 514–596; John M. Doyle, *Pulse Fundamentals* (Englewood Cliffs, N.J.: Prentice-Hall, Inc., 1963), pp. 314–332; 455–458.

but the horizontal time base is replaced with an external signal that is applied to the horizontal amplifier via a preamplifier and the EXT position on the sweep selector. If the X-Y display must represent a true relationship between the horizontal and vertical signals, the two systems must have the same phase delay, deflection factor, and bandpass. These requirements place the horizontal amplifier system in the same class as the vertical amplifier system.

In the more advanced laboratory-type CRO, the input stage is often combined with the sweep generator to form a plug-in time-base unit, with the paraphase and output amplifiers remaining in the mainframe of the CRO.

9-7 CRO PROBES

9-7.1 Introduction

The CRO probe performs the important function of connecting the circuit under investigation to the vertical input terminals of the CRO, without loading or otherwise disturbing the test setup. To meet the requirements of the many general-purpose and special-application CROs, a variety of probes is available, from simple passive voltage probes to sophisticated active probes for special measurement applications. In each case, however, the probe must not degrade the performance of the CRO, and the probe-CRO combination must be properly matched and calibrated as a measurement system to ensure maximum measurement accuracy.

Figure 9-37 is a general block diagram applicable to all CRO probes.

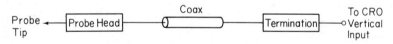

FIGURE 9-37   General block diagram of a CRO probe.

The probe head contains the signal-sensing circuitry. This circuitry may be passive, such as a 10-MΩ resistor shunted by a 7-pF capacitor, or it may be active, such as a FET source follower plus associated elements. A *coaxial cable* (the type of cable depends on the type of probe) is used to couple the probe head to the termination circuitry, which also can be active or passive. The *termination* circuit provides the CRO with the source impedance it requires and terminates the coax cable in its characteristic impedance.

9-7.2 Passive Voltage Probes

The most popular and convenient probe for coupling the signal of interest to the CRO is the passive voltage probe (so called because it contains no active elements).

The simplest passive probe is the nonattenuating, or 1X, probe. This probe consists of a length of coaxial cable with a probe tip at one end and a BNC connector at the other. Although the connection from the test point to the CRO input is a direct one, the shunting capacitance of the cable plays a role and must be taken into account. Typically, the capacitance of a 50-Ω coaxial cable is approximately 30 pF/ft, so that a 5-ft coax adds approximately 150 pF to the CRO input capacitance. The 1X probe is therefore essentially a large shunting capacitance, with an input terminal that is located several feet from the CRO input. Because the 1X probe presents a large load to high-frequency signals, it is normally limited to low-frequency applications, such as the measurement of ac power supply ripple.

One of the most widely used passive voltage probes is the *compensated* 10X probe of Fig. 9-38, designed to provide signal attenuation of ten-to-one over a wide frequency range. In Fig. 9-38 the probe head contains attenuating

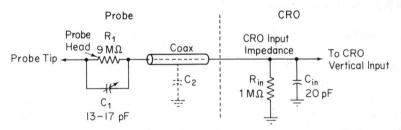

FIGURE 9-38   Compensated 10X probe.

resistor R_1 shunted by a small variable capacitor C_1 for probe compensation. (The reader is referred to Sec. 9-4.3 in which the subject of attenuator compensation is discussed in detail). A length of coaxial cable connects the probe head to the CRO, whose input impedance is represented by resistor R_{in} in parallel with capacitor C_{in}. For a general-purpose laboratory instrument, $R_{in} = 1$ MΩ and $C_{in} = 20$ pF are representative values.

As far as dc voltages are concerned, the probe-CRO combination is a 10-to-1 voltage divider, whose dc transfer characteristic is

$$v_{out} = v_{in}\frac{R_{in}}{R_1 + R_{in}} \tag{9-35}$$

For the circuit values given in Fig. 9-38 this yields

$$v_{out} = 0.1\ v_{in}$$

which is indeed a 10-to-1 voltage division.

To compensate this voltage divider over the frequency range of the CRO, time constant $\tau_1 = R_1 C_1$ of the input network must be equal to time constant $\tau_2 = R_{in}(C_2 + C_{in})$ of the output network. Note that C_2 represents the capacitance of the coaxial cable. Assuming a cable length of 3.5 ft, and a capacitance of 30 pF/ft, the total coax capacitance is $C_2 = 3.5 \times 30$ pF

$= 105 \, \text{pF}$. Hence time constant $\tau_2 = 1 \, \text{M}\Omega \ (105 \, \text{pF} + 20 \, \text{pF}) = 125 \, \mu\text{s}$. For $R_1 = 9 \, \text{M}\Omega$, compensating capacitor C_1 must be adjusted to $C_1 = 125 \, \mu\text{s}/9 \, \text{M}\Omega = 13.88 \, \text{pF}$. Since the input capacitance of a CRO can range from approximately 15 pF to 50 pF, depending on the individual instrument, compensating capacitor C_1 must be adjustable from approximately 13 pF to 17 pF, as indicated in Fig. 9-38.

It is important to realize that when the 10X attenuator probe is first connected to the CRO, compensating capacitor C_1 must be adjusted for optimum frequency response of the probe-CRO combination. This adjustment is most easily done by connecting the probe tip to a 1-kHz square-wave test signal (calibrator voltage) and observing the CRT display for optimum response, while adjusting C_1. Typical response patterns are shown in Fig. 9-19.

9-7.3 Active Voltage Probes

Active voltage probes, designed to provide an efficient method of coupling high-frequency, fast-risetime signals to the CRO input, contain active devices such as diodes, FETs, BJTs, or miniature vacuum tubes. In general, the active probe has a very high input impedance with less attenuation than the passive probe. Because they contain electronic circuitry, active probes are more expensive and more bulky than passive probes, but they extend the measurement capability of the probe-CRO system considerably.

An early version of the active probe is the *cathode-follower (CF) probe* of Fig. 9-39, which uses a miniature vacuum triode as the active element.

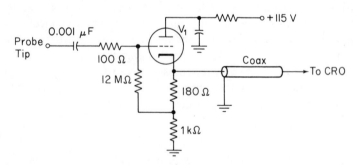

FIGURE 9-39   ac coupled cathode-follower probe (courtesy Tektronix, Inc.).

The entire CF circuitry is contained in the probe head; a coaxial cable connects the CF output to the CRO input terminals. Special provision is made for the high voltage and filament supplies to the vacuum triode by means of a separate cable connection. The input impedance of the cathode-follower circuit is very high, typically on the order of 10 MΩ or more, while the input capacitance is very low (approximately 5 pF). The output impedance of the cathode follower is designed to drive the coax cable that is terminated in its characteristic impedance at the CRO input. The CF probe is limited to input voltages not exceeding a few volts, although its voltage range can be extended by adding a compensated 10-to-1 divider to the CF input by means of an add-on probe tip.

A more sophisticated version of the active voltage probe is the *FET probe* of Fig. 9-40, where a field effect transistor in the source follower con-

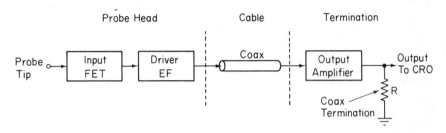

FIGURE 9-40   FET-input active probe.

figuration is used as the active input element. The FET probe, like other probes, consists of three parts: the probe head, the coax, and the termination. The probe head contains the FET source follower plus an EF driver amplifier to drive the coax. The input impedance of the FET circuit is approximately 10 MΩ shunted by 5 pF, and the dynamic signal range of the probe amplifier is limited to approximately ± 500 mV. To extend this limited input voltage range, 10X and 100X attenuators are usually available as add-on devices. The coax cable connects the probe head to the termination box, which in turn is connected to the CRO input. The coax is terminated in its characteristic impedance (for example, $Z_o = 50 \, \Omega$) by active devices contained in the termination box. Additional circuitry designed to improve circuit stability and frequency response, often consisting of active devices and an output amplifier, are also provided in the termination box.

9-7.4   Current Probes

The current probe provides a method of inductively coupling the signal to the CRO input, so that a direct electrical connection to the test circuit is not required. As in the case of the voltage probe, the current probe consists of a sensor, a coax cable, and a termination circuit.

There are various types of current probe. A popular current probe is the so-called *split-core* passive current probe of Fig. 9-41 that can be opened and clipped around a conductor whose current is to be measured. The current-sensing device of this probe is a so-called *current transformer* of split-core

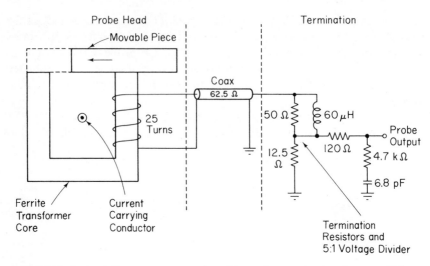

FIGURE 9-41   Split-core current probe with passive termination (courtesy Tektronix, Inc.).

design, consisting of a stationary U-piece and a movable flat piece. A multi-turn coil of approximately 25 turns is wound on one leg of the ferrite core to form the secondary transformer winding. The conductor under test is the single-turn primary.

The input signal to the probe is the current in the conductor under test; the output signal is the voltage developed across the transformer secondary. It is clear that this current probe only senses *changes* in current and therefore can only be used to measure ac signals. When correctly terminated, the sensitivity of this probe is on the order of 10 mA/mV (1 millivolt output signal as a result of a 10-mA change in input current). The transformer output voltage is coupled from the probe head to the termination via a coaxial cable. The termination circuitry can be passive or active, depending on the kind of probe, and generally provides for termination of the coax in its characteristic impedance. Additional circuitry to improve the response characteristics of the probe is also contained in the termination box.

9-7.5 High-voltage Probes

The high-voltage probe is used to apply kilovolt signals to the conventional CRO by providing voltage division ratios of 1,000-to-1 or more. The probe head of the high-voltage probe is made of high-impact strength thermoplastic material and is of special design to protect the user against electrical shock hazards.

Figure 9-42 shows the circuit diagram of a typical 1,000-to-1 high-voltage probe. The probe head contains a 100-MΩ resistor, approximately 4 in. long, whose distributed capacitance is indicated in the schematic. A

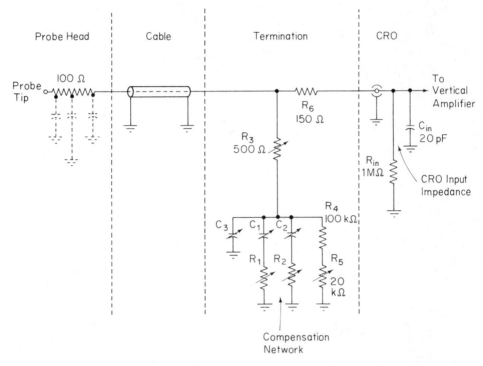

FIGURE 9-42   High-voltage probe (courtesy Tektronix, Inc.).

special probe cable connects the head to the termination box, which can be plugged into the vertical input terminals of the CRO. The 1,000-to-1 attenuation ratio is obtained by adjusting resistor R_5, in series with $R_4 = 100$ kΩ, and with a CRO input resistance of 1 MΩ as shown. The probe is compensated to the input time constant of the CRO by adjusting the network consisting of R_1, C_1, R_2, C_2, and C_3. The probe cable is terminated in its characteristic impedance by resistors R_3 and R_6.

The high-voltage capability of the probe is affected by the shunting capacitance of the input circuit, which can become noticeable at frequencies above 100 kHz. Increased temperatures also derate the high-voltage capability of the probe.

9-8   LISSAJOUS FIGURES

9-8.1   Construction of the Lissajous Figure

Lissajous figures result when sine waves are applied simultaneously to the horizontal and vertical deflection plates of the CRO. The construction of a Lissajous figure is shown graphically in Fig. 9-43. Sine wave e_v represents

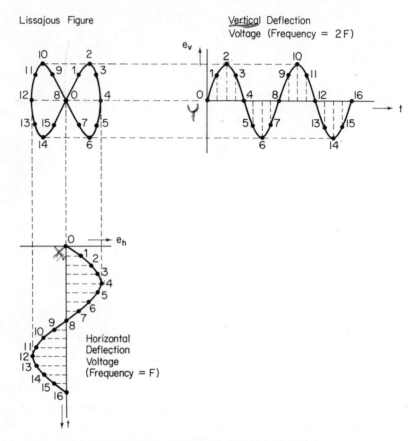

FIGURE 9-43   Construction of a Lissajous figure.

the vertical deflection voltage and sine wave e_h the horizontal deflection voltage. The frequency of the vertical signal is twice that of the horizontal signal, so that the CRT spot travels two complete cycles in the vertical direction against one cycle in the horizontal direction. Figure 9-43 shows that numbers 1 to 16 on both waveforms represent points of corresponding time intervals. Assuming that the spot starts in the center of the CRT screen (point 0), the travel of the spot can be reconstructed in the manner indicated, and the resultant pattern is called a *Lissajous figure*.

Two sine waves of the same frequency produce a Lissajous figure which may be a straight line, an ellipse, or a circle, depending on the phase and amplitude of the two signals. A circle can be formed only when the amplitudes of both signals are equal. If they are not equal and/or out of phase an ellipse is formed, the axes of which are the horizontal and the vertical planes (assuming normal positioning of the CRT). Excluding the consideration of signal amplitude, the property that determines the type of pattern formed when two signals of the same frequency are applied to the deflection plates is the *phase difference* between these signals. Figure 9-44 shows the phase relationships necessary for each of the patterns produced.

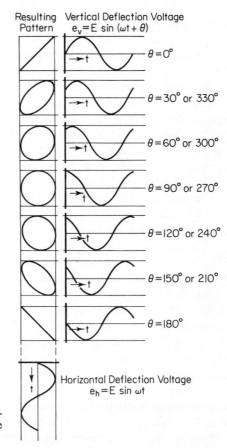

FIGURE 9-44   1/1 Lissajous patterns showing the effect of phase relationships.

A number of conclusions can be drawn from a study of these patterns. For example, a *straight line* results when the two signals are either in phase or 180° out of phase with each other. The angle formed with the horizontal will be exactly 45° when the amplitudes of the signals are equal. An increase in the vertical deflection voltage causes the line to have an angle greater than 45° with the horizontal. Similarly, a reduction in vertical amplifier gain results in a line with an angle smaller than 45° with the horizontal. A *circle* is displayed when the phase difference between the two signals is exactly 90° or 270°, assuming that the two signals are equal in amplitude. If the vertical signal has a larger amplitude, an *ellipse* with a vertical major axis is formed.

When the horizontal signal is larger, the main axis of the ellipse will lie along the horizontal axis. In the case of ellipses resulting from phase differences other than 90°, a change in relationship between the deflection voltages has the same effect.

9-8.2 Frequency Determination

There are many possible configurations for any ratio of applied signals. One consideration is whether the higher or the lower frequency is applied to the horizontal deflection plates. The most significant consideration, however, is the *phase* of the high-frequency signal with respect to the low-frequency signal. The pattern of Fig. 9-43 shows a figure eight, resting on its side, which results when both signals start out together. A tangent drawn against the top edge of the pattern would make contact at two places; a tangent drawn against a vertical side would make contact at one place. Evidently, the number of horizontal tangencies corresponds to the frequency of the vertical deflection voltage, and the number of vertical tangencies to the frequency of the horizontal deflection voltage. Hence the ratio of the vertical deflection frequency to the horizontal deflection frequency is 2/1.

Interesting patterns result when the high-frequency signal and the low-frequency signal do not start at the same time but are shifted in phase. Figure 9-45 shows the intermediate phase relationships between the signals. Figure 9-45(b) shows the situation where the high-frequency signal is shifted ahead by 90°. Here the high-frequency signal is at its maximum when the low-frequency signal is just starting its cycle. When this condition occurs, the resulting pattern forms an inverted parabola. A pattern of this type is commonly referred to as a *double image*, since the electron beam, after reversing its direction, traces out exactly the same path.

When a double image, such as the parabola, is developed, a different method of evaluating the frequency ratio must be used. In this case, a tangent drawn against an open end of the pattern is counted as a half tangency. For example, in Fig. 9-45(d), a tangent drawn against the top makes two contacts which are open ended and therefore count as one-half tangency each, giving a total of one point. Against the vertical side there is only one open contact, giving a count of one-half. The ratio of vertical frequency to horizontal frequency therefore is still 2/1. There are some restrictions on the frequencies that can be applied to the deflection plates. One, obviously, is that the CRO must have the bandwidth required for these frequencies. The other restriction is that the relationship between the two frequencies should not result in a pattern that would be too involved for an accurate determination of the frequency ratio. As a rule, ratios as high as 10/1 and as low as 10/9 can be determined comfortably.

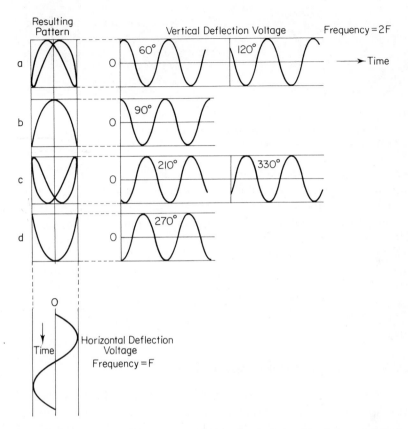

FIGURE 9-45 Lissajous figures for different phase relationships between vertical and horizontal deflection voltages.

In addition to the patterns for integral ratios of the two frequencies, there are many patterns for which the numerator and denominator of the ratio are whole numbers. For example, Fig. 9-46 shows the patterns for a 3/2 ratio and a 5/3 ratio. In every case, the method for determining the ratio of the applied frequencies is the same as described earlier.

Frequency comparison, using Lissajous patterns, is often performed on the CRO. When a Lissajous figure is formed by two equal-amplitude signals, with only a small difference in their frequencies, the pattern appears to drift slowly, according to the phase difference between them. From the beginning, when the two signals are in phase and the pattern is a straight line, as in Fig. 9-44, the line opens to an ellipse, then to a circle, closes to an ellipse, and then to a straight line with inclination opposite to the original. This sequence

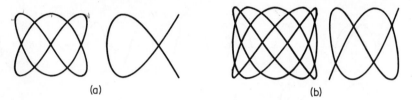

FIGURE 9-46 Frequency determination using Lissajous figures. (a) 3/2 Lissajous figure. (b) 5/3 Lissajous figure.

occurs in a drift of one-half cycle. At the completion of one cycle of difference, the pattern has returned to its original starting position. For example, if a Lissajous figure is used to compare two oscillators, one with a frequency of 1,000 Hz and the other with a frequency of 1,001 Hz, the picture on the CRT screen completes one cycle of change in one second. If the frequency of one oscillator is adjustable so that several seconds are needed for one complete change of the pattern, then the two frequencies are known to be within a fraction of a cycle within each other, which is a very small percentage in terms of the 1,000-Hz frequency of oscillation. When one frequency drifts slightly with respect to the other the pattern will rotate, or *barrel*, from the integral ratio.

9-8.3 Computation of Phase Angles

Regardless of the relative amplitudes of the applied voltages, the ellipse provides a simple means of finding the phase difference between two signals of the same frequency. The method is illustrated in Fig. 9-47. The sine of the phase angle between the two signals is equal to the ratio between the Y-axis intercept represented by Y_1 to the maximum vertical deflection represented by Y_2. We can write

$$\sin \theta = \frac{Y_1}{Y_2} \qquad (9\text{-}36)$$

For convenience, the gains of the vertical and horizontal amplifiers are adjusted so that the ellipse fits exactly within a square as marked by the coordinate lines on the graticule. Figure 9-47 shows how to interpret the phase angle corresponding to the orientation of the ellipse. If the major axis lies in the first and third quadrants, as shown in Fig. 9-47(b), the phase angle is either between 0° and 90° or between 270° and 360°. When the major axis passes through the second and fourth quadrants, the phase angle is between 90° and 180° or between 180° and 270°. In the example of Fig. 9-47 the sine of the phase angle equals 0.5, corresponding to the different values of phase angles indicated.

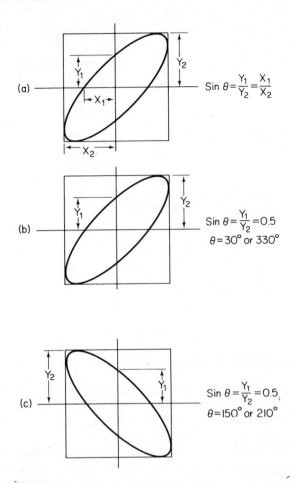

(a) $\sin\theta = \dfrac{Y_1}{Y_2} = \dfrac{X_1}{X_2}$

(b) $\sin\theta = \dfrac{Y_1}{Y_2} = 0.5$
       $\theta = 30°$ or $330°$

(c) $\sin\theta = \dfrac{Y_1}{Y_2} = 0.5$
       $\theta = 150°$ or $210°$

FIGURE 9-47   Computation of the phase angle between two signals of the same frequency.

9-9   SPECIAL-PURPOSE CROS

9-9.1   Dual-trace CRO

The single-trace capability of the conventional CRT can be modified to produce a dual image, or dual-trace display, by means of electronic switching of two separate input signals. The simplified block diagram of Fig. 9-48 shows that the dual-trace CRO has two vertical input circuits, marked channel A and channel B, with identical preamplifiers and delay lines. The A and B preamplifier outputs are fed to an electronic switch that alternately connects

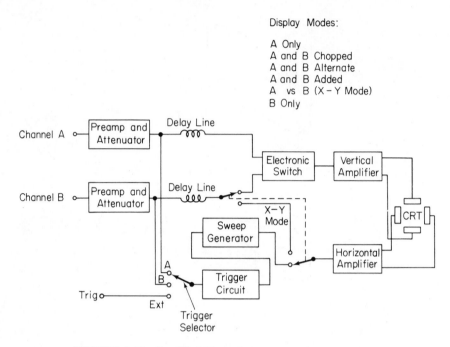

FIGURE 9-48   Simplified block diagram of a dual-trace CRO.

the input of the main vertical amplifier to the two signal inputs. The electronic switch also contains circuitry to select a variety of display modes. Although the display mode selectors are not shown in the block diagram, they are clearly visible in Fig. 9-49 as front panel controls.

When the display mode selector is in the *alternate* position, the electronic switch alternately connects the main vertical amplifier to channel A and to channel B. This switching takes place at the start of each new sweep. The switching rate of the electronic switch is synchronized to the sweep rate, so that the CRT spot traces channel A signal on one sweep and channel B signal on the next sweep. Since each vertical amplifier has a calibrated input attenuator and a vertical position control, the amplitudes of the input signals can be adjusted individually and the two images positioned separately on the screen. This mode of operation is especially useful with relatively fast sweep rates, when the two images appear as one simultaneous and stable display.

Note that the sweep trigger signal is available from channel A or channel B, and that it is picked off before the electronic switch. This arrangement maintains the correct phase relationship between the A and B signals.

In the *chopped* mode of operation the electronic switch is free-running at the rate of 100 to 500 kHz, entirely independent of the frequency of the sweep generator. In this mode the switch successively connects small seg-

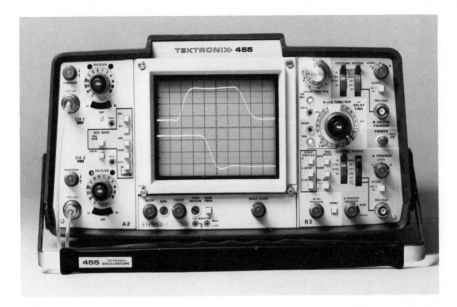

FIGURE 9-49   Front panel layout of a 50-MHz dual-trace portable oscilloscope (courtesy Tektronix, Inc.)

ments of the A and B waveforms to the main vertical amplifier. At the relatively fast chopping rate of 500 kHz, for example, 1-μs segments of each waveform are fed to the CRT for display. If the chopping rate is much faster than the horizontal sweep rate, the individual little segments fed to the main vertical amplifier together reconstitute the original A and B waveforms on the CRT screen, without any visible interruptions in the two images. If the sweep rate approaches the chopping rate, the little segments of the chopped waveforms become visible as individual images, and the continuity of image display is lost. In this case, it would be better to use the *alternate* mode of operation.

In the *added* mode of operation the A and B signals are algebraically added and their summation is displayed as a single image, as a function of time. If the polarity inversion switches are used in both channels, it is possible to display A + B, A − B, B − A, and −A−B.

In the *X-Y* mode of operation the sweep generator is disconnected and channel B is connected to the horizontal amplifier. Since both preamplifiers are identical and have the same delay, truly accurate X-Y measurements can be made.

In addition to these special display modes, the dual-trace CRO can be used as a conventional instrument, displaying either channel A or channel B as a function of time.

9-9.2 Dual-beam CRO

The dual-beam CRO accepts two vertical input signals and displays them as two separate images on the CRT screen. Instead of electronically switching the two signals to a single vertical amplifier, as in the dual-trace CRO of Sec. 9-9.1, the dual-beam oscilloscope uses a special CRT that produces two completely separate electron beams that can be independently deflected in the vertical direction. In some dual-beam CRTs the output of a single electron gun is mechanically split into two separate beams (the so-called *split-beam technique*), while other CRTs contain two separate electron guns, each producing its own beam. The dual-beam CRT has two sets of vertical deflection plates, one set for each channel, and a single set of horizontal deflection plates.

The simplified block diagram of Fig. 9-50 shows that the dual-beam CRO has two identical vertical channels, marked A and B. Each channel

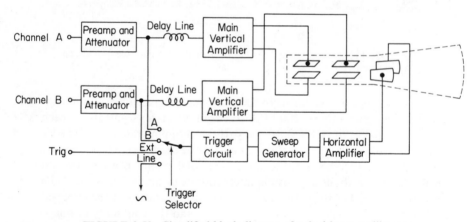

FIGURE 9-50   Simplified block diagram of a dual-beam oscilloscope.

consists of a preamplifier and input attenuator, a delay line, a main vertical amplifier, and the vertical CRT plates. The time-base generator, driving the single set of horizontal CRT plates, sweeps both beams across the screen at the same rate. The sweep generator can be triggered internally from either channel, from an externally applied trigger signal, or from the line voltage.

Since the dual-beam CRO does not have the same number of display modes as the dual-trace instrument, it may appear to be less versatile, but it is ideally suited for simultaneous display of widely different input signals. The photographs of Fig. 9-51 show the versatility of the dual-beam CRO in presenting displays of related phenomena. Figure 9-51(a) shows the input and output waveforms of a pulse-shaping circuit; Fig. 9-51(b) gives an example of typical modulation patterns in a communications circuit.

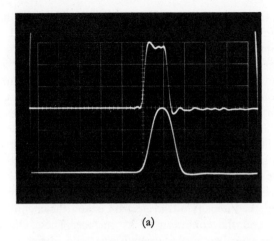

(a)

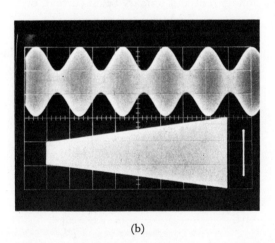

(b)

FIGURE 9-51   Typical dual-beam displays. (a) Input and output waveform of a
pulse-shaping circuit. (b) Modulation patterns (courtesy Tektronix, Inc.).

9-9.3   Storage CRO

In the conventional CRT the persistence of the phosphor ranges from
a few milliseconds to several seconds (see Table 9-1), so that an event that
occurs only once will disappear from the screen after a relatively short period
of time. A storage CRT can retain the display much longer, up to several
hours after the image was first written on the phosphor. This *retention* feature
can also be useful when displaying the waveform of a very low-frequency
signal. In the conventional (non-storage) CRO, the start of such a display
would fade before the end is written.

Storage CRTs can be classified as bistable tubes and halftone tubes. The bistable tube will either store or not store an event and produces only one level of image brightness. The halftone tube can retain an image for varying lengths of time (variable persistence) and at different levels of image brightness. Both the bistable and halftone tubes use the phenomenon of secondary electron emission to build up and store electrostatic charges on the surface of an insulated target. The following discussion applies to either type of tube.

When a target is bombarded by a stream of primary electrons, an energy transfer takes place which separates other electrons from the surface of the target in a process known as *secondary emission*. The number of secondary electrons emitted from the target surface depends on the velocity of the primary electrons, the intensity of the electron beam, the chemical composition of the target, and the condition of its surface. These characteristics are reflected in the so-called *secondary-emission ratio*, defined as the ratio of secondary-emission current and primary beam current, or

$$\delta = I_s/I_p \qquad\qquad (9\text{-}37)$$

The simple experimental circuit of Fig. 9-52 can be used to demonstrate how the secondary-emission ratio varies as a function of the target voltage

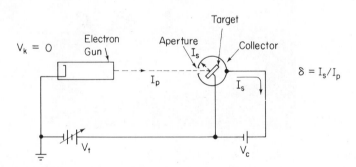

FIGURE 9-52 Experimental circuit used to demonstrate secondary-electron emission.

V_t. The electron gun in Fig. 9-52 emits a focused beam of high-velocity electrons in much the same way as does the gun in a conventional CRT. This electron beam is directed at the surface of a metal target which will emit secondary electrons under favorable conditions. The collector, which completely surrounds the target except for a small aperture to pass the primary beam, collects all secondary-emission electrons. This constitutes the secondary current I_s. The target voltage is adjustable over a wide range (from 0 to +3,000 V), while the collector is held a few volts above the target by battery V_c.

The bombarding energy of a primary electron is directly related to the potential difference between the electron source (the cathode) and the target. When the target voltage is zero, the energy of the bombarding electron is zero and there is no secondary emission. Hence $\delta = 0$. When the target voltage is increased from zero, the bombarding energy increases and causes some secondary-electron emission. Hence δ increases from zero, as shown in the secondary-emission curve of Fig. 9-53. At some positive target voltage

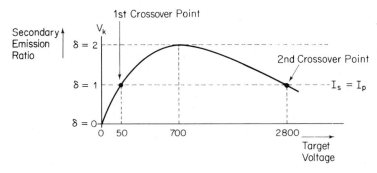

FIGURE 9-53   Typical secondary-emission curve.

($+50$ V in Fig. 9-53), the number of secondary emission electrons equals the number of primary beam electrons, so that $I_s = I_p$ and $\delta = 1$. This point on the curve is known as the *first crossover point*. When the target voltage is increased beyond this crossover point, the secondary-emission ratio initially increases to some maximum value ($\delta = 2$ in Fig. 9-53), and then it decreases again until $I_s = I_p$ and $\delta = 1$. This point on the curve is the *second crossover point*.

Figure 9-54(a) is a modification of the previous circuit and shows the collector voltage fixed at $+200$ V. The target voltage is adjustable over a wide range, as before. The fixed collector voltage drastically modifies the secondary-emission ratio, as indicated in Fig. 9-54(b). When the target voltage is larger than the collector voltage, the secondary electrons emitted from the target enter the retarding field of the collector and are reflected back into the target. Hence the target collects the total primary beam current I_p and the collector current I_s is zero. The *effective* secondary-emission ratio, defined by Eq. (9-37) as $\delta = I_s/I_p$, is therefore zero, and the curve is modified as in Fig. 9-54(b). The other change occurs when the target voltage is approximately 0 V. When the target is slightly negative, the primary electrons cannot reach the target but are deflected to the collector. Although there can be no secondary-electron emission, the collector current equals the primary beam current, and the target has an *apparent*, or *effective*, secondary-emission ratio of one. As the target voltage is increased from the negative side and approaches

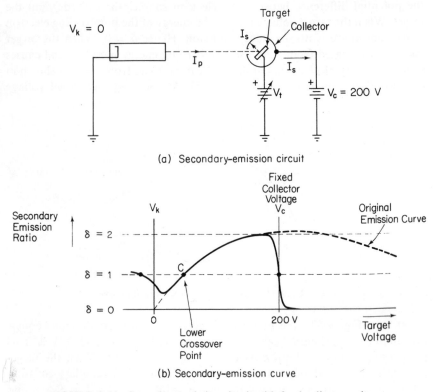

(a) Secondary-emission circuit

(b) Secondary-emission curve

FIGURE 9-54   Secondary-emission circuit with fixed collector voltage.

zero, the target no longer repels the primary beam, so that actual target bombardment takes place and real secondary emission results. These effects are shown in the modified curve of Fig. 9-54(b).

A further modification of the basic circuit is shown in Fig. 9-55(a). The collector voltage is again fixed at $+200$ V, but the target can be disconnected by switch S to become a so-called *floating target*. This CRT with floating target is capable of simple storage effects. Note that the secondary-emission curve for this tube, shown in Fig. 9-55(b), is similar to the one shown for the previous circuit.

Switch S is initially closed and the target voltage is set at some low value, say $+20$ V. At this point, the secondary-emission ratio is typically on the order of 0.5, so that the current in the collector circuit is one-half the primary beam current, or $I_s = \frac{1}{2}I_p$. The other half of the primary current is simply collected by the target and returned to the target battery. Hence target current $I_t = \frac{1}{2}I_p$. When switch S is now opened, the current in the target lead is interrupted and the primary beam current charges the target in the

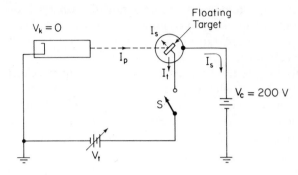

(a) Secondary-emission circuit with floating target

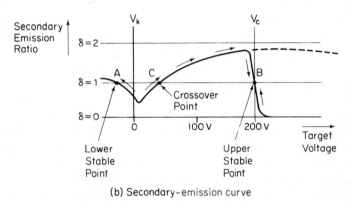

(b) Secondary-emission curve

FIGURE 9-55   Secondary-emission circuit with fixed collector voltage and floating target. The target voltage always assumes one of the stable conditions A or B.

negative direction. The target voltage therefore decreases (becomes less positive), and the secondary emission ratio changes, following the curve of Fig. 9-55(b). The rate of charge decreases as the target voltage approaches point A on the curve. At this point, the secondary-emission current equals the primary beam current, and the net charging rate is zero. At point A, the target voltage is slightly negative, the secondary emission ratio is one, and the target has reached a stable condition. Point A is called the *lower stable point*, and the target is considered to be in the *erased* condition.

If the initial or starting voltage of the target is to the right of crossover point C, say at $+100$ V in Fig. 9-55(b), the secondary emission ratio is greater than one. This means that I_s is greater than I_p and there must therefore be a net electron flow leaving the target surface. When switch S is now opened, the target continues to emit secondary electrons, so that it discharges and becomes more positive. Hence the secondary emission ratio moves up along the curve to point B where the rate of discharge is once again zero and the target obtains a stable condition. At this so-called *upper stable point* the secondary emission ratio is one, and the target is considered to be in the *written* condition.

As long as the primary gun is *on* and primary electrons bombard the target, the target will always be at a stable point, upper or lower, depending on the initial voltage of the target. Crossover point C on the curve is uniquely unstable in the sense that the target voltage will always move up to point B or down to point A, depending on which way the target voltage is first shifted by noise.

The CRT of Fig. 9-55 is an elementary bistable storage device. Its condition can be interrogated by measuring the target voltage. If the target voltage is "high," the target is written; if the target voltage is "low," the target is erased. The tube therefore has an electrical readout and its storage condition is not visible.

Figure 9-56(a) shows the principle of a bistable storage tube capable of writing, storing, and erasing an image. This storage tube differs from the one in Fig. 9-55(a) in two aspects: It has a multiple-target area, and it has a second electron gun. The second electron gun is called the *flood gun*; it emits low-velocity primary electrons that flood the entire target area. The distinguishing feature of the flood gun is that it floods the target at all times and not just intermittently as does the writing gun. The cathode of the flood gun is at ground potential, so that the target voltage will follow the secondary-emission curve indicated in Fig. 9-56(b). The lower stable point of the target is a few volts negative with respect to the flood gun cathode, and the upper stable point is at +200 V, the collector voltage. The cathode of the writing gun, however, is at −2,000 V, and its secondary-emission curve is superimposed on the flood gun curve. It is found that the combined effect of writing gun and flood gun is simply the sum of the individual effects of each electron beam by itself.

The flood gun is on at all times. Assume that the target is at its lower stable point, the erased condition. When the writing gun is gated on, its primary electrons arrive at the target with a potential of 2,000 V, which causes high secondary emission from the target. The target voltage therefore leaves the lower stable point and starts to increase. The flood gun electrons, however, attempt to maintain the target in its stable condition and oppose the increase in target voltage. If the writing gun is switched on long enough to carry the target past the crossover point, the flood gun electrons will aid the writing gun electrons and carry the target all the way to the upper stable point, so that the target is written. Even if the writing gun is now switched off, the target will be held in its upper stable condition by the flood gun electrons, thereby *storing* the information delivered by the writing gun. When the writing gun is not switched on long enough to carry the target past the

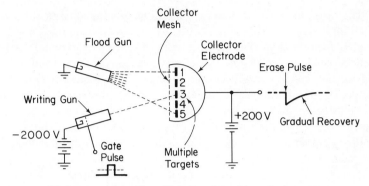

(a) Storage tube with multiple targets and two electron guns

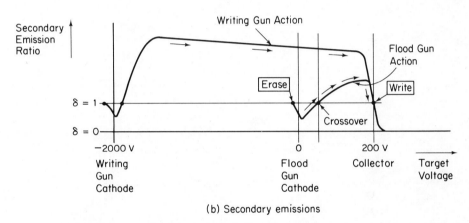

(b) Secondary emissions

FIGURE 9-56   Storage CRT with multiple targets and two electron guns.

crossover point, the flood gun electrons will simply move the target back to its lower stable condition, and storage does not occur.

Erasing the target simply means restoring the target voltage to the lower stable point. This can be accomplished by pulsing the collector negative, so that it momentarily repels the secondary-emission electrons and reflects them back into the target. This reduces the collector current I_s and the secondary-emission ratio drops below one. The target then collects primary electrons from the flood gun (remember that the writing gun is off) and charges negative. The target voltage decreases until it reaches the lower stable point where the charging ceases, and the target is in the erased condition. After erasure, the collector must be returned to its original positive voltage ($+200$ V in this case), and the erase pulse must therefore be returned to zero. As indicated in Fig. 9-56(a), this must happen gradually, so that the target is not accidentally driven past the crossover point and becomes written again.

The target area of the storage tube in Fig. 9-56(a) consists of a number of small individual metal targets electrically separated from one another and numbered from 1 to 5. The flood gun is of simple construction, without deflection plates, and it emits low-velocity electrons that cover all the individual targets. When the writing gun is gated on, a focused beam of high-velocity electrons is directed at one small target (number 3 in this case). This one target then charges positive and is written to the upper stable point. When the writing gun is turned off again, the flood electrons hold target 3 at its upper stable point (store). All the other targets are held at their lower stable points (erase).

The last step in our development of the bistable direct-viewing storage tube consists of replacing the individual metal targets with a single dielectric sheet, as in the typical tube of Fig. 9-57. This dielectric storage sheet consists

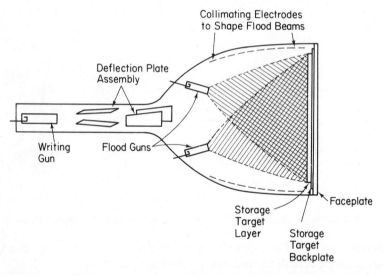

FIGURE 9-57   Schematic view of a bistable storage tube (courtesy Tektronix, Inc.).

of a layer of scattered phosphor particles capable of having any portion of its surface area written and held positive or erased and held negative without affecting the adjacent areas on the surface of the sheet. This dielectric sheet is deposited on a conductive-coated glass faceplate. The conductive coating is called the *storage target backplate*, and it is the collector of secondary-emission electrons. In addition to the writing gun and its deflection plate assembly, this storage CRT has two flood guns and a number of collimation electrodes that form an electron lens to distribute the flood electrons evenly over the entire surface area of the storage target.

After the write gun has written a charge image on the storage target, the flood guns will store the image. The written portions of the target are being bombarded by flood electrons that transfer energy to the phosphor layer in the form of visible light. This light pattern can be viewed through the glass faceplate. Since the storage target areas are either positive or negative, the light output produced by the flood electrons is either at full brightness or at minimum brightness. There is no gray scale in between.

9-9.4 Sampling CRO

When the frequency of the vertical deflection signal increases, the writing speed of the electron beam increases. The immediate result of higher writing speed is a reduction in image intensity on the CRT screen. In order to obtain sufficient image brilliance, the electron beam must be accelerated to a higher velocity so that more kinetic energy is available for transfer to the screen and normal image brightness is maintained. An increase in electron beam velocity is easily achieved by raising the voltage on the accelerating anodes. A beam with higher velocity also needs a greater deflection potential to maintain the deflection sensitivity. This immediately places higher demands on the vertical amplifier.

One way to improve the deflection system at higher frequencies has been successfully developed by the CRO manufacturers and consists of the *traveling-wave* type CRT. Figure 9-58 shows such a CRT, in which a series of

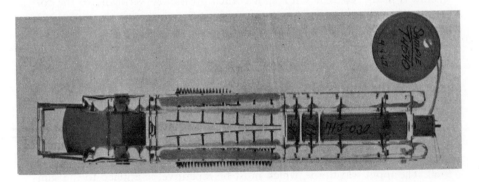

FIGURE 9-58 Traveling-wave type CRT. A special high-frequency CRT with a series of vertical deflection plates (courtesy Tektronix, Inc.).

deflection plates, mounted inside the tube, is so shaped and spaced that an electron traveling between them will receive from each plate an additional deflecting force in the proper time sequence. The vertical deflection signal is applied to each plate through a delay line which is so designed that the time delays correspond exactly to the transit times of the electrons traveling down the CRT toward the screen. The electron speed must be controlled very accurately in order to avoid distortion in the trace.

In addition to this special high-frequency CRT, new fluorescent materials have been developed to increase the image brightness at the higher frequencies. Further improvement in the vertical deflection system must be found in the vertical amplifiers themselves.

The *sampling* CRO uses a different approach to improve high-frequency performance. In the sampling CRO the input waveform is reconstructed from many samples taken during recurrent cycles of the input waveform and so circumvents the bandwidth limitations of conventional CRTs and amplifiers. The technique is illustrated by the waveforms indicated in Fig. 9-59.

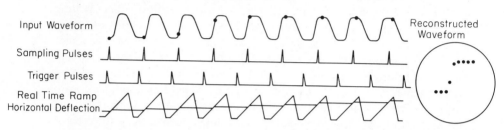

FIGURE 9-59   Waveforms pertinent to the operation of the sampling oscilloscope.

In reconstructing the waveform, the sampling pulse turns the sampling circuit on for an extremely short time interval. The waveform voltage at that instant is measured. The CRT spot is then positioned vertically to the corresponding voltage input. The next sample is taken during a subsequent cycle of the input waveform at a slightly later position. The CRT spot is moved horizontally over a very short distance and is repositioned vertically to the new value of the input voltage. In this way the CRO plots the waveform *point by point*, using as many as 1,000 samples to reconstruct the original waveform. The sample frequency may be as low as one-hundredth of the input signal frequency. If the input signal has a frequency of 1,000 MHz, the required bandwidth of the amplifier would be only 10 MHz, a very reasonable figure.

A simplified block diagram of the sampling circuitry is given in Fig. 9-60. The input waveform, which must be repetitive, is applied to the sampling gate. Sampling pulses momentarily bias the diodes of the balanced sampling gate in the forward direction, thereby briefly connecting the gate input capacitance to the test point. These capacitances are slightly charged toward the voltage level of the input circuit. The capacitor voltage is amplified by the vertical amplifier and applied to the vertical deflection plates. Since the sampling must be synchronized with the input signal frequency, the signal is delayed in the vertical amplifier, allowing the sweep triggering to be done by the input signal. When a trigger pulse is received, the avalanche blocking oscillator (so called because it uses avalanche transistors) starts an exactly

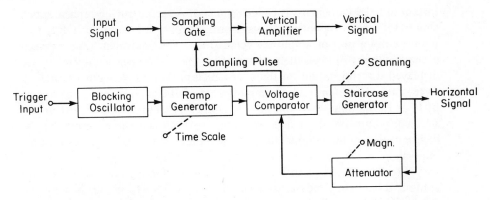

FIGURE 9-60 Simplified block diagram of the sampling circuitry (courtesy Hewlett-Packard Company).

linear ramp voltage, which is applied to a voltage comparator. The voltage comparator compares the ramp voltage to the output voltage of a staircase generator. When the two voltages are equal in amplitude, the staircase generator is allowed to advance one step and simultaneously a sampling pulse is applied to the sampling gate. At this moment, a sample of the input voltage is taken, amplified, and applied to the verical deflection plates.

The real-time horizontal sweep is shown in Fig. 9-59, indicating the horizontal deflection rate of the beam. Notice that the horizontal displacement of the beam is *synchronized* with the trigger pulses which also determine the moment of sampling. The resolution of the final image on the CRT screen is determined by the size of the steps of the staircase generator. The larger these steps, the greater the horizontal distance between the CRT spots that reconstitute the trace.

9-9.5 Digital Readout CRO

The *digital readout CRO* introduces the concept of providing digital readout of signal information, such as voltage or time, in addition to the conventional CRT display. The digital readout CRO basically consists of a conventional high-speed laboratory CRO plus an electronic counter, both contained in one cabinet. The circuitry of both units is connected by means of a display-logic control, allowing measurements to be made with great speed and accuracy. The digital readout CRO gives readout of rise time, amplitude, time difference, depending on the positions of the various operating controls, such as TIME/DIV, AMPLITUDE/DIV, and PROGRAM. The input waveform is sampled by means of a sampling unit. With each repetition of the input signal, the sampling unit strobes one point at a time, a little later than the previous sample. (The process of advancing the sampling instant in fixed increments is called *strobing*). A reconstructed replica, much slower than the original input waveform, is reproduced on the CRT as a point-by-point plot of amplitude versus time. The equivalent time between each sample depends on the number of samples taken per centimeter of the displayed waveform, and on the sweep time per centimeter. For example, a sweep rate of 1 ns/cm and a sampling rate of 100 samples/cm gives a time of 10 ps per sample. By counting the number of samples taken between two selected points on the waveform, the time between these points can be determined. Countercircuits, driving nixie tubes, display the measured time interval.

Figure 9-61 shows in block diagram form the operation of the digital readout scope when measuring time. The input waveform, strobed by the sampling unit, is applied to the CRT and is displayed in point-by-point form.

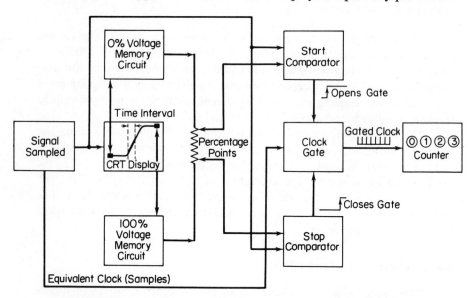

FIGURE 9-61   Block diagram of the digital readout oscilloscope when it is measuring time.

Two intensified portions of the CRT trace identify 0-per cent and 10-per cent zones. Each zone can be positioned to cover any portion of the CRT display. The input waveform amplitudes, corresponding to the intensified zones on the display, are stored in voltage-memory circuits. Voltage divider taps between the 0 per cent and 100 per cent memory voltages are set for start-and-stop timing at selected percentage points of the waveform to be measured. Coincidence of the input waveform amplitude with the selected reference amplitude (percentage selected) is sensed by voltage comparators that open and close the clock-gate to the digital counter. The number of clock pulses are read out digitally in nanoseconds, microseconds, milliseconds, or seconds on the nixie display tubes. In the time measurement application, the clock pulses consist of the actual samples taken.

Figure 9-62 shows in block diagram form the operation of the digital readout CRO when measuring voltage. The input waveform again is strobed

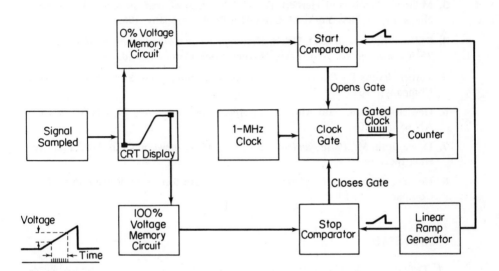

Voltage-to- Digital
Conversion. Gated
Clock to Counter.
Number of Clock
Pulses is Proportional
to Voltage.

FIGURE 9-62   Block diagram of the digital readout oscilloscope when it is measuring voltage.

by the sampling unit and displayed on the CRT. The voltage-memory circuits provide intensified portions on the CRT display and identify the 0-per cent and 100-per cent reference voltages. The 0-per cent reference voltage is applied to the start comparator and the 100-per cent reference voltage is applied to the stop comparator. A linear ramp generator is connected to both comparators. For the period of time that the linear ramp voltage is at values between the 0-per cent and 100-per cent amplitudes, as set by the voltage-memory circuits, 1-MHz clock pulses are gated to the digital countercircuit. The number of clock pulses is directly proportional to the voltage between the selected measurement points and is read out in millivolts or volts by the nixie tube display.

REFERENCES

1. Branson, Lane K., *Introduction to Electronics*, chaps. 8, 9. Englewood Cliffs, N.J.: Prentice-Hall, Inc., 1967.

2. Stout, Melville B., *Basic Electrical Measurements*, 2nd ed., chap. 17. Englewood Cliffs, N.J.: Prentice-Hall, Inc., 1960.

3. Millman, Jacob, and Herbert Taub, *Pulse, Digital, and Switching Waveforms*, chaps. 14, 15. New York: McGraw-Hill Book Company, 1965.

4. Rider, John F., and S. D. Uslan, *Encyclopedia on Cathode Ray Oscilloscopes and Their Uses*, 2nd ed. New York: Hayden Book Company, n.d.

5. Turner, Rufus P., *Practical Oscilloscope Handbook*. New York: Hayden Book Company, 1964.

6. Hewlett-Packard, Palo Alto, Calif., *Application Notes* AN 29, AN 36, AN 62, AN 65, AN 108.

7. Doyle, John M., *Pulse Fundamentals*, chap. 16. Englewood Cliffs, N.J.: Prentice-Hall, Inc., 1963.

8. Herrick, Clyde N., *Oscilloscope Handbook*. Reston, Va.: Reston Publishing Company, Inc., 1974.

PROBLEMS

1. Define the following terms: fluorescence, phosphorescence, persistence, luminescence.

2. A laboratory-type CRO generally uses an "unblanking cathode follower" in its CRT circuit. Explain its function and describe its operation.

3. Discuss the relationship between wide-band performance and high sensitivity of a general-purpose CRO. Make suggestions about steps to be taken to improve the gain-bandwidth performance of a CRO.

4. A simple RC time-base generator generally delivers a nonlinear ramp voltage which may be unsuitable for the time base of a laboratory-type CRO. Suggest several methods which may be used to improve the time-base linearity, and explain the principles involved in these methods of linearization.

5. What is the reason for using a delay line in the vertical deflection system of a laboratory-type CRO?

6. The input attenuator in the vertical amplifier of a general-purpose CRO is generally followed by an emitter-follower or cathode-follower circuit. Suggest three possible reasons for using this circuit.

7. Draw the block diagram of a general-purpose CRO. Label all the blocks and show the waveforms entering and leaving each block (when applicable) assuming a sinusoidal voltage applied to the input of the vertical amplifier.

8. The gain, frequency response, and phase shift of a 10-W audio amplifier (frequency range 20 Hz–20 kHz) are to be measured, using a CRO as the basic measuring tool. An audio oscillator and several types of ac and dc voltage and current meters are available. Suggest a measurement technique, indicating the equipment required to perform each measurement. The result of each measurement is to be presented in graphical form. Suggest a suitable way of presenting the measurement results, and approximately sketch the expected shape of each graph.

9. The input attenuator of Fig. 9-18 is used in a CRO that requires a time constant $\tau = 4 \ \mu s$. Calculate the values of C_a, C_i, R_a, and R_i, if the sum of R_a and R_i is 2 MΩ.

10. The calibrated time base of a laboratory-type CRO is set at 0.2 mV/cm. The horizontal display switch is in the "5 × magnified" position. A sinusoidal waveform of unknown frequency is applied to the input terminals of the vertical amplifier and produces $3\frac{1}{2}$ cycles over a sweep width of 10 cm. Determine the frequency of the input voltage.

11. A certain Lissajous pattern is produced by applying sinusoidal voltages to the vertical and horizontal input terminals of a CRO. The pattern makes five tangencies with the vertical and three with the horizontal. Calculate the frequency of the signal applied to the vertical amplifier if the frequency of the horizontal input voltages is 3 kHz.

12. Voltage V_1 is applied to the vertical input and voltage V_2 to the horizontal input of a CRO. The Lissajous pattern is symmetrical about the vertical and horizontal axes, with V_1 and V_2 having the same frequency. The slope of the major axis is positive, with a maximum vertical value of 2.5 divisions. The point where the figure crosses the vertical axis is 1.2 divisions high. Calculate the possible phase angles of V_2 with respect to V_1.

13. The transit time of an electron through the deflection plates is one of the factors determining the frequency limits of a CRO. Assuming that this transit time should be kept below 0.1 cycle, determine the upper frequency limit of an electrostatic deflection system with plates of 1-cm length if electrons enter with velocities corresponding to a kinetic energy of 1,000 eV.

14. Calculate the deflection sensitivity, S, for the CRO of Prob. 13 if L in Fig. 9-13 is 20 cm and the distance d between the plates is 5 mm.

15. Which factor(s) can be changed if the upper frequency limit of the CRO of Prob. 13 is to be doubled without affecting the deflection sensitivity calculated in Prob. 14?

16. The accelerating voltage of a CRT is 1,000 V. A sinusoidal voltage is applied to a set of deflecting plates whose axial length is 1 cm. Calculate (a) the maximum frequency of the sinusoidal voltage if the electrons are not to remain between the plates more than one-half cycle; (b) the time, in μs, that an electron remains in the region of the deflection plates if the frequency of the applied voltage is 60 Hz.

17. The calibrated time base of a laboratory-type CRO is set to 0.1 ms/cm. The sweep width is 10 cm. Assuming that the sweep voltage is an ideal ramp with zero retrace time, sketch the waveform patterns resulting from applying the following signals to the vertical amplifier input terminals:
    (a) sine wave with a frequency of 5 kHz
    (b) sine wave with a period of 0.5 ms
    (c) cosine wave with a period of 2 ms
    (d) square wave with a frequency of 10 kHz
    (e) pulse with a repetition rate of 2,000 cycles per second and a duty cycle of 25 per cent

18. Explain the function of each of the following CRO controls and indicate in which section of the CRO circuitry these controls are found:
    (a) focus   (b) horizontal position
    (c) sweep vernier   (d) external horizontal input
    (e) Z-axis modulation

19. The horizontal amplifier section of a CRO usually provides several connections to trigger the time base. Give one example of application for each of the following trigger input settings:
    (a) internal   (b) line
    (c) external

20. Use simple waveform sketches to explain the to effect of excessive sync amplitude on the displayed waveform pattern.

10

Electronic Instruments for the Measurement of Voltage, Current, Resistance, and Other Circuit Parameters

10-1 INTRODUCTION

Electronic voltmeters, ammeters, and ohmmeters use amplifiers, rectifiers, and other circuits to generate a current proportional to the quantity being measured. This current then drives a conventional meter movement of the type discussed in Chapter 4. It is interesting to note that many modern electronic voltmeters use taut-band suspension movements (Sec. 4-3.3) instead of the more conventional pivot-and-jewel mechanisms. Instruments that use meter movements to indicate the magnitude of the quantity under measurement on a continuous scale are sometimes called *analog* instruments.

When the result of the measurement is displayed in *discrete* intervals or numerals (instead of by a pointer deflection on a continuous scale), we speak of a *digital* indication. Direct numeral readout reduces human error and tedium, eliminates parallax and other reading errors, and increases reading speed. Additional features in modern digital instruments, such as automatic polarity and range-changing facilities, further reduce measurement error and possible instrument damage due to accidental overloading.

Digital instruments are available to measure dc and ac voltage, current, and resistance. Other physical variables can be measured by using suitable transducers. Many digital instruments have auxiliary output provisions to make permanent records of measurement results using printers, card or tape

punches, or magnetic tape equipment. With data already in digital form, they
may then be processed without loss of accuracy.

This chapter presents some of the common analog instruments and also
analyzes a few of the major digital devices in block diagram form. The last part
of the chapter deals with some special-purpose instruments used for measuring circuit or component parameters.

10-2 ELECTRONIC VOLTMETERS

10-2.1 dc Voltmeter with Direct-coupled Amplifier

The dc electronic voltmeter represents a straightforward application of
electronics to measuring instruments. The instrument usually consists of an
ordinary dc meter movement preceded by a dc amplifier of one or more
stages. The dc amplifiers used in electronic voltmeters can be classified into
two groups: (a) *direct-coupled dc amplifiers;* (b) *chopper-type dc amplifiers.*

Direct-coupled dc amplifiers are attractive because they are economical;
they are commonly found in lower-priced dc voltmeters. Figure 10-1 shows a

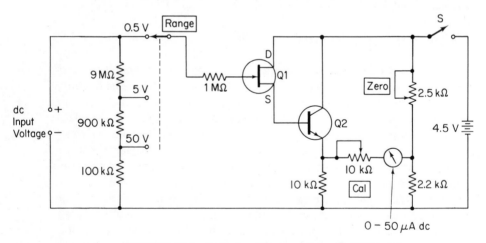

FIGURE 10-1   Basic dc voltmeter circuit with FET input.

schematic diagram of a FET-input direct-coupled dc amplifier with an
indicating meter. The dc input voltage is applied to the input attenuator, a
calibrated front panel control marked RANGE. This input voltage divider
permits a maximum voltage of 0.5 V to be applied to the gate of the n-channel
FET, without causing distortion. The FET is connected as a source follower
and is directly coupled to npn transistor Q_2, an emitter follower. Q_2 is one
arm of a bridge circuit whose remaining arms consist of the 10-kΩ emitter

resistor of Q_2 and the 2.5-kΩ potentiometer in series with the 2.2-kΩ resistor. Bridge balance, or zero meter current, is obtained by adjusting the ZERO SET potentiometer. Full-scale calibration is adjusted by the 10-kΩ potentiometer marked CALIBRATION, in series with the 50-μA meter movement. The input impedance of this voltmeter is 10 MΩ, which is sufficiently high to ignore any possible loading effects on the circuit under measurement.

10-2.2   dc Voltmeter with Chopper-type Amplifier

Instruments in the microvolt range of measurement require a high-gain dc amplifier in order to supply sufficient current to drive the meter movement. To avoid the drift problems usually associated with direct-coupled dc amplifiers, high-sensitivity voltmeters often use the *chopper-type dc amplifier*. In the chopper amplifier the dc input voltage is converted into an ac voltage, amplified by an ac amplifier and then converted back into a dc voltage, proportional to the original input signal.

The block diagram of Fig. 10-2 illustrates the operation of the chopper-type amplifier. Photodiodes are used as nonmechanical choppers for modulation (conversion from dc into ac) and demodulation (conversion from ac back into dc). A photoconductor has a low resistance, from a few hundred to a few thousand ohms, when it is illuminated by a neon or incandescent lamp. The photoconductor resistance increases sharply, usually to several megohms, when it is not illuminated. In the circuit of Fig. 10-2 an oscillator drives two neon lamps into illumination on alternate half-cycles of oscillation. Each neon lamp illuminates one photoconductor in the input circuit of the amplifier and one in the output circuit. The two photodiodes in the input circuit form a series-shunt half-wave *modulator*, or *chopper*. Together they act like a switch across the input of the amplifier, alternately opening and closing at a rate determined by the frequency of the neon oscillator.

The input to the amplifier is a square-wave voltage with an amplitude proportional to the level of the input voltage and a frequency equal to the oscillator frequency. This frequency is limited to a few hundred Hertz because the transition time between the high- and low-resistance states of the photodiodes limits the chopping rate. The ac amplifier delivers an amplified square wave at its output terminals. The two photodiodes in the amplifier output circuit, operating in antisynchronism with the input chopper, recover the dc signal by their demodulating action and the output capacitor becomes charged to the peak of the ac output voltage. This dc output voltage is then passed through a low-pass filter to remove any residual ac components and is finally applied to the meter movement.

The input impedance of the chopper-amplifier dc voltmeter is usually on the order of 10 MΩ or more, except on the very low input ranges. To eliminate measurement errors caused by high source impedances a *nulling* feature is

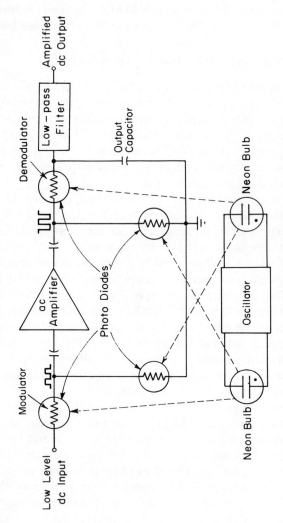

FIGURE 10-2 Nonmechanical photoconductive chopper.

sometimes included in the meter circuitry. This extremely useful addition places a *bucking* voltage in series with the input. A front panel control allows the user to null the input voltage with the bucking voltage. When a null is indicated on the meter, the bucking voltage equals the input voltage and no current is drawn from the source. The meter therefore represents an infinite input impedance and eliminates any loading effect. The function switch then allows the input to be disconnected from the meter circuit, and the bucking voltage (which is equal to the input voltage) is displayed on the meter.

10-2.3 ac Voltmeter Using Rectifiers

Electronic ac voltmeters are basically identical to dc voltmeters except that the ac input voltage must be *rectified* before it can be applied to the dc meter circuit. In some instances, rectification takes place *before* amplification, in which case a simple diode rectifier circuit precedes the amplifier and meter, as in Fig. 10-3(a). This approach ideally requires a dc amplifier with zero drift characteristics and unity voltage gain, and a dc meter movement with adequate sensitivity.

In another approach the ac signal is rectified *after* amplification, as in Fig. 10-3(b) where full-wave rectification takes place in the meter circuit

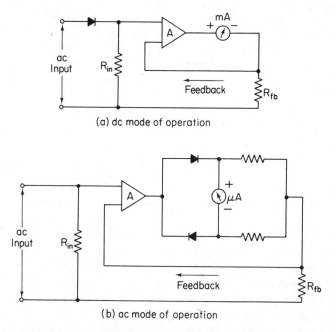

(a) dc mode of operation

(b) ac mode of operation

FIGURE 10-3 Basic ac voltmeter circuits. (a) The ac input signal is first *rectified* and then applied to a dc amplifier and meter movement. (b) The ac input signal is first *amplified* and then applied to a full-wave rectifier in the meter circuit.

connected to the output terminals of the ac amplifier. This approach generally requires an ac amplifier with high open-loop gain and large amounts of negative feedback to overcome the nonlinearity of the rectifier diodes.

Ac voltmeters are usually of the *average-responding* type, with the meter scale *calibrated* in terms of the rms value of a sine wave. Since so many waveforms in electronics are sinusoidal, this is an entirely satisfactory solution and certainly much less expensive than a true rms-responding voltmeter. Non-sinusoidal waveforms, however, will cause this type of meter to read high or low, depending on the form factor of the waveform.

A few basic rectifier circuits are shown in Fig. 10-4. The series-connected

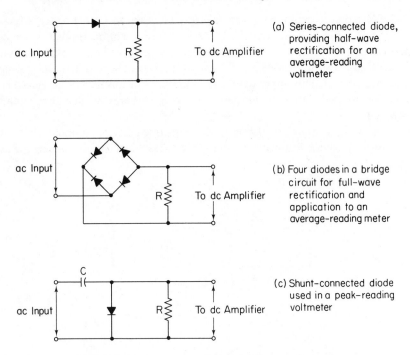

(a) Series-connected diode, providing half-wave rectification for an average-reading voltmeter

(b) Four diodes in a bridge circuit for full-wave rectification and application to an average-reading meter

(c) Shunt-connected diode used in a peak-reading voltmeter

FIGURE 10-4   Rectifier circuits used in ac voltmeters.

diode of Fig. 10-4(a) provides half-wave rectification, and the average value of the half-wave voltage is developed across the resistor and applied to the input terminals of the dc amplifier. Full-wave rectification can be obtained by the bridge circuit of Fig. 10-4(b), where the average value of the sine wave is applied to the amplifier and meter circuit. In some cases, there may be a requirement to measure the peak value of a waveform instead of the average value; the circuit of Fig. 10-4(c) may then be used. In this circuit the rectifier diode charges the small capacitor to the peak of the applied input voltage and the meter will therefore indicate the peak voltage. In most cases, the meter scale is calibrated in terms of both the rms and peak values of the sinusoidal input waveform.

The rms value of a voltage wave that has equal positive and negative excursions is related to the average value by the *form factor*. The form factor, as the ratio of the rms value to the average value of this waveform, can be expressed as

$$k = \frac{\sqrt{\dfrac{1}{T} \int_0^T e^2 \, dt}}{\dfrac{2}{T} \int_0^{T/2} e \, dt} \tag{10-1}$$

If the waveform is sinusoidal, the form factor equals

$$k_{\text{sinusoidal}} = \frac{E_{\text{rms}}}{E_{\text{av}}} = \frac{0.707 E_m}{0.636 E_m} = 1.11 \tag{10-2}$$

Therefore when an average-responding voltmeter has scale markings corresponding to the rms value of the applied sinusoidal input waveform, those markings are actually corrected by a factor of 1.11 from the true (average) value of applied voltage.

Nonsinusoidal waveforms, when applied to this voltmeter, will cause the meter to read either high or low, depending on the form factor of the waveform. An illustration of the effect of nonsinusoidal waveforms on ac voltmeters is given in Examples 10-1 and 10-2.

Example 10-1: The symmetrical square-wave voltage of Fig. 10-5(a) is applied to an average-responding ac voltmeter with a scale calibrated in terms of the rms value of a sine wave. Calculate (a) the form factor of the square-wave voltage, (b) the error in the meter indication.

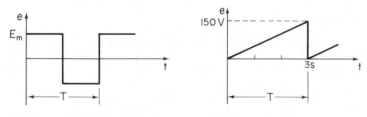

(a) Symmetrical square wave   (b) Sawtooth waveform

FIGURE 10-5   Waveforms used in Examples 10-1 and 10-2.

SOLUTION:

(a) The rms value of the square-wave voltage is

$$E_{\text{rms}} = \sqrt{\frac{1}{T} \int_0^T e^2 \, dt} = E_m$$

and the average value is

$$E_{\text{av}} = \frac{2}{T} \int_0^{T/2} e\, dt = E_m$$

so that the form factor equals, by definition,

$$k = \frac{E_{\text{rms}}}{E_{\text{av}}} = 1$$

(b) The meter scale is calibrated in terms of the rms value of a sine-wave voltage, where $E_{\text{rms}} = k \times E_{\text{av}} = 1.11\, E_{\text{av}}$. For the square-wave voltage, $E_{\text{rms}} = E_{\text{av}}$, since $k = 1$. Therefore the meter indication for the square-wave voltage is *high* by a factor $k_{\text{sine wave}}/k_{\text{square wave}} = 1.11$. The percentage error equals

$$\frac{1.11 - 1}{1} \times 100\% = 11\%$$

Example 10-2: Repeat Example 10-1 if the voltage applied to the meter consists of a sawtooth waveform with a peak value of 150 V and a period of 3 s as shown in Fig. 10-5(b).

SOLUTION:

(a) The analytical expression for the sawtooth waveform between the limits of $t = 0$ and $t = T = 3$ s is $e = 50t$ V. Therefore

$$E_{\text{rms}} = \sqrt{\frac{1}{T} \int_0^T e^2\, dt} = \sqrt{\frac{1}{3} \int_0^3 (50\, t)^2\, dt} = 50\sqrt{3}\ \text{V}$$

$$E_{\text{av}} = \frac{1}{T} \int_0^T e\, dt = \frac{1}{3} \int_0^3 50\, t\, dt = 75\ \text{V}$$

$$\text{Form factor, } k = \frac{50\sqrt{3}}{75} = 1.155$$

(b) The ratio of the two form factors is

$$\frac{k_{\text{sine wave}}}{k_{\text{sawtooth}}} = \frac{1.11}{1.155} = 0.961$$

The meter indication is *low* by a factor of 0.961. The percentage error equals

$$\frac{0.961 - 1}{1} \times 100\% = -3.9\%$$

Examples 10-1 and 10-2 point out that any departure from a true sinusoidal waveform may cause an appreciable error in the result of the measurement.

10-2.4   True rms-responding Voltmeter

Complex waveforms are most accurately measured with an *rms-responding voltmeter*. This instrument produces a meter indication by sensing waveform *heating power*, which is proportional to the square of the rms value of the voltage. This heating power can be measured by feeding an amplified version of the input waveform to the heater element of a *thermocouple* whose output voltage is then proportional to E_{rms}^2.

One difficulty with this technique is that the thermocouple is often non-linear in its behavior. This difficulty is overcome in some instruments by placing two thermocouples in the same thermal environment, as shown in the block diagram of the true rms-responding voltmeter of Fig. 10-6. The effect of

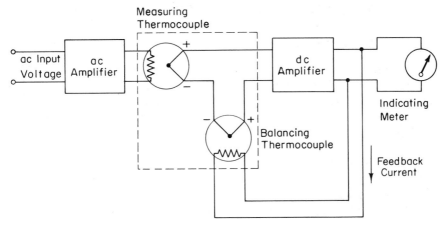

FIGURE 10-6   Block diagram of a true rms-reading voltmeter. The measuring and balancing thermocouples are located in the same thermal environment.

the nonlinear behavior of the couple in the input circuit (the *measuring* thermocouple) is canceled by similar nonlinear effects of the couple in the feedback circuit (the *balancing* thermocouple). The two couple elements form part of a bridge in the input circuit of a dc amplifier. The unknown ac input voltage is amplified and applied to the heating element of the measuring thermocouple. The application of heat produces an output voltage that upsets the balance of the bridge. The unbalance voltage is amplified by the dc amplifier and fed back to the heating element of the balancing thermocouple. Bridge balance will be reestablished when the feedback current delivers sufficient heat to the balancing thermocouple, so that the voltage outputs of both couples are the same. At this point the dc current in the heating element of the feedback couple is equal to the ac current in the input couple. This dc current is therefore directly proportional to the effective, or rms, value of the input voltage and is indicated on the meter movement in the output circuit of the dc amplifier. The true rms value is measured independently of the waveform of the ac signal, provided that the peak excursions of the waveform do not exceed the dynamic range of the ac amplifier.

A typical laboratory-type rms-responding voltmeter provides accurate rms readings of complex waveforms having a *crest factor* (ratio of peak value to rms value) of 10/1. At 10 per cent of full-scale meter deflection, where there is less chance of amplifier saturation, waveforms with crest factors as high as 100/1 could be accommodated. Voltages throughout a range of 100 μV to 300 V within a frequency range of 10 Hz to 10 MHz may be measured with most good instruments.

10-3   ELECTRONIC MULTIMETER

10-3.1   Basic Circuit

One of the most versatile general-purpose shop instruments capable of measuring dc and ac voltages as well as current and resistance is the solid-state electronic multimeter or VOM. Although circuit details will vary from one instrument to the next, an electronic multimeter generally contains the following elements:

(a) Balanced-bridge dc amplifier and indicating meter
(b) Input attenuator or RANGE switch, to limit the magnitude of the input voltage to the desired value
(c) Rectifier section, to convert an ac input voltage to a proportional dc value
(d) Internal battery and additional circuitry, to provide the capability of resistance measurement
(e) FUNCTION switch, to select the various measurement functions of the instrument

In addition, the instrument generally has a built-in power supply for ac line operation and, in most cases, one or more batteries for operation as a portable test instrument.

Figure 10-7 shows the schematic diagram of a balanced-bridge dc amplifier using field effect transistors or FETs. This circuit also applies to a bridge amplifier with ordinary bipolar transistors or BJTs. The circuit shown here consists of two FETs which should be reasonably well matched for current gain to ensure thermal stability of the circuit. The two FETs form the upper arms of a bridge circuit. Source resistors R_1 and R_2, together with ZERO adjust resistor R_3, form the lower bridge arms. The meter movement is connected between the source terminals of the FETs, representing two opposite corners of the bridge.

Without an input signal, the gate terminals of the FETs are at ground potential and the transistors operate under identical quiescent conditions. In

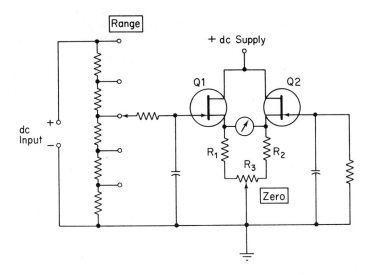

FIGURE 10-7   Balanced-bridge dc amplifier with input attenuator and indicating meter.

this case, the bridge is balanced and the meter indication is zero. In practice, however, small differences in the operating characteristics of the transistors, and slight tolerance differences in the various resistors, cause a certain amount of unbalance in the drain currents, and the meter shows a small deflection from zero. To return the meter to zero, the circuit is balanced by ZERO adjust control R_3 for a true null indication.

When a positive voltage is applied to the gate of input transistor Q_1, its drain current increases which causes the voltage at the source terminal to rise. The resulting unbalance between the Q_1 and Q_2 source voltages is indicated by the meter movement, whose scale is calibrated to agree with the magnitude of the applied input voltage.

The maximum voltage that can be applied to the gate of Q_1 is determined by the operating range of the FET and is usually on the order of a few volts. The range of input voltages can easily be extended by an input attenuator or RANGE switch, as shown in Fig. 10-8. The unknown dc input voltage is applied through a large resistor in the probe body to a resistive voltage divider. Thus, with the RANGE switch in the 3-V position as shown, the voltage at the gate of the input FET is developed across 8 MΩ of the total resistance of 11.3 MΩ and the circuit is so arranged that the meter deflects full scale with 3 V applied to the tip of the probe. With the RANGE switch in the 12-V position, the gate voltage is developed across 2 MΩ of the total divider resistance of 11.3 MΩ and an input voltage of 12 V is required to cause the same full-scale meter deflection.

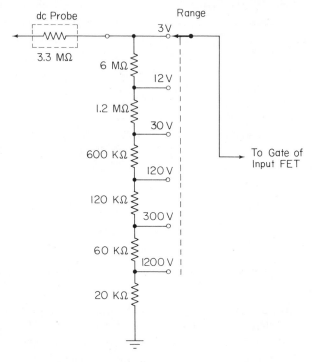

FIGURE 10-8   Typical input voltage attenuator for a VOM. The RANGE switch on the front panel of the VOM allows selection of the desired voltage range.

10-3.2   Resistance Ranges

When the function switch of the multimeter is placed in the OHMS position, the unknown resistor is connected in series with an internal battery, and the meter simply measures the voltage drop across the unknown. A typical circuit is shown in Fig. 10-9, where a separate divider network, used only for resistance measurements, provides for a number of different resis-

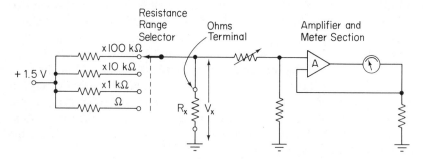

FIGURE 10-9   Resistance range selector circuit of a VOM.

tance ranges. When unknown resistor R_x is connected to the OHMS terminals of the multimeter, the 1.5-V battery supplies current through one of the range resistors and the unknown resistor to ground. Voltage drop V_x across R_x is applied to the input of the bridge amplifier and causes a deflection on the meter. Since the voltage drop across R_x is directly proportional to its resistance, the meter scale can be calibrated in terms of resistance.

Note that the resistance scale of the multimeter reads increasing resistance from left to right, opposite to the way resistance scales read on conventional multimeters (Sec. 4-11). This can be expected because the electronic multimeter reads a larger resistance as a higher voltage, whereas the ordinary multimeter indicates a higher resistance as a smaller current.

10-3.3 Commercial Multimeter

The simplified metering circuit of a commercial solid-state VOM is given in Fig. 10-10. The dc voltage from the input voltage divider (Fig. 10-8) is

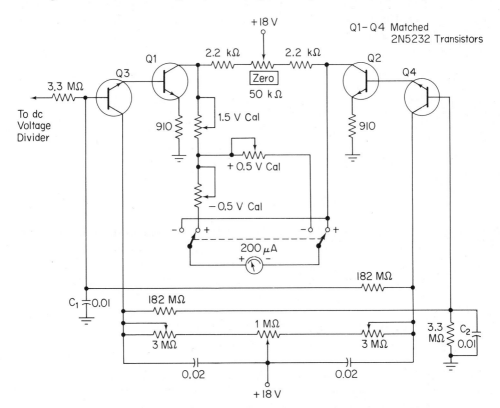

FIGURE 10-10 Typical metering circuit of a solid-state VOM (adapted from RCA WV-510A circuit).

applied to the bases of bridge preamplifier transistors Q_3 and Q_4. These emitter followers provide nearly infinite input impedance and therefore present a minimum load to the high-resistance input voltage divider. Preamplifier transistors Q_3 and Q_4 drive the bases of bridge amplifier transistors Q_1 and Q_2, respectively. The input impedances of Q_1 and Q_2 are very high because of their unbypassed emitter resistors, which prevent loading of the Q_3 and Q_4 emitters. The output voltage of the bridge amplifier is indicated on the 200-μA meter, connected between the collectors of Q_1 and Q_2. The front panel ZERO control balances the meter amplifier output with zero input signal. Internal adjustments allow for meter calibration with two accurate test voltages of 0.5 V and 1.5 V, respectively. Also note that bypass capacitors C_1 and C_2 prevent ac signals from reaching the amplifier and affecting the meter reading.

Ac voltages being measured are applied to a full-wave peak-to-peak rectifier that charges a capacitor to the peak-to-peak value of the ac signal. A circuit of this type is also known as a *voltage doubler* and is shown in Fig. 10-11. The rectified ac voltage is then fed to the amplifier through the regular RANGE voltage divider.

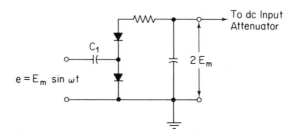

FIGURE 10-11   Full-wave peak-to-peak rectifier, also known as a voltage doubler.

When resistance is being measured, 1.5 V dc is applied to the unknown resistor through one of the resistance range resistors, as shown in Fig. 10-9. The known and the unknown resistances form a voltage divider whose output is fed to the amplifier and read on the meter in terms of resistance.

10-4 CONSIDERATIONS IN CHOOSING
AN ANALOG VOLTMETER

The most appropriate instrument for a particular voltage measurement depends on the performance required in a given situation. Some important considerations in choosing a voltmeter are summarized below.

10-4.1   Input Impedance

To avoid loading effects, the input resistance or impedance of the volt-meter should be at least an order of magnitude higher than the impedance of the circuit under measurement. For example, when a voltmeter with a 10-MΩ input resistance is used to measure the voltage across a 100-kΩ resistor, the circuit is hardly disturbed and the loading effect of the meter on the circuit is negligible. The same meter placed across a 10-MΩ resistor, however, seriously loads the circuit and causes an error in measurement of approximately 50 per cent.

The input impedance of the voltmeter is a function of the inevitable shunt capacitance across the input terminals. The loading effect of the meter is particularly noticeable at the higher frequencies, when the input shunt capacitance greatly reduces the input impedance.

In some applications, a passive voltage-divider probe can be used to reduce the input capacitance at the point of measurement at the sacrifice of perhaps 20 dB of sensitivity. With such a probe, measurements can be easily made at random points without disturbing the circuit under test.

10-4.2   Voltage Ranges

The voltage ranges on the meter scale may be in the 1-3-10 sequence with 10 dB of separation, or in the 1.5-5-15 sequence, or in a single scale calibrated in decibels. In any case, the scale divisions should be compatible with the accuracy of the instrument. For example, a linear meter with a 1-per cent full-scale accuracy should have 100 divisions on the 1.0-V scale so that 1 per cent can be easily resolved. An instrument with an accuracy of 1 per cent or less should also have a mirror-backed scale to reduce parallax and improve accuracy.

10-4.3   Decibels

Use of the decibel scale can be very effective in measurements that cover a *wide range* of voltages. A measurement of this kind is found, for example, in the frequency response curve of an amplifier or a filter, where the output voltage is measured as a function of the frequency of the applied input voltage. Almost all voltmeters with dB scales are calibrated in dBm, referenced to some particular impedance. The 0-dBm reference for a 600-Ω system is 0.7746 V; for a 50-Ω system it is 0.2236 V. In many applications only a 0-dB reference is needed. In this case, 0 dBv (relative to 1 V) can be used for any impedance system.

10-4.4   Sensitivity Versus Bandwidth

Noise is a function of bandwidth. A voltmeter with a broad bandwidth will pick up and generate more noise than one operating over a narrow range of frequencies. In general, an instrument with a bandwidth of 10 Hz to 10 MHz has a sensitivity of 1 mV. A voltmeter whose bandwidth extends only to 5 MHz could have a sensitivity of 100 μV.

10-4.5   Battery Operation

For field work, a voltmeter powered by an internal battery is essential. If an area contains troublesome groundloops, a battery-powered instrument is preferred over a mains-powered voltmeter to remove the groundpaths.

10-4.6   ac Current Measurements

Current measurements can be made by a sensitive ac voltmeter and a series resistance. In the usual case, however, an ac *current probe* is used which enables the operator to measure an ac current without disturbing the circuit under test. The current probe simply clips around the wire carrying the unknown current and in effect makes the wire the one-turn primary of a transformer formed by a ferrite core and a many-turn secondary within the current-probe body. The signal induced in the secondary winding is amplified and the output voltage of the amplifier is applied to a suitable ac voltmeter for measurement. Normally, the amplifier is designed so that 1 mA in the wire being measured produces 1 mV at the amplifier output. The current is then read directly on the voltmeter, using the same scale as for voltage measurements.

In *summarizing* the preceding considerations, the following general guidelines can be stated:

(a) For measurements involving dc applications, select the meter with the broadest capability meeting the circuit's requirements.
(b) For ac measurements involving sine waves with only modest amounts of distortion ($<$ 10 per cent), the average-responding voltmeter provides the best accuracy and most sensitivity per dollar investment.
(c) For high-frequency measurements ($>$ 10 MHz), the peak-responding voltmeter with a diode-probe input is the most economical choice. Peak-responding circuits are acceptable if the inaccuracies caused by distortion in the input waveform can be tolerated.
(d) For measurements where it is important to determine the effective power of waveforms which depart from the true sinusoidal form, the rms-responding·voltmeter is the appropriate choice.

10-5   DIFFERENTIAL VOLTMETERS

10-5.1   Basic Differential Measurement

One of the most accurate methods of measuring an unknown voltage is the *differential voltmeter* technique, where the voltmeter is used to indicate the difference between the unknown voltage and a known voltage. The principle of operation of the differential voltmeter is similar to that of the potentiometer discussed in Chapter 6 and, for this reason, the instrument is sometimes called a *potentiometric* voltmeter.

The classic differential voltage measurement is shown in basic form in the circuit of Fig. 10-12. In this circuit, the null meter is connected between

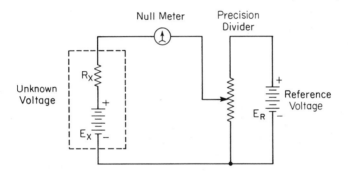

FIGURE 10-12   Basic differential voltage measurement.

the unknown voltage source and the output terminals of a precision voltage divider, and hence indicates the difference between the two. This voltage divider is connected across a *reference* voltage source and can be adjusted to provide accurately known fractions of the reference voltage.

In performing a measurement, the divider is adjusted until its output voltage equals the unknown voltage. The null meter, which is connected between the unknown source and the divider-output terminals, indicates zero volts when the two voltages are equal. In this *null condition* neither the source nor the reference supplies current to the meter, and the differential voltmeter therefore presents an *infinite impedance* to the source under test. Note that the null meter serves to indicate only the residual differential between the known voltage and the unknown voltage. To detect small differences in unbalance potentials, a sensitive meter movement is required; accuracy of the meter is of secondary importance since the meter is not used to indicate the absolute value of the unknown voltage.

The reference source usually consists of a low-voltage dc standard, such as a 1-V dc laboratory reference standard or a low-voltage Zener-controlled precision supply.

When measuring high voltages, a high-voltage reference supply can be used. In the usual case, however, a voltage divider is placed across the unknown source to reduce the voltage to a sufficiently low value for direct comparison against the usual low-voltage dc standard. The main drawback of this system is that a differential voltmeter with input divider has a relatively low input resistance, especially for unknown voltages much higher than the reference standard. This low input resistance is undesirable because of its loading effect. A differential voltmeter offers an input resistance approaching infinity only at the null condition and then *only* if an input divider is *not* used.

The *ac differential voltmeter* is a modification of the dc instrument and incorporates a precision *rectifier* circuit. The unknown ac voltage is applied to the rectifier for conversion to a dc voltage equivalent to the average value of the ac. The resulting dc is then applied to the potentiometric voltmeter in the usual manner. The simplified block diagram of an ac differential voltmeter shown in Fig. 10-13 is self-explanatory.

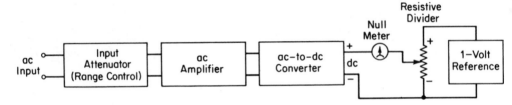

FIGURE 10-13   Simplified block diagram of an ac differential voltmeter.

10-5.2 dc Standard/Differential Voltmeter

The differential voltmeter requires a reference source to make a measurement and a meter circuit to detect an unbalance between the unknown and the known voltage. Some manufacturers combine the various elements in a multifunction laboratory instrument called a dc standard/differential voltmeter. This instrument has three modes of operation: as a *dc voltage standard*, as a *dc differential voltmeter*, and as a *dc voltmeter*. Function switching allows many of the same basic circuits to be used in each mode of operation.

The block diagram of Fig. 10-14 illustrates the *standard* mode of operation, where the instrument generates precision output voltages from 0 V to 1,000 V as a reference source for many laboratory applications. A temperature-controlled reference supply generates a very stable $+1$ V dc, which is applied to a decimal divider network. The divider ratios are controlled by a set of front panel switches allowing the reference supply to be adjusted in $1\text{-}\mu\text{V}$ steps from 0 V to 1 V. This reference output voltage is applied to a high-gain dc amplifier with degenerative voltage feedback to obtain precisely controlled gain characteristics.

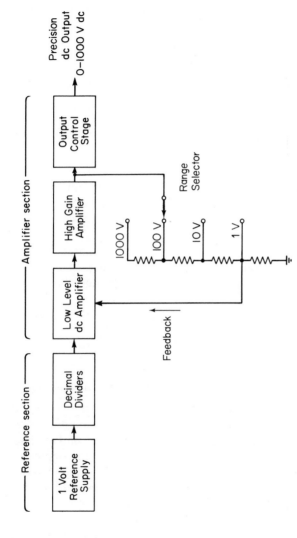

FIGURE 10-14 Block diagram of the dc standard/differential voltmeter in the *standard* mode of operation. The reference section in conjunction with the dc amplifier section provides precision dc output voltages from 0 V to 1,000 V.

The dc amplifier, consisting of several stages in cascade, provides an open-loop gain of 10^8 or higher. The feedback network monitors the actual output voltage and feeds a controlled fraction of the output back to the amplifier input. The closed-loop gain of the feedback amplifier can be expressed by the relationship

$$G = \frac{A}{1 + \beta A} \tag{10-3}$$

where G = closed-loop gain (voltage gain with feedback)

A = open-loop gain (voltage gain without feedback)

β = fraction of the output voltage used as degenerative feedback

If the open-loop gain is very high (ideally infinite), Eq. (10-3) reduces to

$$G = \frac{1}{\beta} \tag{10-4}$$

showing that the gain of the amplifier depends only on the amount of degenerative feedback. The accuracy of the closed-loop gain therefore depends only on the accuracy of the voltage divider which determines β. The feedback divider, shown in the block diagram of Fig. 10-14, is made from stable, wire-wound precision resistors, enabling the amplifier to have precisely controlled closed-loop gain characteristics. The output terminals of the instrument in the *standard* mode of operation provide the following precision voltage ranges:

0–1 V in 1-μV steps   (1-V range)

0–10 V in 10-μV steps   (10-V range)

0–100 V in 100-μV steps (100-V range)

0–1,000 V in 1-mV steps (1,000-V range)

In the second mode of operation, the instrument is connected as a *differential voltmeter*, using the same building blocks as before, together with a meter circuit. This is illustrated in the block diagram of Fig. 10-15. The unknown dc voltage is applied to the input terminals of the amplifier section. One half of the divider network feeds a fraction of the output voltage back to the input stage and thereby controls the closed-loop gain of the amplifier. The second section of the divider network applies a fraction of the output voltage to the differential input of the meter amplifier.

The meter circuit measures the difference between the feedback voltage and the reference voltage, and indicates a null when the two voltages are equal. The range selector on the front panel of the instrument controls both the feedback voltage and the voltage that is applied in opposition to the reference divider output in such a way that the 1-V capability of the reference supply is never exceeded.

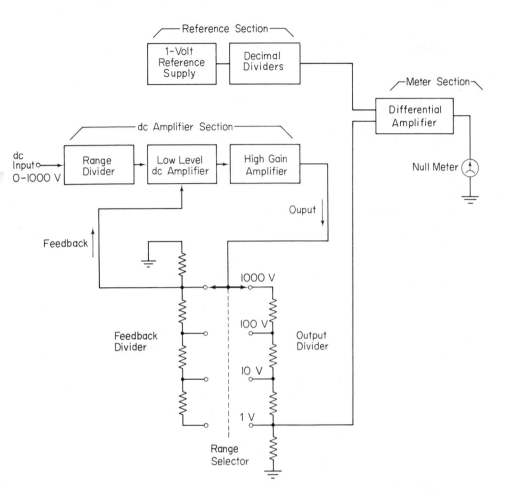

FIGURE 10-15 Block diagram of the dc standard/differential voltmeter in the *differential* mode of operation. The meter section indicates the voltage balance between the reference section and the dc amplifier section.

In the third mode of operation, the instrument is connected as a *voltmeter* and the dc amplifier acts as a buffer stage to provide high-input impedance to the unknown voltage source. The input voltage is amplified, and the dc output voltage is applied directly to the meter circuit. The meter circuit incorporates a feedback-controlled amplifier and allows selection of its sensitivity by adjustment of the feedback loop through a front panel control, marked *sensitivity*. This feature provides for extreme sensitivity of the meter circuit, often on the order of 1 μV full-scale deflection. Meaningful measurements at the very high sensitivities, however, are difficult to make because of the problems of noise generation and pickup.

An ac-to-dc converter can be incorporated in the instrument to provide the capability of ac voltage measurement by potentiometric methods.

10-6 DIGITAL VOLTMETERS

10-6.1 General Characteristics

The digital voltmeter (DVM) displays measurements of dc or ac voltages as discrete numerals instead of a pointer deflection on a continuous scale as in analog devices. Numerical readout is advantageous in many applications because it reduces human reading and interpolation errors, eliminates parallax error, increases reading speed, and often provides outputs in digital form suitable for further processing or recording.

The DVM is a versatile and accurate instrument that can be used in many laboratory measurement applications. Since the development and perfection of integrated circuit (IC) modules, the size, power requirements, and cost of the DVM have been drastically reduced so that DVMs can actively compete with conventional analog instruments, both in portability and price.

The DVM's outstanding qualities can best be illustrated by quoting some typical operating and performance characteristics. The following specifications do not all apply to one particular instrument, but they do represent valid information on the present state of the art:

(a) Input range: from ± 1.000000 V to $\pm 1,000.000$ V, with automatic range selection and overload indication
(b) Absolute accuracy: as high as ± 0.005 per cent of the reading
(c) Stability: short-term, 0.002 per cent of the reading for a 24-hr period; long-term, 0.008 per cent of the reading for a 6-month period
(d) Resolution: 1 part in 10^6 (1 μV can be read on the 1-V input range)
(e) Input characteristics: input resistance typically 10 MΩ; input capacitance typically 40 pF
(f) Calibration: internal calibration standard allows calibration independent of the measuring circuit; derived from stabilized reference source
(g) Output signals: print command allows output to printer; BCD (binary-coded-decimal) output for digital processing or recording

Optional features may include additional circuitry to measure current, resistance, and voltage ratios. Other physical variables may be measured by using suitable transducers.

Digital voltmeters can be classified according to the following broad categories:

(a) Ramp-type DVM
(b) Integrating DVM
(c) Continuous-balance DVM
(d) Successive-approximation DVM

10-6.2   Ramp-type DVM

The operating principle of the ramp-type DVM is based on the measurement of the time it takes for a linear ramp voltage to rise from 0 V to the level of the input voltage, or to decrease from the level of the input voltage to zero. This time interval is measured with an electronic time-interval counter, and the count is displayed as a number of digits on electronic indicating tubes.

Conversion from a voltage to a time interval is illustrated by the waveform diagram of Fig. 10-16. At the start of the measurement cycle, a ramp

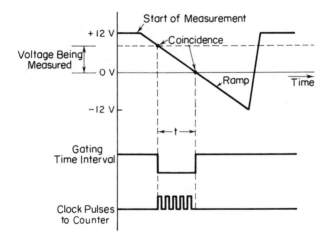

FIGURE 10-16   Voltage-to-time conversion using gated clock pulses.

voltage is initiated; this voltage can be positive-going or negative-going. The negative-going ramp, shown in Fig. 10-16, is continuously compared with the unknown input voltage. At the instant that the ramp voltage equals the unknown voltage, a coincidence circuit, or comparator, generates a pulse which opens a gate. This gate is shown in the block diagram of Fig. 10-17. The ramp voltage continues to decrease with time until it finally reaches 0 V (or ground potential) and a second comparator generates an output pulse which closes the gate.

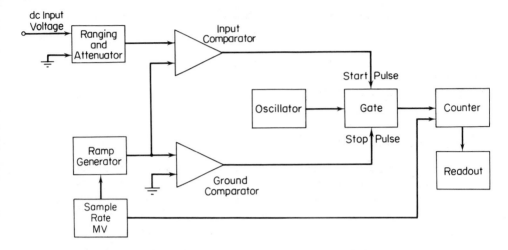

FIGURE 10-17   Block diagram of a ramp-type digital voltmeter.

An oscillator generates clock pulses which are allowed to pass through the gate to a number of decade counting units (DCUs) which totalize the number of pulses passed through the gate. The decimal number, displayed by the indicator tubes associated with the DCUs, is a measure of the magnitude of the input voltage.

The sample-rate multivibrator determines the rate at which the measurement cycles are initiated. The oscillation of this multivibrator can usually be adjusted by a front panel control, marked *rate*, from a few cycles per second to as high as 1,000 or more. The sample-rate circuit provides an initiating pulse for the ramp generator to start its next ramp voltage. At the same time, a reset pulse is generated which returns all the DCUs to their 0 state, removing the display momentarily from the indicator tubes.

10-6.3   Staircase-ramp DVM

The staircase-ramp DVM is given in block diagram form in Fig. 10-18. It is a variation of the ramp-type DVM but is somewhat simpler in overall design, resulting in a moderately priced general-purpose instrument that can be used in the laboratory, on production test-stands, in repair shops, and at inspection stations.

This DVM makes voltage measurements by comparing the input voltage to an internally generated staircase-ramp voltage. The instrument shown in Fig. 10-18 contains a 10-MΩ input attenuator, providing five input ranges from 100 mV to 1,000 V full scale. The dc amplifier, with a fixed gain of 100, delivers 10 V to the comparator at any of the full-scale voltage settings of the input divider. The comparator senses coincidence between the amplified

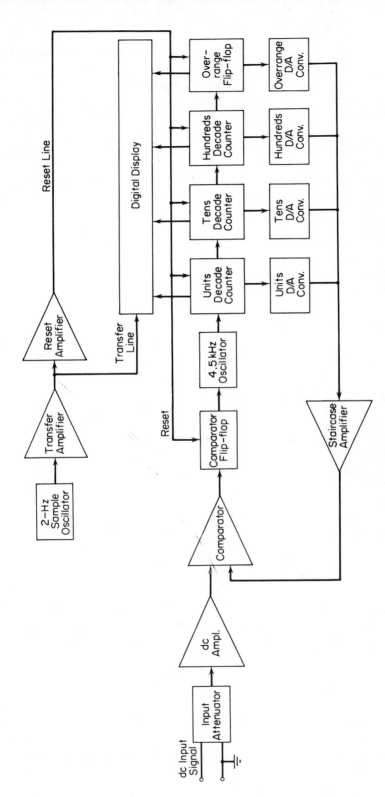

FIGURE 10-18   Block diagram of a staircase-ramp digital voltmeter.

input voltage and the staircase-ramp voltage which is generated as the measurement proceeds through its cycle.

When the measurement cycle is first initiated, the clock (a 4.5-kHz relaxation oscillator) provides pulses to three DCUs in cascade. The *units* counter provides a carry pulse to the *tens* decade at every tenth input pulse. The *tens* decade counts the carry pulses from the *units* decade and provides its own carry pulse after it has counted ten carry pulses. This carry pulse is fed to the *hundreds* decade which provides a carry pulse to an overrange circuit. The overrange circuit causes a front panel indicator to light up, warning the operator that the input capacity of the instrument has been exceeded. The operator should then switch to the next higher setting on the input attenuator.

Each decade counter unit is connected to a digital-to-analog (D/A) converter. The outputs of the D/A converters are connected in parallel and provide an output current proportional to the current count of the DCUs. The staircase amplifier converts the D/A current into a staircase voltage which is applied to the comparator. When the comparator senses coincidence of the input voltage and the staircase voltage, it provides a trigger pulse to stop the oscillator. The current content of the counter is then proportional to the magnitude of the input voltage.

The sample rate is controlled by a simple relaxation oscillator. This oscillator triggers and resets the transfer amplifier at a rate of two samples per second. The transfer amplifier provides a pulse that transfers the information stored in the decade counters to the front panel display unit. The trailing edge of this pulse triggers the reset amplifier which sets the three decade counters to zero and initiates a new measurement cycle by starting the master oscillator or clock.

The display circuits store each reading until a new reading is completed, eliminating any blinking or counting during the computation.

10-6.4 Integrating DVM

The integrating DVM measures the *true average* of the input voltage over a fixed measuring period, in contrast to the ramp-type DVM which *samples* the voltage at the end of a measuring cycle. A widely used technique to accomplish integration employs a voltage-to-frequency (V/F) converter. The V/F converter functions as a feedback control system that governs the rate of pulse generation in proportion to the magnitude of the input voltage.

The simplified block diagram of an integrating DVM is given in Fig. 10-19. The dc voltage under test is applied to the input stage which isolates the meter circuitry from the test circuit and provides the necessary input attenuation. The attenuated input signal is applied to the V/F converter. This circuit consists of an integrating amplifier, a level detector (*comparator circuit*),

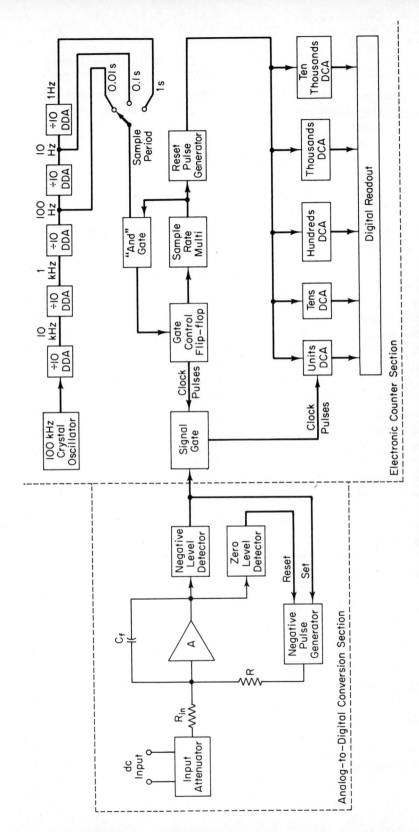

FIGURE 10-19   Block diagram of an integrating digital voltmeter.

and a pulse generator. The integrating amplifier produces an output voltage proportional to the input voltage, related to the input and feedback elements by the equation

$$V_{out} = -\frac{1}{C} \int i \, dt$$

$$= -\frac{1}{RC} \int V_{in} \, dt \qquad \textbf{(10-5)}$$

If the input voltage is constant, the output is a linear ramp following the equation

$$V_{out} = -V_{in}\frac{t}{RC} \qquad \textbf{(10-6)}$$

When the ramp reaches a certain negative voltage level, the level detector triggers the pulse generator, which applies a negative voltage step to the summing junction of the integrating amplifier. The sum of the input voltage and the pulse voltage is negative, causing the ramp to reverse its direction. This "retrace" is very rapid since the pulse is large in amplitude compared to the input voltage. When the now positive-going ramp reaches 0 V, the level detector generates a reset trigger to the pulse generator. The negative pulse is removed from the summing junction of the integrating amplifier and only the original input voltage is left. The amplifier then produces a negative-going ramp again and the procedure repeats.

The rate of pulse generation is governed by the magnitude of the dc input voltage. A larger input voltage causes a steeper ramp and therefore a higher pulse repetition rate (PRR).

The major advantage of this system of A/D conversion is its ability to measure accurately in the presence of large amounts of superimposed noise, since the input is integrated.

The level-detector output pulse controls the signal gate allowing the decimal counters to accumulate a count provided by the crystal oscillator circuitry. The remainder of the circuit is essentially identical to any conventional counter and needs no further elaboration.

10-6.5   Continuous-balance DVM

The continuous-balance DVM is a low-cost instrument that provides excellent performance. The accuracy of this voltmeter is usually on the order of 0.1 per cent of its input range. It has an input impedance of about 10 MΩ and acceptable resolution.

The block diagram of a servo-driven continuous-balance DVM is given in Fig. 10-20. The dc input voltage is applied to an input attenuator that provides suitable range switching. The input attenuator is a front panel control

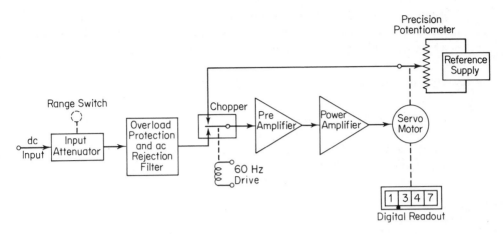

FIGURE 10-20   Functional block diagram of a servo-balancing potentiometer-type digital voltmeter.

that also causes a decimal point indicator to move on the display area in accordance with the input range selected. After passing through an over-voltage protection circuit and ac rejection filter, the input voltage is applied to one side of a mechanical *chopper comparator*. The other side of the comparator is connected to the wiper arm of the motor-driven precision poten-tiometer, connected across a reference supply. The output of the chopper comparator, which is driven by the line voltage and vibrates at the line frequency rate, is a square-wave signal. The amplitude of the square wave is a function of the difference in magnitude and polarity of the dc voltages connected to the opposite sides of the chopper. The square-wave signal is amplified by a high-impedance, low-noise preamplifier and fed to a power amplifier. This amplifier has special damping to minimize overshoot and hunting at the null position. The servo motor, on receiving the amplified square-wave difference signal, drives the arm of the precision potentiometer in the direction required to *cancel* the difference voltage across the chopper comparator. The servo motor also drives a drum-type mechanical indicator that has the digits 0 to 9 imprinted about the periphery of its drum seg-ments. The position of the servo motor shaft corresponds to the amount of feedback voltage required to null the chopper input, and this position is indicated by the drum-type indicator. The position of the shaft therefore is an indication of the magnitude of the input voltage.

It is clear that this instrument does not "sample" the unknown dc voltage at regular intervals, as is the case with more sophisticated instruments, but continuously seeks to *balance* the input voltage against the internally generated reference. Because of the different mechanical movements involved in the mechanism, such as the positioning of the potentiometer arm and the rotation of the indicator mechanism, the average reading time is approximately 2 s. Simplicity of design and low cost, however, make this instrument a very attractive choice when extreme accuracy is not required.

10-6.6   Successive-approximation DVM

Digital voltmeters capable of 1,000 readings per second or more are now commercially available. These instruments generally use successive-approximation converters to perform the *digitization* (analog-to-digital conversion). A simplified block diagram of such a DVM is shown in Fig. 10-21.

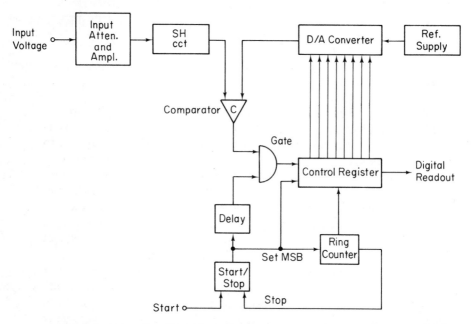

FIGURE 10-21   Simplified block diagram of a successive-approximation digital voltmeter.

At the beginning of the measurement cycle, a start pulse is applied to the start-stop multivibrator. This sets a 1 in the most significant bit (MSB) of the control register and a 0 in all bits of less significance. Assuming an 8-bit control register, its reading would then be 1000 0000. This initial setting of the control register causes the output of the D/A converter to be one-half the reference supply voltage ($\frac{1}{2}$ V). The converter output is compared to the unknown input by the comparator. If the input voltage is larger than the converter reference voltage, the comparator produces an output that causes the control register to *retain* the 1 setting in its MSB, and the converter continues to supply its reference output voltage of $\frac{1}{2}$ V.

The ring counter next advances one count, shifting a 1 in the second MSB of the control register, and its reading becomes 1100 0000. This causes the D/A converter to increase its reference output by one increment, to $\frac{1}{2}$ V + $\frac{1}{4}$ V, and another comparison with the unknown input voltage takes place. If in this case, the accumulated reference voltage *exceeds* the unknown voltage, the comparator produces an output that causes the control register to *reset* its second MSB to 0. The converter output then returns to its previous level of $\frac{1}{2}$ V and awaits another input from the control register for the next approximation. When the ring counter advances another count, the third MSB of the control register is set to 1 and the converter output rises by the next increment, to $\frac{1}{2}$ V + $\frac{1}{8}$ V. The measurement cycle thus proceeds through a series of *successive approximations*, as shown in Fig. 10-22, retaining or

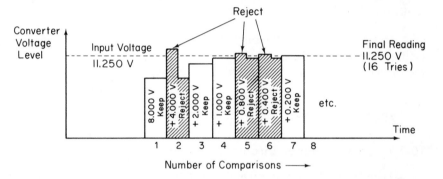

FIGURE 10-22   Successive approximations are used to make an analog-to-digital conversion. Converter reference voltages are switched to the comparator in an 8-4-2-1 sequence and are rejected if the accumulated converter output exceeds the input voltage.

rejecting the converter output in the manner described. Finally, when the ring counter reaches its last count, the measurement cycle stops, and the digital output of the control register represents the final approximation of the unknown input voltage.

With input voltages other than dc, the input level changes during digitization and decisions made during conversion are not consistent. To avoid this *conversion error*, a sample-and-hold (SH) circuit is placed in the input, directly following the input attenuator and amplifier, as shown in Fig. 10-21. In its simplest form, the SH circuit can be represented by a switch and a capacitor, as in Fig. 10-23. In the *sample* mode the switch is closed and the capacitor charges to the instantaneous value of the input voltage. In the *hold* mode the switch is opened and the capacitor holds the voltage that it had at the instant the switch was opened. If the switch drive is synchronous with the ring counter pulse, the actual measurement and conversion take place when the SH circuit is in the hold mode.

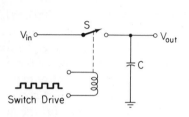

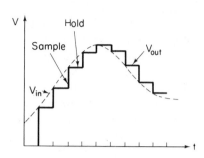

(a) Simple sample-and-hold circuit   (b) Reconstruction of the waveform

FIGURE 10-23   A sample-and-hold circuit freezes the input voltage during digitization, so that voltage levels do not change during the successive approximation process.

In a practical circuit the simple switch of Fig. 10-23 is replaced by fast-acting transistor switches, and an operational amplifier is added to increase the charging current into the capacitor.

10-7   Q METER

10-7.1   Basic Q-meter Circuit

The Q meter is an instrument designed to measure some of the electrical properties of coils and capacitors. The operation of this useful laboratory instrument is based on the familiar characteristics of a series-resonant circuit, namely, that the voltage across the coil or the capacitor is equal to the applied voltage times the Q of the circuit. If a fixed voltage is applied to the circuit, a voltmeter across the capacitor can be calibrated to read Q directly.

The voltage and current relationships of a series-resonant circuit are shown in Fig. 10-24. At resonance, the following conditions are valid:

$$X_C = X_L$$
$$E_C = IX_C = IX_L$$
$$E = IR$$

where E = applied voltage

I = circuit current

E_C = voltage across the capacitor

X_C = capacitive reactance

X_L = inductive reactance

R = coil resistance

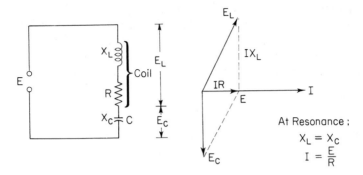

FIGURE 10-24   Series-resonant circuit.

The magnification of the *circuit*, by definition, is Q, where

$$Q = \frac{X_L}{R} = \frac{X_C}{R} = \frac{E_C}{E} \tag{10-7}$$

Therefore if E is maintained at a constant and known level, a voltmeter connected across the capacitor can be calibrated directly in terms of the circuit Q.

A practical Q-meter circuit is shown in Fig. 10-25. The wide-range oscillator with a frequency range from 50 kHz to 50 MHz delivers current to

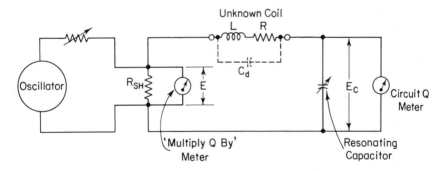

FIGURE 10-25   Basic Q-meter circuit.

a low-value shunt resistance R_{SH}. The value of this shunt is very low, typically on the order of 0.02 Ω. It introduces almost no resistance into the oscillatory circuit and it therefore represents a voltage source of magnitude E with a very small (in most cases negligible) internal resistance. The voltage E across the shunt, corresponding to E in Fig. 10-24, is measured with a thermocouple meter, marked "Multiply Q by." The voltage across the variable capacitor, corresponding to E_C in Fig. 10-24, is measured with an electronic voltmeter whose scale is calibrated directly in Q values.

To make a measurement, the unknown coil is connected to the test terminals of the instrument, and the circuit is tuned to resonance either by setting the oscillator to a given frequency and varying the internal resonating capacitor or by presetting the capacitor to a desired value and adjusting the frequency of the oscillator. The Q reading on the output meter must be multiplied by the index setting of the "Multiply Q by" meter to obtain the actual Q value.

The *indicated* Q (which is the resonant reading on the "circuit Q" meter) is called the *circuit* Q because the losses of the resonating capacitor, voltmeter, and insertion resistor are all included in the measuring circuit. The *effective* Q of the measured coil will be somewhat greater than the indicated Q. This difference can generally be neglected, except in certain cases where the resistance of the coil is relatively small in comparison with the value of the insertion resistor. (This problem is discussed in Example 10-6.)

The inductance of the coil can be calculated from the known values of frequency (f) and resonating capacitance (C), since

$$X_L = X_C \quad \text{and} \quad L = \frac{1}{(2\pi f)^2 C} \text{ henry} \tag{10-8}$$

10-7.2 Measurement Methods

There are three methods for connecting unknown components to the test terminals of a Q meter. The type of component and its size determine the method of connection.

(*a*) *The Direct Connection*   Most coils can be connected directly across the test terminals, exactly as shown in the basic Q-meter circuit of Fig. 10-25. The circuit is resonated by adjusting either the oscillator frequency or the resonating capacitor. The indicated Q is read directly from the "Circuit Q" meter, modified by the setting of the "Multiply Q by" meter. When the last meter is set at the unity mark, the "Circuit Q" meter reads the correct value of Q directly.

(*b*) *The Series Connection*   Low-impedance components, such as low-value resistors, small coils, and large capacitors, are measured *in series* with the measuring circuit. Figure 10-26 shows the connections. The component to be measured, here indicated by [Z], is placed in series with a stable *work coil* across the test terminals. (The work coil is usually supplied with the instrument.) Two measurements are made: In the first measurement the unknown is short-circuited by a small *shorting strap* and the circuit is resonated, establishing a reference condition. The values of the tuning capacitor (C_1) and the indicated Q (Q_1) are noted. In the second measurement the shorting strap is removed and the circuit is retuned, giving a new value for the tuning capacitor (C_2) and a change in the Q value from Q_1 to Q_2.

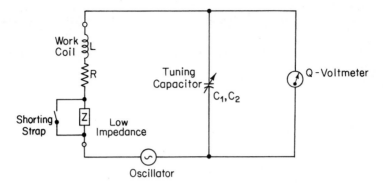

FIGURE 10-26 *Q*-meter measurement of a low-impedance component in the series connection.

For the reference condition,

$$X_{C_1} = X_L \quad \text{or} \quad \frac{1}{\omega C_1} = \omega L \tag{10-9}$$

and neglecting the resistance of the measuring circuit,

$$Q_1 = \frac{\omega L}{R} = \frac{1}{\omega C_1 R} \tag{10-10}$$

For the second measurement, the reactance of the unknown can be expressed in terms of the *new* value of the tuning capacitor (C_2) and the *in-circuit* value of the inductor (L). This yields

$$X_S = X_{C_2} - X_L \quad \text{or} \quad X_S = \frac{1}{\omega C_2} - \frac{1}{\omega C_1} \tag{10-11}$$

so that

$$X_S = \frac{C_1 - C_2}{\omega C_1 C_2} \tag{10-12}$$

X_S is inductive if $C_1 > C_2$ and capacitive if $C_1 < C_2$. The resistive component of the unknown impedance can be found in terms of reactance X_S and the indicated values of circuit Q, since

$$R_1 = \frac{X_1}{Q_1} \quad \text{and} \quad R_2 = \frac{X_2}{Q_2}$$

Also,

$$R_S = R_2 - R_1 = \frac{1}{\omega C_2 Q_2} - \frac{1}{\omega C_1 Q_1}$$

so that

$$R_S = \frac{C_1 Q_1 - C_2 Q_2}{\omega C_1 C_2 Q_1 Q_2} \tag{10-13}$$

If the unknown is *purely resistive*, the setting of the tuning capacitor would not have changed in the measuring process, and $C_1 = C_2$. The equation for resistance reduces to

$$R_S = \frac{Q_1 - Q_2}{\omega C_1 Q_1 Q_2} = \frac{\Delta Q}{\omega C_1 Q_1 Q_2} \tag{10-14}$$

If the unknown is a *small inductor*, the value of the inductance is found from Eq. (10-12) and equals

$$L_S = \frac{C_1 - C_2}{\omega^2 C_1 C_2} \tag{10-15}$$

The Q of the coil is found from Eq. (10-12) and Eq. (10-13) since, by definition,

$$Q_S = \frac{X_S}{R_S}$$

and

$$Q_S = \frac{(C_1 - C_2)(Q_1 Q_2)}{C_1 Q_1 - C_2 Q_2} \tag{10-16}$$

If the unknown is a *large capacitor*, its value is determined from Eq. (10-12), and

$$C_S = \frac{C_1 C_2}{C_2 - C_1} \tag{10-17}$$

The Q of the capacitor may be found by using Eq. (10-16).

(c) *The Parallel Connection*   High-impedance components, such as high-value resistors, certain inductors, and small capacitors, are measured by connecting them *in parallel* with the measuring circuit. Figure 10-27 shows the

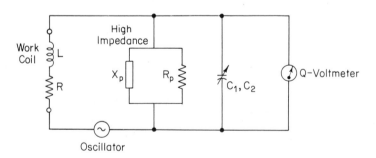

FIGURE 10-27 Q-meter measurement of a high-impedance component in the parallel connection.

connections. Before the unknown is connected, the circuit is resonated, by using a suitable work coil, to establish reference values for Q and C (Q_1 and C_1). Then, when the component under test is connected to the circuit, the capacitor is readjusted for resonance, and a new value for the tuning capacitance (C_2) is obtained and a change in the value of circuit Q (ΔQ) from Q_1 to Q_2.

In a parallel circuit, computation of the unknown impedance is best approached in terms of its parallel components X_p and R_p, as indicated in Fig. 10-27. At the initial resonance condition, when the unknown is not yet connected into the circuit, the working coil (L) is tuned by the capacitor (C_1). Therefore

$$\omega L = \frac{1}{\omega C_1} \tag{10-18}$$

and
$$Q_1 = \frac{\omega L}{R} = \frac{1}{\omega C_1 R} \tag{10-19}$$

When the unknown impedance is now connected into the circuit and the capacitor is tuned for resonance, the reactance of the working coil (X_L) equals the parallel reactances of the tuning capacitor (X_{C_2}) and the unknown (X_p). Therefore

$$X_L = \frac{(X_{C_2})(X_p)}{X_{C_2} + X_p}$$

which reduces to

$$X_p = \frac{1}{\omega(C_1 - C_2)} \tag{10-20}$$

If the unknown is *inductive*, $X_p = \omega L_p$, and Eq. (10-20) yields the value of the unknown impedance:

$$L_p = \frac{1}{\omega^2(C_1 - C_2)} \tag{10-21}$$

If the unknown is *capacitive*, $X_p = 1/\omega C_p$ and Eq. (10-20) yields the value of the unknown capacitor:

$$C_p = C_1 - C_2 \tag{10-22}$$

In a parallel resonant circuit the total resistance at resonance is equal to the product of the circuit Q and the reactance of the coil. Therefore

$$R_T = Q_2 X_L$$

or by substitution of Eq. (10-18),

$$R_T = Q_2 X_{C_1} = \frac{Q_2}{\omega C_1} \tag{10-23}$$

The resistance (R_p) of the unknown impedance is most easily found by computing the *conductances* in the circuit of Fig. 10-27.

Let G_T = total conductance of the resonant circuit

G_p = conductance of the unknown impedance

G_L = conductance of the working coil

Then

$$G_T = G_p + G_L \quad \text{or} \quad G_p = G_T - G_L \tag{10-24}$$

From Eq. (10-23),

$$G_T = \frac{1}{R_T} = \frac{\omega C_1}{Q_2}$$

Therefore

$$\frac{1}{R_p} = \frac{\omega C_1}{Q_2} - \frac{R}{R^2 + \omega^2 L^2}$$

$$= \frac{\omega C_1}{Q_2} - \left(\frac{1}{R}\right)\left(\frac{1}{1 + \omega^2 L^2/R^2}\right)$$

$$= \frac{\omega C_1}{Q_2} - \frac{1}{RQ_1^2}$$

Substituting Eq. (10-19) in the foregoing expression, we obtain

$$\frac{1}{R_p} = \frac{\omega C_1}{Q_2} - \frac{\omega C_1}{Q_1}$$

and after simplifying, we obtain

$$R_p = \frac{Q_1 Q_2}{\omega C_1 (Q_1 - Q_2)} = \frac{Q_1 Q_2}{\omega C_1 \Delta Q} \qquad \text{(10-25)}$$

The Q of the unknown is then found by using Eq. (10-20) and Eq. (10-25) so that

$$Q_p = \frac{R_p}{X_p} = \frac{(C_1 - C_2)(Q_1 Q_2)}{C_1(Q_1 - Q_2)} = \frac{(C_1 - C_2)(Q_1 Q_2)}{C_1 \Delta Q} \qquad \text{(10-26)}$$

10-7.3 Sources of Error

Probably the most important factor affecting measurement accuracy, and the most often overlooked, is the *distributed capacitance* or *self-capacitance* of the measuring circuit. The presence of distributed capacitance in a coil modifies the *actual* or *effective* Q and the inductance of the coil. At the frequency at which the self-capacitance and the inductance of the coil are resonant, the circuit exhibits a purely resistive impedance. This characteristic may be used for measuring the distributed capacitance.

One simple method of finding the distributed capacitance (C_d) of a coil involves making two measurements at different frequencies. The coil under test is connected directly to the test terminals of the Q meter, as shown in the circuit of Fig. 10-28. The tuning capacitor is set to a high value, preferably to its maximum position, and the circuit is resonated by adjusting the oscillator frequency. Resonance is indicated by maximum deflection on the "Circuit Q" meter. The values of the tuning capacitor (C_1) and the oscillator frequency (f_1) are noted. The frequency is then increased to twice its original value $(f_2 = 2f_1)$ and the circuit is retuned by adjusting the resonating capacitor (C_2).

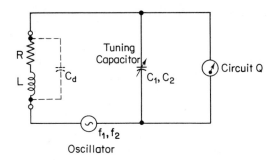

FIGURE 10-28   Determination of the distributed capacitance of an inductor.

The resonant frequency of an *LC*-circuit is given by the well-known equation

$$f = \frac{1}{2\pi\sqrt{LC}} \tag{10-27}$$

At the initial resonance condition, the capacitance of the circuit equals $C_1 + C_d$, and the resonant frequency equals

$$f_1 = \frac{1}{2\pi\sqrt{L(C_1 + C_d)}} \tag{10-28}$$

After the oscillator and the tuning capacitor are adjusted, the capacitance of the circuit is $C_2 + C_d$, and the resonant frequency equals

$$f_2 = \frac{1}{2\pi\sqrt{L(C_2 + C_d)}} \tag{10-29}$$

Since $f_2 = 2f_1$, Eqs. (10-28) and (10-29) are related so that

$$\frac{1}{2\pi\sqrt{L(C_1 + C_d)}} = \frac{2}{2\pi\sqrt{L(C_1 + C_d)}}$$

and

$$\frac{1}{C_2 + C_d} = \frac{4}{C_1 + C_d}$$

Solving for the distributed capacitance yields

$$C_d = \frac{C_1 - 4C_2}{3} \tag{10-30}$$

Example 10-3: The self-capacitance of a coil is to be measured by using the procedure just outlined. The first measurement is at $f_1 = 2\,\text{MHz}$ and $C_1 = 460\,\text{pF}$. The second measurement, at $f_2 = 4\,\text{MHz}$, yields a new value of tuning capacitor, $C_2 = 100\,\text{pF}$. Find the distributed capacitance, C_d.

SOLUTION: Using Eq. (10-30), we obtain

$$C_d = \frac{C_1 - 4C_2}{3} = \frac{460 - 400}{3} = 20 \text{ pF}$$

Example 10-4: Compute the value of self-capacitance of a coil when the following measurements are made: At frequency $f_1 = 2$ MHz, the tuning capacitor is set at 450 pF. When the frequency is increased to 5 MHz, the tuning capacitor is tuned at 60 pF.

SOLUTION: Since $f_2 = 2.5 \, f_1$, Eqs. (10-28) and (10-29) are related as follows:

$$\frac{1}{2\pi\sqrt{L(C_2 + C_d)}} = \frac{2.5}{2\pi\sqrt{L(C_1 + C_d)}}$$

This reduces to

$$\frac{1}{C_2 + C_d} = \frac{6.25}{C_1 + C_d}$$

Solving for C_d, we obtain

$$C_d = \frac{C_1 - 6.25C_2}{5.25}$$

Substituting the values for $C_1 = 450$ pF and $C_2 = 60$ pF, we see that the value of the distributed capacitance is $C_d = 14.3$ pF.

The effective Q of a coil with distributed capacitance is less than the true Q by a factor that depends on the value of the self-capacitance and the resonating capacitor. It can be shown that

$$\text{true } Q = Q_e \left(\frac{C + C_d}{C} \right) \tag{10-31}$$

where Q_e = effective Q of the coil

   C = resonating capacitance

   C_d = distributed capacitance

The *effective* Q can usually be considered the *indicated* Q.

For many measurements, the *residual* or *insertion* resistance (R_{SH}) of the Q-meter circuit of Fig. 10-25 is sufficiently small to be considered negligible. Under certain circumstances, it can contribute an error to the measurement of Q. The effect of the insertion resistor on the measurement depends on the magnitude of the unknown impedance and, of course, on the size of the insertion resistor. For instance, the 0.02 Ω of insertion resistance may be neglected in comparison with a coil resistance of 10 Ω, but it assumes importance when compared to a coil resistance of 0.1 Ω. The effect of the 0.02-Ω insertion resistance is illustrated by Examples 10-5 and 10-6.

Example 10-5: A coil with a resistance of 10 Ω is connected in the "direct-measurement" mode. Resonance occurs when the oscillator frequency is 1.0 MHz and the resonating capacitor is set at 65 pF. Calculate the percentage error introduced in the calculated value of Q by the 0.02-Ω insertion resistance.

SOLUTION: The *effective* Q of the coil equals

$$Q_e = \frac{1}{\omega CR} = \frac{1}{(2\pi)(10^6)(65 \times 10^{-12})(10)} = 245$$

The *indicated* Q of the coil equals

$$Q_i = \frac{1}{\omega C(R + 0.02)} = 244.5$$

The percentage error is then $\dfrac{245 - 244.5}{245} \times 100\% = 0.2\%$

Example 10-6: Repeat the problem of Example 10-5 for the following conditions:

   The coil resistance is 0.1 Ω
   The frequency at resonance is 40 MHz.
   The tuning capacitor is set at 135 pF.

SOLUTION: The *effective* Q of the coil is

$$Q^e = \frac{1}{\omega CR} = \frac{1}{2\pi \times 40 \times 10^6 \times 135 \times 10^{-12} \times 0.1} = 295$$

The *indicated* Q of the coil is

$$Q_i = \frac{1}{\omega C(R + 0.02)} = 245$$

The percentage error equals

$$\frac{295 - 245}{295} \times 100\% = 17\%$$

Other sources of error include the *residual inductance* of the instrument, which is usually in the order of 0.015 μH and affects the measurement of only very small inductors ($< 0.5\ \mu$H). The *conductance* of the Q voltmeter has a slight shunting effect on the tuning capacitor at the higher frequencies, but this effect can usually be neglected.

10-8   VECTOR IMPEDANCE METER

Impedance measurements are concerned with both the magnitude (Z) and the phase angle (θ) of a component. At frequencies below 100 MHz, measurement of voltage and current is usually sufficient to determine the magnitude of the impedance. The phase difference between the voltage wave-form and the current waveform indicates whether the component is inductive or capacitive. If the phase angle can be determined, for example, by using a CRO displaying a Lissajous pattern, the reactance can be calculated. If a component must be fully specified, its properties should be determined at several different frequencies, and many measurements may be required. Especially at the higher frequencies, these measurements become rather elaborate and time consuming, and many steps may be required to obtain the desired information.

The development of such instruments as the *vector impedance meter* makes impedance measurements over a wide frequency range possible. *Sweep-frequency plots* of impedance and phase angle versus frequency, providing complete coverage within the frequency band of interest, can also be made.

The vector impedance meter, shown in Fig. 10-29, makes simultaneous measurements of impedance and phase angle over a frequency range of from

FIGURE 10-29   Vector impedance meter (courtesy Hewlett-Packard Co.).

5 Hz to 500 kHz. The unknown component is simply connected across the input terminals of the instrument, the desired frequency is selected by turning the front panel controls, and the two front panel meters indicate the magnitude of the impedance and the phase angle.

The operation of the vector impedance meter is best understood by referring to the block diagram of Fig. 10-30. Two measurements take place: (1) The magnitude of the impedance is determined by measuring the current through the unknown component when a known voltage is applied across it, or by measuring the voltage across the component when a known current is passed through it; (2) the phase angle is found by determining the phase difference between the voltage across the component and the current through the component.

The block diagram of Fig. 10-30 shows that the instrument contains a *signal source* (Wien bridge oscillator) with two front panel controls to select the frequency range and to continuously adjust the selected frequency. The oscillator output is fed to an *AGC amplifier* which allows accurate gain adjustment by means of its feedback voltage. This gain adjustment is an internal control actuated by the setting of the *impedance range* switch, to which the AGC amplifier output is connected. The impedance range switch is a precision attenuator network controlling the oscillator output voltage and at

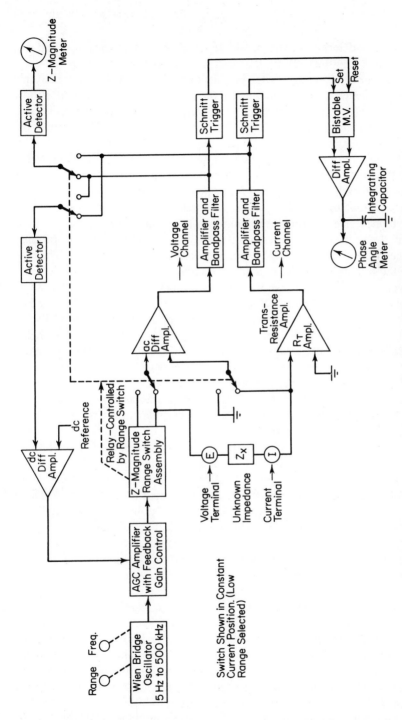

FIGURE 10-30 Block diagram of the vector impedance meter (courtesy Hewlett-Packard Co.).

the same time determining the manner in which the unknown component will be connected into the circuitry that follows the range switch.

The impedance range switch permits operation of the instrument in two modes: the *constant-current* mode and the *constant-voltage* mode. The three lower ranges ($\times$ 1, $\times$ 10, and $\times$ 100) operate in the constant-current mode and the four higher ranges ($\times$ 1k, $\times$ 10k, $\times$ 100k, and $\times$ 1M) operate in the constant-voltage mode.

In the constant-current mode the unknown component is connected across the input of the ac differential amplifier. The current supplied to the unknown depends on the setting of the impedance range switch. This current is held constant by the action of the transresistance or R_T amplifier, which converts the current through the unknown to a voltage output equal to the current times its feedback resistance. The R_T amplifier is an operational amplifier whose output voltage is proportional to its input current. The output of the R_T amplifier is fed to a detector circuit and compared to a dc reference voltage. The resulting control voltage regulates the gain of the AGC amplifier and hence the voltage applied to the impedance range switch. The output of the ac differential amplifier is applied to an amplifier and filter section consisting of high- and low-band filters that are changed with the frequency range to restrict the amplifier bandwidth. The output of the band-pass filter is connected, when selected, to a detector that drives the *Z-magnitude* meter. Since the current through the unknown is held constant by the R_T amplifier, the Z-magnitude meter, which measures the voltage across the unknown, deflects in proportion to the magnitude of the unknown impedance and is calibrated accordingly.

In the constant-voltage mode the two inputs to the differential amplifier are switched. The terminal that was connected to the input of the transresistance amplifier in the constant-current mode is now grounded. The other input of the differential amplifier that was connected to the voltage terminal of the unknown component is now connected to a point on the Z-magnitude range switch which is held at a constant potential. The voltage terminal of the unknown is connected to this same point of constant potential, or depending on the setting of the Z-magnitude range switch, to a decimal fraction of this voltage. In any case, the voltage across the unknown is held at a constant level. The current through the unknown is applied to the transresistance amplifier which again produces an output voltage proportional to its input current.

The roles of the ac differential amplifier and the transresistance amplifier are now reversed. The voltage output of the R_T amplifier is applied to the detector and then to the Z-magnitude meter. The output voltage of the differential amplifier controls the gain of the AGC amplifier in the same manner that the R_T amplifier did in the constant-current mode.

Phase-angle measurements are carried out simultaneously. The outputs of both the voltage channel and the current channel are amplified and each output is connected to a Schmitt trigger circuit. The Schmitt trigger circuits produce a positive-going spike every time the input sine wave goes through a zero crossing. These positive spikes are applied to a *binary phase detector* circuit. The phase detector consists of a bistable multivibrator, a differential amplifier, and an integrating capacitor. The positive-going pulse from the constant-current channel sets the multivibrator, and the pulse from the constant-voltage channel resets the multivibrator. The "set" time of the MV is therefore determined by the zero crossings of the voltage and current waveforms. The "set" and "reset" outputs of the MV are applied to the differential amplifier, which applies the difference voltage to an integrating capacitor. The capacitor voltage is directly proportional to the zero-crossing time interval and is applied to the *phase-angle* meter which then indicates the phase difference, in degrees, between the voltage and current waveforms.

Calibration of the vector impedance meter is usually performed by connecting standard components to the input terminals. These components may be standard resistors or capacitors. An electronic counter is needed to accurately determine the period of the applied test frequency. When the value of the component under test and the frequency of the test signal are both known accurately, the impedance or reactance can be calculated and compared to the indication on the Z-magnitude meter. With a standard resistor connected to the input terminals, the phase-angle meter should read 0°.

10-9   VECTOR VOLTMETER

A *vector voltmeter* measures the amplitude of a signal at two points in a circuit and simultaneously measures the phase difference between the voltage waveforms at these two points. This instrument can be used in a wide variety of applications, especially in situations where other methods are very difficult or time consuming. The vector voltmeter is useful in VHF applications and can be used successfully in such measurements as:

(a) Amplifier gain and phase shift
(b) Complex insertion loss
(c) Filter transfer functions
(d) Two-port network parameters

The vector voltmeter basically converts two RF signals of the same fundamental frequency (from 1 MHz to 1 GHz) to two IF signals with 20-kHz fundamental frequencies. These IF signals have the same amplitudes, waveforms and phase relationships as the original RF signals. Consequently, the fundamental components of the IF signals have the same amplitude and phase relationships as the fundamental components of the RF signals. These fundamental components are filtered from the IF signals and are measured by a voltmeter and a phase meter.

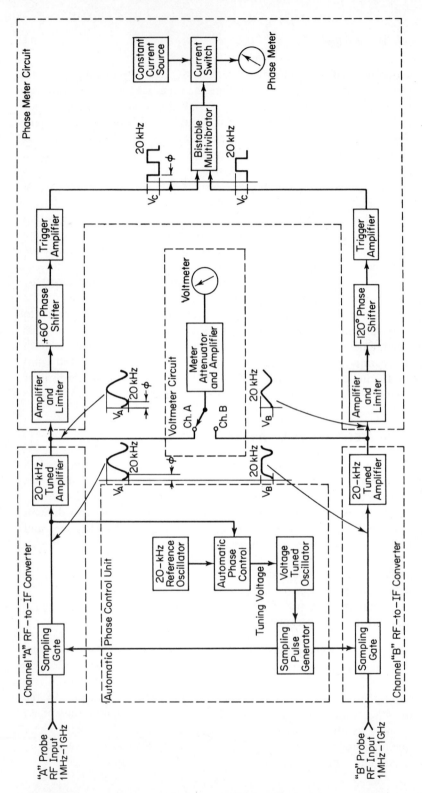

FIGURE 10-31 Block diagram of the vector voltmeter (courtesy Hewlett-Packard Co.).

The block diagram of Fig. 10-31 shows that the instrument consists of five major sections as follows: two RF-to-IF converters, an automatic phase control section, a phase meter circuit, and a voltmeter circuit. The RF-to-IF converters and the phase control section produce two 20-kHz sine waves with the same amplitudes and the same phase relationship as the fundamental components of the RF signals applied to channels A and B. The phase meter section continuously monitors these two 20-kHz sine waves and indicates the phase angle between them. The voltmeter section can be switched to channel A or channel B to provide a meter display of the amplitude.

Each RF-to-IF converter consists of a sampler and a tuned amplifier. The sampler produces a 20-kHz replica of the RF input waveform, and the tuned amplifier extracts the 20-kHz fundamental component from this waveform replica. *Sampling* is a time-stretching process, with which a high-frequency repetitive signal is duplicated at a much lower frequency. The process is illustrated in the diagram of Fig. 10-32. An electronic switch is

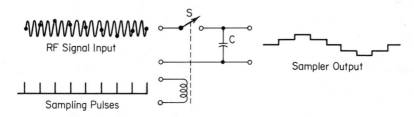

FIGURE 10-32   Simplified diagram of a sampling circuit.

connected between the RF input waveform and a storage capacitor. Each time the switch is momentarily closed, the capacitor is charged to the instantaneous value of the input voltage and holds this until the next switch closure. With appropriate timing, samples are taken at progressively later points on the RF waveform. Provided that the RF waveform is repetitive, the samples reconstruct the original waveform at a much lower frequency. Each input channel has a sampler consisting of a sampling gate and a storage capacitor. The sampling gates are controlled by pulses from the same pulse generator. Samples are taken in each channel at exactly the same instant, and the phase relationship of the input signals is therefore preserved in the IF signals.

The phase control unit is a rather sophisticated circuit that generates the sampling pulses for both RF-to-IF converters and automatically controls the pulse rate to produce 20-kHz IF signals. The sampling pulse rate is controlled by a voltage-tuned oscillator (VTO) for which the tuning voltage is supplied by the automatic phase control section. This section locks the IF signal of channel A to a 20-kHz reference oscillator. To get initial locking, the phase control section applies a ramp voltage to the VTO. This ramp voltage sweeps the sampling rate until channel A IF is 20 kHz and in phase with the reference oscillator. Then the sweep stops and channel A IF is held in phase with the reference oscillator.

The tuned amplifier passes only the 20-kHz fundamental component of the IF signal of each channel. The output of each tuned amplifier then consists of a signal that has retained its original phase relationship with respect to the signal in the other channel and also its correct amplitude relationship. The two filtered IF signals can be connected to the voltmeter circuit by a front panel switch, marked *channel A* and *channel B*. The voltmeter circuit contains an input attenuator to provide the appropriate meter range. This attenuator is also a front panel control, marked *amplitude range*. The meter amplifier consists of a stable fixed-gain feedback amplifier, followed by a rectifier and a filter section. The rectified signal is applied to a dc voltmeter.

To determine the *phase difference* between the two IF signals, the tuned amplifiers are followed by the phase meter circuit. Each channel is first amplified and then limited, resulting in square-wave signals at the inputs to the IF phase-shifting circuits. The circuit in channel A shifts the phase of the square-wave signal by $+60°$; the circuit in channel B shifts the phase of its signal by $-120°$. Both phase shifts are accomplished by a combination of capacitive networks and inverting and noninverting amplifiers whose vector-sum outputs provide the desired phase shift. The outputs of the phase-shift circuits are amplified and clipped, producing square waveforms, and applied to the trigger amplifiers. These circuits convert the square-wave input signals to positive spikes with very fast rise times. The bistable multivibrator is triggered by pulses from both channels. Channel A is connected to the *set* input of the MV; channel B is connected to the *reset* input of the MV. If the initial phase shift between the RF signals at the probes was $0°$, the trigger pulses into the multivibrator are $180°$ out of phase owing to the action of the phase-shift circuits. The MV then produces a square-wave output voltage which is symmetrical about zero. Any phase shift at the RF probes carries through the entire system and varies the trigger pulses from their $180°$ relationship, producing an asymmetrical waveform.

The (asymmetrical) square wave controls the current switch, which is a transistor switched into conduction by the negative portion of the square wave. The switch connects the constant current supply to the *phase meter*. At $0°$ phase shift at the RF input, the switch is turned off and on for equal amounts of time and the current supply is adjusted to cause the meter to read $0°$ or center scale. Any RF phase shift results in an asymmetrical waveform and allows either more or less current to the phase meter, depending on whether the phase shift caused the negative half-cycle of the square wave to be larger or smaller. An input phase shift of $180°$ would cause the square wave to collapse into either a positive or a negative dc voltage and the switch would then allow no current or maximum current to the phase meter. These maximum deviations from the center reading of $0°$ are marked on the meter face as $+180°$ and $-180°$. The *phase range* can be selected by a front panel switch that places a shunt across the phase meter and changes its sensitivity.

The instrument contains a power supply section, which is not shown on the block diagram of Fig. 10-31. The power supply generates all the necessary supply voltages for the various sections of the instrument.

Calibration procedures and the testing of performance specifications vary from one instrument to the next. Complete descriptions of the various tests are given in the manual of the instrument and usually include the procedure and instrumentation needed for such tests.

REFERENCES

1. Thomas, Harry E., and Carole A. Clarke, *Handbook of Electronic Instruments and Measurement Techniques*. Englewood Cliffs, N.J.: Prentice-Hall, Inc., 1967.

2. Hewlett-Packard, Palo Alto, Calif., *Application Notes*.

3. Stout, Melville B., *Basic Electrical Measurements*, 2nd ed., chap. 10. Englewood Cliffs, N.J.: Prentice-Hall, Inc., 1960.

4. Malvino, Albert Paul, *Electronic Instrumentation Fundamentals*, chap. 11. New York: McGraw-Hill Book Company, 1967.

5. Lenk, John D., *Handbook of Electronic Test Equipment*, chap. 2. Englewood Cliffs, N.J.: Prentice-Hall, Inc., 1971.

PROBLEMS

1. An average-responding electronic ac voltmeter with an rms-calibrated scale is used to measure the following nonsinusoidal voltages:
   (a) dc voltage of 10 V;
   (b) square-wave voltage with an amplitude of 10 V and a duty cycle of 75%;
   (c) triangular voltage of symmetrical waveform and a peak value of 10 V.
      Calculate the percentage error indicated by the ac voltmeter for each of the voltages.

2. A 25-mA current meter with an internal resistance of 100 Ω is available for constructing an ac voltmeter with a voltage range of 200 V rms. Using four diodes in a bridge arrangement, where each diode has a forward resistance of 500 Ω and infinite reverse resistance, calculate the necessary series-limiting resistance for the 200-V rms voltage range.

3. In checking the distributed capacitance of a certain coil using the Q-meter circuit of Fig. 10-28, initial resonance is obtained with the resonating capacitor set to 450 pF. Resonance at twice the initial frequency is obtained with the tuning capacitor at 11 pF. Calculate the value of the distributed capacitance of the coil.

4. A coil with a resistance of 3 Ω is connected to the terminals of the Q meter of Fig. 10-25. Resonance occurs at an oscillator frequency of 5 MHz and resonating capacitance of 100 pF. Calculate the percentage error introduced by the insertion resistance $R_{sh} = 0.1$ Ω.

5. The following laboratory test equipment is available for calibrating multimeters:
    (a) an accurate differential voltmeter;
    (b) a stable dc power supply;
    (c) a number of assorted precision resistors. With the aid of diagrams, suggest a measurement procedure for the calibration of the current and voltage ranges of a multimeter.

6. Design a range switch for the dc volt section of a balanced-bridge voltmeter (see Fig. 10-8). The total resistance of the attenuator should be 11 MΩ. The attenuation should be so arranged that input voltages from 3 V to 1,000 V can be accommodated in the customary 1-3-10 sequence. The bridge circuit (see Fig. 10-7) requires 1 V at the gate of the input FET to cause full-scale meter deflection.

7. The differential voltage measurement of Fig. 10-12 uses a reference source with an internal resistance of 200 Ω and a terminal voltage of 3.0 V. The galvanometer has a current sensitivity of 1 mm/μA and internal resistance of 100 Ω. Neglect the internal resistance of the unknown source and calculate its terminal voltage if the galvanometer deflection is 250 mm.

8. For the measurement conditions given in Prob. 7, calculate the resolution of the measurement setup if the galvanometer deflection can be read to 1 mm.

9. For the differential voltage measurement of Prob. 7, a second galvanometer has become available. This galvanometer has a current sensitivity of 5 mm/μA and internal resistance of 1,000 Ω. Calculate which of the two galvanometers provides the greater sensitivity to unbalance. Express the results in millimeters per microvolt.

11

Instruments for Generating and Analyzing Waveforms

11-1 BASIC OSCILLATOR CIRCUITS

11-1.1 Introduction

A test oscillator or a signal generator is an indispensable tool in electronic service work, in the research laboratory, or on the production line. It provides the technician with various kinds of test signal to perform a wide range of operations, such as measuring the frequency response of an amplifier, the bandpass characteristics of a filter, the alignment of a radio receiver or television set, or simply tracing faults in a piece of electronic equipment.

The various signal sources are described by different names: test oscillator, audio oscillator, signal generator, sweep generator, function generator, and so on, depending on the characteristics of the instrument and its intended use. Generally speaking, an *oscillator* is a device that provides a sine-wave signal of a certain frequency and a certain amplitude, while a *signal generator* has the added capacity of amplitude modulation of the output signal, and a wide tuning range. The oscillator circuit itself, however, is the basic element common to all signal sources.

There are various oscillator circuits, with circuit designs that depend on the frequency they are required to produce. Low-frequency oscillators, operating roughly in the 1-Hz to 1-MHz range, are often based on the familiar Wien bridge circuit. The Wien bridge provides a combination of

319

variable frequency range and stable output in a reliable, inexpensive instrument. High-frequency oscillators, covering a frequency range of 100 kHz to 500 MHz or higher, are generally based on some variation of the well-known LC tank circuit. The following sections describe some of these basic oscillator circuits.

11-1.2   LC Tank Circuit Oscillators

The principle of operation of LC tank circuits is simple and almost identical for the many variations of the basic circuit, and can be described as follows: A parallel *LC* combination is excited into oscillation and the ac voltage across this LC tank circuit is amplified by a transistor amplifier. Part of the amplified ac voltage is fed back to the tank circuit by inductive or capacitive coupling to compensate for the power losses in the tank circuit. This *regenerative* feedback causes an output voltage of constant amplitude at the resonant frequency of the tank circuit, described by the equation

$$f = \frac{1}{2\pi\sqrt{LC}} \qquad\qquad \textbf{(11-1)}$$

LC tank circuit oscillators can operate at very high frequencies, up to several hundred megahertz. Specially designed tubes, such as klystrons and magnetrons, extend the frequency range in the gigahertz range.

The *Armstrong oscillator* of Fig. 11-1 is one of the early RF oscillator circuits. When supply voltage V_{cc} is first switched on, the transistor conducts and collector current flows. Since coil L_2 is inductively coupled to coil L_1, the increasing collector current induces a voltage across L_1 in such a direction that the base of the transistor is driven positive: the top of the coil has a

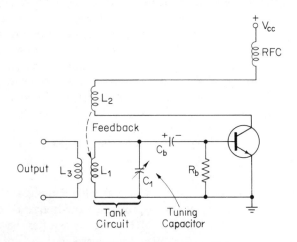

FIGURE 11-1   Armstrong oscillator.

positive polarity. This causes the collector current to increase at a faster rate and the induced voltage increases further. As a result, a high positive voltage is built up across the tank circuit, and capacitor C_1 charges with a positive polarity on its top plate. Since, at the same time, the base is driven positive, base current charges capacitor C_b to the peak value of the induced voltage with the polarity as shown. As the transistor begins to saturate, the rate of rise in collector current decreases and therefore the induced voltage decreases. C_b must therefore discharge and does this through R_b, making the base of the transistor negative. This causes a chain reaction.

The collector current starts to decrease from its maximum (saturation) value and the magnetic field of L_2 collapses. This induces a *negative* voltage across coil L_1. In the meantime, C_b is still discharging and this drives the transistor into cutoff so that collector current ceases abruptly. The induced voltage across L_1 causes C_1 to discharge and then charge up again to the peak value of the negative voltage. Now the transistor is cut off and C_1 is charged with a negative polarity on its top plate. During the next half-cycle of action the tank circuit brings the transistor out of its cutoff condition. C_1 starts to discharge through L_1 and the base of the transistor is raised in potential until conduction starts again. As soon as the transistor conducts, energy is transferred from the collector circuit to the tank circuit, and C_1 charges up again to the peak value of the now positive voltage. The entire cycle repeats in the manner described. The frequency of oscillation is governed by the charge and discharge characteristics of the tank circuit and is given by Eq. (11-1).

The *Hartley oscillator* of Fig. 11-2 uses only one coil with a tap which corresponds to the common ac ground of the Armstrong circuit. Tuning capacitor C_1 is shunted across the entire coil ($L_1 + L_2$). Since the tap of the coil is connected to ground, the rotor of the variable capacitor can no longer be grounded. The output signal is available through the *RC* coupling circuit

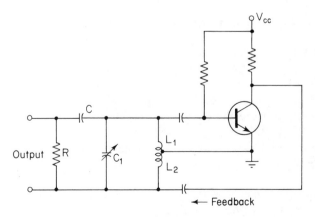

FIGURE 11-2   Hartley oscillator.

instead of by inductive coupling as in the case of the Armstrong oscillator. This has no bearing on the operation of the circuit; either type of output coupling can be used.

The *Colpitts oscillator* of Fig. 11-3 is another variation of the basic

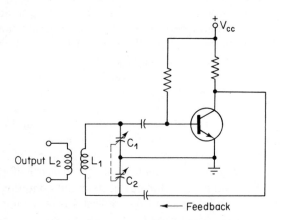

FIGURE 11-3   Colpitts oscillator.

Armstrong circuit. The tank circuit now consists of one inductor, L_1, and two capacitors in series (C_1 and C_2). Notice that, except for the method of tapping in the tank circuit, this circuit is identical to the circuit of the Hartley oscillator. The amount of feedback in the Colpitts circuit depends on the relative values of capacitors C_1 and C_2. The smaller C_1 is, the greater the feedback. As the tuning is varied, both capacitor values increase or decrease simultaneously, but the ratio of the two values remains fixed. The output signal is available through an extra winding on the tank circuit coil.

11-1.3   Wien Bridge Oscillator

The Wien bridge oscillator is one of the standard circuits used to generate sine-wave signals in the audiofrequency range. The oscillator is of simple design, has a relatively pure waveform, and enjoys excellent frequency stability. It is essentially a feedback amplifier with a Wien bridge as the feedback network between the output and input terminals of the amplifier, as shown in Fig. 11-4. This amplifier will oscillate when two basic conditions, known as the *Barkhausen criteria* for oscillation, are met. These two basic requirements for oscillation can be stated as follows:

(a) The voltage gain around the amplifier and feedback loop, called the *loop gain*, must be unity, or $A_v\beta = 1$.
(b) The phase shift between input voltage v_i and feedback voltage v_f, called the *loop phase shift*, must be zero.

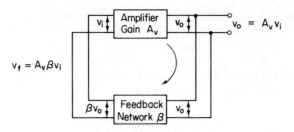

FIGURE 11-4   Block diagram of a feedback oscillator.

If these conditions are satisfied, the feedback amplifier of Fig. 11-4 will generate a sinusoidal output waveform.

The Wien bridge was first introduced in Sec. 8-6, where it was shown as a frequency-selective network consisting of resistive and capacitive elements. The circuit is shown again in Fig. 11-5. The bridge has a series RC combina-

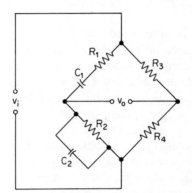

FIGURE 11-5   Wien bridge circuit.

tion in one arm and a parallel RC combination in the adjoining arm. The other arms are purely resistive. The equations for bridge balance yield

$$\mathbf{Z_1 Z_4 = Z_2 Z_3} \quad \text{or} \quad \mathbf{Z_3 = Z_1 Z_4 Y_2} \tag{11-2}$$

where $\mathbf{Z_1} = R_1 - j/\omega C_1$, $\mathbf{Y_2} = 1/R_2 + j\omega C_2$, $\mathbf{Z_3} = R_3$, and $\mathbf{Z_4} = R_4$. Substituting the appropriate expressions into Eq. (11-2), we obtain

$$R_3 = \left(R_1 - \frac{j}{\omega C_1}\right) R_4 \left(\frac{1}{R_2} + j\omega C_2\right) \tag{11-3}$$

and expanding, we obtain

$$R_3 = \frac{R_1 R_4}{R_2} + j\omega C_2 R_1 R_4 - \frac{j R_4}{\omega C_1 R_2} + \frac{R_4 C_2}{C_1} \tag{11-4}$$

At bridge balance, both the real terms and the imaginary terms must equate. Separating them in Eq. (11-4) yields, for the real terms,

$$\frac{R_3}{R_4} = \frac{R_1}{R_2} + \frac{C_2}{C_1} \tag{11-5}$$

and for the imaginary terms,

$$\omega C_2 R_1 = \frac{1}{\omega C_1 R_2} \tag{11-6}$$

where $\omega = 2\pi f$. Equation (11-6) can be solved to obtain an expression for the frequency of the input voltage and we find that

$$f = \frac{1}{2\pi\sqrt{C_1 C_2 R_1 R_2}} \tag{11-7}$$

In the usual case, the bridge components are selected so that $R_1 = R_2 = R$ and $C_1 = C_2 = C$. Equation (11-5) then yields

$$\frac{R_3}{R_4} = 2 \tag{11-8}$$

while the balancing, or resonant, bridge frequency is

$$f = \frac{1}{2\pi RC} \tag{11-9}$$

In other words, the bridge is balanced (*zero* output voltage) when the resistance ratio of the nonreactive arms satisfies Eq. (11-8) and when the excitation voltage has a frequency given by Eq. (11-9). When the Wien bridge is used as the feedback network in an oscillator, as in Fig. 11-4, the circuit must be slightly modified. This follows from considering the Barkhausen criteria. The voltage gain of the amplifier of Fig. 11-4 is a finite quantity (say $A_v = 100$). The Barkhausen condition of unity loop gain ($A_v \beta = 1$) implies that the feedback voltage derived from the Wien bridge must also be a finite quantity and cannot be zero. The bridge must therefore be modified so that it does provide an output voltage at the resonant frequency, while still maintaining its zero phase shift.

Consider now the modified Wien bridge of Fig. 11-6. The impedances of the reactive arms at the resonant frequency ($f = 1/2\pi RC$) can be written as

$$\mathbf{Z}_1 = R - \frac{j}{\omega C} = (1 - j)R \tag{11-10}$$

and

$$\mathbf{Z}_2 = \frac{1}{1/R + j\omega C} = \frac{(1 - j)R}{2} \tag{11-11}$$

Hence voltage drop v_a across $\mathbf{Z}_2$ equals

$$v_a = \frac{\mathbf{Z}_2}{\mathbf{Z}_1 + \mathbf{Z}_2} v_i = \frac{v_i}{3} \tag{11-12}$$

and voltage drop v_b across R_2 is

$$v_b = \frac{R_4}{R_3 + R_4} v_i \tag{11-13}$$

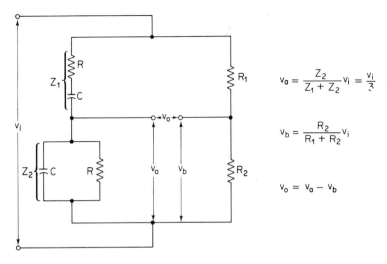

$$v_a = \frac{Z_2}{Z_1 + Z_2} v_i = \frac{v_i}{3}$$

$$v_b = \frac{R_2}{R_1 + R_2} v_i$$

$$v_o = v_a - v_b$$

FIGURE 11-6   Voltages in the Wien bridge.

The output voltage of the bridge is

$$v_o = v_a - v_b \tag{11-14}$$

If a null is desired, the output voltage must be zero and $v_a = v_b$. To achieve this result, R_3 and R_4 must be chosen so that $v_b = 1/3 v_i$. Then $R_4/(R_3 + R_4)$ $= 1/3$ or $R_3 = 2R_4$. In the case under discussion, however, the output voltage must *not* be zero and hence the ratio $R_4/(R_3 + R_4)$ should be *smaller* than 1/3. Let, for example,

$$\frac{v_b}{v_i} = \frac{R_4}{R_3 + R_4} = \frac{1}{3} - \frac{1}{\delta} \tag{11-15}$$

where δ is a number larger than 3. Then

$$\beta = \frac{v_o}{v_i} = \frac{v_a - v_b}{v_i} = \frac{v_a}{v_i} - \left(\frac{1}{3} - \frac{1}{\delta}\right) \tag{11-16}$$

Hence to produce an output voltage at the resonant bridge frequency f_o, and thereby provide the necessary feedback voltage for oscillation, $v_a/v_i = 1/3$ and $\beta = 1/\delta$. The Barkhausen criterion for oscillation, unity loop gain or $A\beta = 1$, is then satisfied by making the amplifier gain $A = \delta$.

Two important observations can be made at this point: (1) The frequency of oscillation is exactly the null frequency of the balanced bridge, that is, $f_o = 1/2\pi RC$. (2) At any other frequency, v_a is not in phase with v_i and therefore v_o is not in phase with v_i, so that the condition of unity loop gain can only be satisfied at the resonant frequency.

The circuit diagram of a simple but practical Wien bridge oscillator is given in Fig. 11-7. The bridge consists of R and C in series, R and C in parallel,

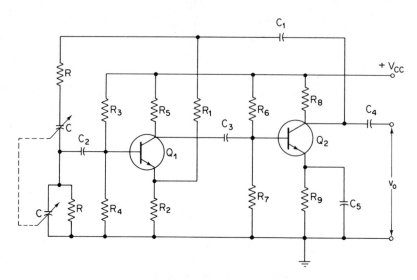

FIGURE 11-7   Two-stage Wien bridge oscillator.

R_1 and R_2. Feedback is applied from the collector of Q_2 through coupling capacitor C_1 to the top of the bridge circuit. C_1 is large enough to introduce no phase shift at the lowest frequency of oscillation. Resistance R_2 serves the dual purpose of emitter resistor of Q_1 and element of the Wien bridge.

The amplitude of oscillation is determined by the extent to which βA is greater than unity. If β is fixed, the amplitude is determined by A, increasing as A increases until further increase is limited by the nonlinear behavior of the transistors. Regulation of the amplitude is provided by resistor R_2, which provides a variable β. Resistor R_2 can be a tungsten-filament light bulb, acting as a variable resistance element. If the output of the amplifier tends to increase, the increased current through R_2 raises its temperature and increases its resistance. Then, from Eq. (11-16), β would decrease and would tend to keep the product $A\beta$ constant, thereby regulating the amplifier output to a constant level. The *thermal lag* of the filament of the tungsten lamp causes its resistance to remain almost constant during the course of an alternating cycle of output voltage or current. At very low frequencies, however, the thermal lag may not be large enough and the resistance of the lamp may change during the cycle. In this case a *thermistor* may be used, which has sufficient bulk to provide adequate thermal lag. Since the thermal coefficient of the thermistor is negative (its resistance decreases with an increase in temperature), the thermistor would have to be placed in the other bridge arm (R_1 instead of R_2).

The frequency of the oscillator can be continuously varied by the two variable air capacitors C, mounted on a common shaft. Different frequency ranges can be provided by switching in different values for the two identical resistors, *R*.

The Wien bridge oscillator is capable of stable oscillations with low distortion output. With the addition of a power amplifier to isolate the oscillator from the load, the circuit is used to provide test signals for a variety of applications. The upper frequency of the Wien bridge oscillator is limited by the amplitude and phase-shift characteristics of the amplifier and is usually on the order of 100 kHz. Above this frequency the well-known RF oscillator circuits are often used.

11-1.4 Phase-shift Oscillator*

Figure 11-8 shows a simple *RC* phase-shift oscillator capable of generating sinusoidal output voltages at frequencies up to several hundred kilohertz.

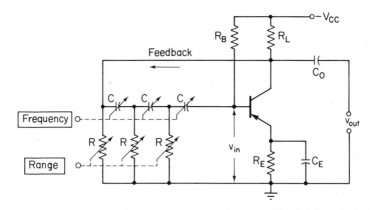

FIGURE 11-8 *RC* phase-shift oscillator, using three cascaded *RC* sections to provide 180° phase shift between collector output and base input. The three ganged air capacitors provide continuous frequency adjustment, and the three resistors are switched in decade steps to provide the frequency range control.

The circuit consists of a single transistor as the amplifier stage, and three cascaded *RC* sections to provide feedback from the output of the amplifier back to the input.

The single transistor in the circuit shifts by 180° the phase of any voltage appearing at its base. The *RC* network provides an additional amount of phase shift. At some particular frequency, the phase shift of the three *RC* network sections will be exactly 180°, and at this frequency the total phase shift from the base of the transistor, around the circuit, back to the base will

---

*Jacob Millman and Herbert Taub, *Pulse, Digital, and Switching Waveforms* (New York: McGraw-Hill Book Company, Inc., 1965), pp. 476–477.

be exactly 360°. Provided, then, that the amplification of the transistor is sufficiently large, the circuit will oscillate at that frequency.

The feedback factor β, defined as the ratio of output voltage, v_o, to input voltage, v_i, can be found by applying conventional network theory to the RC combination. This analysis yields

$$\beta = \frac{v_o}{v_i} = \frac{1}{1 - 5\alpha^2 - j(6\alpha - \alpha^3)} \qquad (11\text{-}17)$$

where $\alpha = \dfrac{1}{\omega RC}$.

The phase shift between v_o and v_i will be 180° when the imaginary portion of the denominator of Eq. (11-17) equals 0, or when $\alpha^2 = 6$. The frequency corresponding to this situation equals

$$f = \frac{1}{2\pi RC\sqrt{6}} \qquad (11\text{-}18)$$

At this frequency of oscillation, the feedback factor is $\beta = -\frac{1}{29}$. To satisfy the condition for oscillation that the product βA shall not be less than unity, it is necessary for A to be *at least* 29. Therefore the transistor chosen as the amplifier in this circuit should be able to supply this amplification.

The phase-shift oscillator is suited to a wide frequency range, from a few hertz to several hundred kilohertz. The upper frequency is limited because the impedance of the RC phase-shifting network may become so small that it loads the amplifier so heavily that corrective measures are required, making the circuit unnecessarily complicated. Frequencies on the order of 1 Hz are easily obtained by using commercially available large values for R and C. The phase-shift oscillator has a decided advantage over LC-tuned oscillators in the low-frequency range since the large inductors required for the LC oscillators would be impractical.

11-1.5   Performance Characteristics

In selecting an oscillator to perform a certain function in a measurement situation, the user should be interested in the *performance characteristics* of the instrument, summarized as follows:

(a) *Frequency range*   The oscillator should be able to supply both the the lowest and the highest frequency of interest. The range of frequencies covered by most laboratory instruments is from 0.00005 Hz to 30 MHz or even higher.

(b) *Available output power or output voltage*   Some measurements require large amounts of power; others merely require sufficient voltage output.

(c) *Output impedance*   Some oscillators have a low output impedance that can be converted to almost any desired impedance by the use of a resistive divider. Other instruments have a transformer-coupled output providing a balanced and isolated output circuit. Since many audio-range oscillators are used with 600-Ω input impedance systems, these oscillators are generally provided with a 600-Ω output attenuator.

(d) *Dial resolution and accuracy*   In the ideal case, the user should be able to set the tuning dial of the oscillator to a particular frequency with the assurance that the instrument will deliver that frequency at all times. On laboratory instruments the tuning dial may be precisely set by a vernier control. The accuracy with which the frequency tracks the tuning dial enters into the overall accuracy figure.

(e) *Frequency stability*   The frequency stability of the oscillator determines its ability to maintain the selected frequency over a period of time. Component aging, temperature changes, and power supply variations all affect stability. Frequency stability can be improved in some instances by using large amounts of negative feedback and carefully selected components.

(f) *Amplitude stability*   Amplitude stability is important in some applications. The frequency response (amplitude variation as the frequency is changed) is of particular interest when the oscillator is used for response measurements over a wide range of frequencies.

(g) *Distortion*   Distortion in the oscillator output signal is an inverse measure of the purity of the waveform. Distortion is undesirable because a harmonic of the test signal may enter the circuits under test and generate a false indication at the output. If the oscillator is used for distortion measurements, the amount of distortion that it contributes to the measurement should be far less than the distortion contributed by the circuits under test.

11-2 PULSE AND SQUARE-WAVE GENERATORS

11-2.1 Introduction

Pulse and square-wave generators are often used with a CRO as the measuring device. The waveforms portrayed by the CRO either at the output or at pertinent points in the system under test provide both qualitative and quantitative information of the system or device under test.

The fundamental difference between a pulse generator and a square-wave generator concerns the *duty cycle*. Duty cycle is defined as the ratio of the average value of the pulse over one cycle to the peak value of the pulse. Since the average value and the peak value are inversely related to their time

duration, the duty cycle can be defined in terms of the *pulsewidth* and the *period* or *pulse repetition time:*

$$\text{duty cycle} = \frac{\text{pulsewidth}}{\text{period}}$$

Square-wave generators produce an output voltage with equal *on* and *off* times, so that their duty cycle equals 0.5, or 50 per cent. The duty cycle remains at 50 per cent as the frequency of oscillation is varied.

The duty cycle of a *pulse generator* may vary; very short duration pulses give a low duty cycle, and the pulse generator generally can supply more power during its *on* period than a square-wave generator can. Short duration pulses reduce the power dissipation in the component under test. For instance, measurements of transistor gain can be made with pulses short enough to prevent junction heating, and the resulting effect of heat on the gain of the transistor is then greatly minimized.

Square-wave generators are used whenever the low-frequency characteristics of a system are being investigated: testing audio systems, for instance. Square waves are also preferable to short-duration pulses if the transient response of a system requires some time to settle down.

11-2.2   Pulse Characteristics and Terminology

In selecting a pulse generator or square-wave generator, the quality of the pulse is of primary importance. A test pulse of high quality insures that any degradation of the displayed pulse may be attributed to the circuit under test and not to the test instrument itself.

The pertinent characteristics of a pulse are shown in Fig. 11-9. The specifications describing these characteristics are usually given in the instrument manual or manufacturer's specification sheets.

The time required for the pulse to increase from 10 per cent to 90 per cent of its normal amplitude is called the *risetime* (t_r). Similarly, the time required for the pulse to decrease from 90 per cent to 10 per cent of its maximum amplitude is called the *falltime* (t_f). In general, the risetime and the falltime of the pulse should be significantly faster than the circuit or component under test.

When the initial amplitude rise exceeds the correct value, *overshoot* occurs. The overshoot may be visible as a single *pip*, or *ringing* may occur. When the maximum amplitude of the pulse is not constant but decreases slowly, the pulse is said to *droop* or *sag*. Any overshoot, ringing, or sag in the test pulse should be known to avoid confusion with similar phenomena caused by the test circuit.

The maximum pulse *amplitude* is of prime concern if appreciable input power is required by the tested circuit, such as, for example, a magnetic core memory. At the same time, the attenuation range of the instrument should

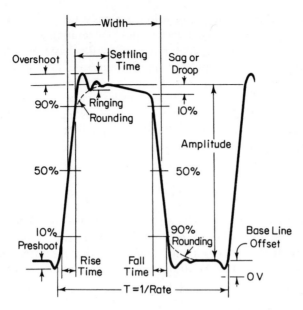

FIGURE 11-9   Characteristics of a pulse.

be adequate to prevent overdriving the test circuit as well as to simulate actual operating conditions.

The range of the frequency control or *pulse repetition rate* (PRR) is of concern if the tested circuit can operate only within a certain range of pulse rates or if a variation in the rate is needed. Some of the more sophisticated pulse generators are capable of repetition rates of up to 100 MHz for testing "fast" circuits; others have a *pulse-burst* feature that allows a train of pulses rather than a continuous output to be used to check a system.

Some pulse generators can be *triggered* by externally applied trigger signals, similar to the trigger features found in laboratory CROs. Conversely, the output of the pulse generator or square-wave generator may be used to provide trigger pulses for operating external circuits. The output trigger circuitry of the pulse generator then allows the trigger pulse to occur either before or after the main output pulse.

The *output impedance* of the pulse generator is another important consideration in fast pulse systems. This is so because the generator, which has a source impedance matched to the connecting cable, will absorb reflections resulting from impedance mismatches in the external circuitry. Without this generator-to-cable match, the reflections would be rereflected by the generator, resulting in spurious pulses or perturbations on the main pulse.

DC coupling of the output circuit is necessary when retention of the dc bias levels in the test circuit is desired, in spite of variations in pulsewidth, pulse amplitude, or PRR.

Circuits used in pulse generation generally fall into two categories: *passive*, or pulse-shaping, and *active*, or pulse-generating circuits. In passive-type circuits, a sine-wave oscillator is used as the basic generator and its output is passed through a pulse-shaping circuit to obtain the desired waveform. For instance, an approximate square waveform may be obtained by first amplifying and then clipping a sine wave. Active generators are usually of the *relaxation* type. The relaxation oscillator uses the charge-and-discharge action of a capacitor to control the conduction of a vacuum tube or a transistor. Some common forms of relaxation oscillator are the *multivibrators* and the *blocking oscillators*.

11-2.3   Astable Multivibrator

The *astable* or *free-running multivibrator* is widely used for the generation of pulses. It can be made to produce either square waves or pulses, depending on the choice of circuit components. A typical free-running multivibrator is shown in Fig. 11-10. Essentially, the circuit consists of a two-

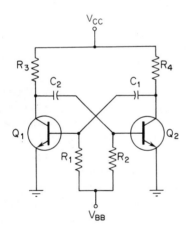

FIGURE 11-10 Astable or free-running multivibrator.

stage *RC*-coupled amplifier, with the output of the second stage (Q_2) coupled back to the input of the first stage (Q_1) via capacitor C_1. Similarly, the output of Q_1 is coupled via C_2 to the input of Q_2. Since the coupling between the two transistors is taken from the collectors, the circuit is known as a *collector-coupled* astable multivibrator.

The usual qualitative analysis of the circuit proceeds as follows: When the power is first applied to the circuit, both transistors start conducting. Because of small differences in their operating characteristics, one of the transistors will conduct slightly more than the other. This starts a series of events. Assume that Q_1 initially conducts more than Q_2. This means that the collector voltage of $Q_1(e_{c1})$ drops more rapidly than the collector voltage of

$Q_2(e_{c2})$. The decrease in e_{c1} is applied to the R_2C_2 network, and because the charge on C_2 cannot change instantaneously, the full negative-going change appears across R_2. This decreases the forward bias on Q_2 which in turn decreases the collector current of $Q_2(i_{c2})$, and the collector voltage of Q_2 rises. This rise in Q_2 collector voltage is applied via the R_1C_1 network to the base of Q_1, increasing its forward bias. Q_1 therefore conducts even more heavily and its collector voltage drops still more rapidly. This negative-going change is coupled to the base of Q_2, further decreasing its collector current. The entire process is cumulative until Q_2 is entirely cut off and Q_1 conducts heavily (*bottoms*).

With Q_2 cut off, its collector voltage practically equals the supply voltage, V_{CC}, and capacitor C_1 charges rapidly to V_{CC} through the low-resistance path from emitter to base of the conducting transistor Q_1. When the circuit action turns Q_1 fully on, its collector potential drops to approximately 0 V, and since the charge on C_2 cannot change instantaneously, the base of Q_2 is at $-V_{CC}$ potential, driving Q_2 deep into cutoff.

The switching action now begins. C_2 begins to discharge exponentially through R_2. When the charge on C_2 reaches 0 V, C_2 attempts to charge up to the value of $+V_{BB}$, the base supply voltage. But this action immediately places a forward bias on Q_2 and this transistor starts to conduct. As soon as Q_2 starts conducting, its collector current causes a decrease in collector voltage e_{c2}. This negative-going change is coupled to the base of Q_1 which starts to conduct less, i.e., it comes out of saturation. This cumulative action repeats until Q_1 finally cuts off and Q_2 conducts heavily. At this instant, the collector voltage of Q_1 reaches its maximum value of V_{CC}. Capacitor C_2 charges to the full value of V_{CC}, and a full cycle of operation has been completed.

The waveforms appearing at the base and the collector of each transistor are the result of a *symmetrical* or *balanced* operation: The time constants R_1C_1 and R_2C_2, the transistors themselves, and the supply voltages are all identical. The conducting and nonconducting periods are therefore of almost the same duration. The waveforms for each of the two transistors are given in the waveform diagram of Fig. 11-11.

Assume that at time $t = 1$, transistor Q_1 is fully turned on and transistor Q_2 is cut off. This makes the collector voltage e_{c1} of Q_1 a minimum (practically 0 V) and the collector voltage e_{c2} of Q_2 a maximum (V_{CC}). Capacitor C_1 is charging through the emitter-to-base resistance of Q_1 toward the supply voltage V_{CC} and reaches its full charge rapidly (low emitter-to-base resistance). Since e_{c1} is 0 V, capacitor C_2 begins to charge exponentially through R_2 toward the base supply voltage V_{BB} with a time constant equal to R_2C_2. Since the early part of the exponential charging curve is almost linear, the increase in Q_2 base voltage (e_{b2}) is indicated by a linear slope on the graph of Fig. 11-11.

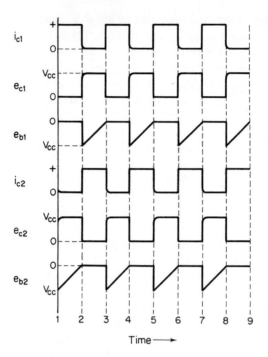

FIGURE 11-11   Waveforms for the astable multivibrator of Fig. 11-10.

At time $t = 2$, e_{c2} reaches a value of approximately 0 V, placing a forward bias on the base of Q_2 which then starts to conduct. Within a very short time, the collector current of Q_2 reaches its maximum and collector voltage e_{c2} drops to 0 V. When Q_2 starts to draw current, the base of Q_1 becomes negative and Q_1 is quickly driven into cutoff. Its collector voltage, e_{c1}, reaches the V_{CC} value and collector current i_{c1} becomes zero. Within a very small fraction of the total Q_2 conducting time, capacitor C_2 is fully charged to V_{CC} through the low-resistance emitter-to-base path of Q_2.

Between times $t = 2$ and $t = 3$, transistor Q_1 is cut off, and its collector current and voltage remain constant. Similarly, the collector voltage and current for Q_2 remain constant. Only capacitor C_1 is charging and the base voltage e_{b1} of Q_1 is rising exponentially toward V_{BB}. At time $t = 3$, the base voltage of Q_1 exceeds the cutoff value (approximately 0 V) and Q_1 starts conducting again. Obviously, one complete cycle of operation, from time $t = 1$ to time $t = 3$, depends on the time required for the base voltage of the cutoff transistor to reach the forward bias value. This time depends on two things: the magnitude of the reverse bias $(-V_{CC})$ and the time constant of the capacitor charging circuit involved, namely, R_1C_1 or R_2C_2.

The *analytical* evaluation of circuit operation proceeds as follows:

During its nonconducting period, the collector voltage of Q_1 equals

$$e_{c1} = V_{CC}(1 - e^{-t/\tau_3}) \tag{11-19}$$

where $\tau_3 = R_3 C_2$.

When Q_1 switches on, its collector voltage is at ground potential and the base voltage of Q_2 becomes $-V_{CC}$ with respect to ground. The subsequent rise in Q_2 base voltage, through the $R_2 C_2$ charging circuit, is described by

$$e_{b2} = (V_{BB} + V_{CC})(1 - e^{-t/\tau_2}) - V_{CC} \tag{11-20}$$

where $\tau_2 = R_2 C_2$.

Q_2 remains cut off until e_{b2} reaches the value of 0 V (by good approximation) and the *off* time interval T_2 of Q_2 can be determined by setting e_{b2} in Eq. (11-20) to zero and solving for t, so that

$$0 = (V_{BB} + V_{CC})(1 - e^{-t/\tau_2}) - V_{CC} \tag{11-21}$$

and
$$T_2 = \tau_2 \ln\left(\frac{V_{BB}}{V_{BB} + V_{CC}}\right) \tag{11-22}$$

Similarly, when Q_2 is off and Q_1 is bottomed, the collector voltage of Q_2 can be described by

$$e_{c2} = V_{CC}(1 - e^{-t/\tau_4}) \tag{11-23}$$

where $\tau_4 = R_4 C_1$.

When now Q_2 is switched on, its collector voltage drops to 0 V and the base voltage of Q_1 is given by

$$e_{b1} = (V_{BB} + V_{CC})(1 - e^{-t/\tau_1}) \tag{11-24}$$

where $\tau_1 = R_1 C_1$.

Solving for the *off* time interval T_1 of Q_1 by setting e_{b1} in Eq. (11-24) to zero, we obtain

$$0 = (V_{BB} + V_{CC})(1 - e^{-t/\tau_1}) - V_{CC} \tag{11-25}$$

and
$$T_1 = \tau_1 \ln\left(\frac{V_{BB}}{V_{BB} + V_{CC}}\right) \tag{11-26}$$

The total period of oscillation is given by

$$T = T_1 + T_2 \tag{11-27}$$

In the case of *symmetrical* operation, when time constants $R_1 C_1$ and $R_2 C_2$ are equal, the waveform is a symmetrical square wave. By making time constant $R_1 C_1$ larger than time constant $R_2 C_2$, the output waveform becomes a pulse train because the *off* time of Q_1 will be larger than the *off* time of Q_2.

11-2.4   Blocking Oscillator

The blocking oscillator is a circuit of considerable practical importance and can be used to generate a single pulse (monostable operation) or a pulse train (astable operation). In either configuration the blocking oscillator consists of an amplifier whose output is coupled back to its input through a *pulse transformer*. If the winding polarities of this transformer are correctly chosen, the feedback will be *regenerative* (positive feedback) and the circuit will produce a pulse. The addition of an *RC* timing network gives the circuit the astable or free-running characteristics and allows it to produce a pulse train. The pulsewidth is determined by the characteristics of the pulse transformer, and to some extent by the other circuit parameters, and can be in the nanosecond to millisecond range.

A typical blocking oscillator circuit is shown in Fig. 11-12. The pulse transformer is connected in the circuit to provide polarity inversion between

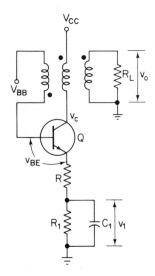

FIGURE 11-12 Astable blocking oscillator with $R_1 C_1$ timing circuit in the emitter.

collector and base voltages, as indicated by the polarity dots on the transformer windings. A third winding connects the load. Its winding direction is arbitrary and may be chosen to obtain either a positive or a negative output pulse.

Qualitatively, the operation of the circuit can be explained as follows: Assume, initially, that voltage v_1 across timing capacitor C_1 is larger than the base-emitter voltage v_{BE}, which keeps the transistor cut off. Capacitor C_1, however, discharges toward ground with a time constant $R_1 C_1$, and v_1 decreases exponentially. When v_1 reaches the value of v_{BE}, the base starts to draw current and the transistor is moved out of its nonconducting or *off* state into the *on* state. Collector current starts to flow, which lowers the collector voltage and, through inductive coupling of the pulse transformer, raises the base voltage. As a result, more collector current flows, causing a further drop in collector voltage and a corresponding rise in base voltage. If the loop gain of the circuit is greater than unity, this regenerative action quickly drives the transistor into saturation and the collector current rises to its maximum value. As the collector current increases, however, the collector-to-emitter voltage decreases and reduces the transformer voltage, which in turn reduces the base current. At this point the transistor leaves the saturation region and enters the active region again. By regenerative action the transistor quickly returns to the *off* state and the pulse ends. During the pulse duration, capacitor C_1 is recharged and the voltage v_1 across it is larger than it was at the beginning of the pulse. The transistor remains in the *off* state until C_1 has discharged to the voltage at which the transistor enters its active region again. At this point the cycle repeats.

The waveform across the R_1C_1 combination and the collector waveform, which is also the output pulse, are shown in Fig. 11-13. The *overshoot*

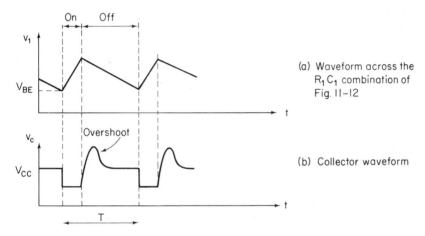

FIGURE 11-13   Blocking oscillator waveforms.

in the collector waveform is caused by the action of the inductance and distributed capacitance of the pulse transformer. In the usual equivalent circuit these two effects are represented by a parallel LC network. When the current in the transformer winding is suddenly interrupted, this LC circuit is excited into oscillation at its resonant frequency and produces the overshoot shown in Fig. 11-13(b). In a practical circuit this oscillation is damped out by using a transformer with low-Q windings and by connecting an appropriate load to the output terminals.

11-2.5 Laboratory Square-wave and Pulse Generator

The block diagram of a typical general-purpose generator providing negative pulses of variable frequency, duty cycle, and amplitude is given in Fig. 11-14. The frequency range of the instrument is covered in seven decade

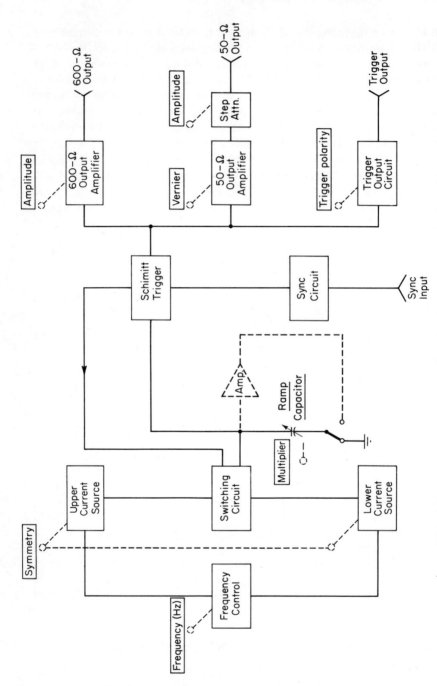

FIGURE 11-14   Block diagram of a pulse generator (courtesy Hewlett-Packard Co.).

steps from 1 Hz to 10 MHz, with a linearly-calibrated dial for continuous adjustment on all ranges. The duty cycle can be varied from 25 per cent to 75 per cent. Two independent outputs are available: a 50-Ω source that supplies pulses with rise and fall times of 5 ns at 5-V peak amplitude, and a 600-Ω source that supplies pulses with rise and fall times of 70 ns at 30-V peak amplitude. The instrument can be operated as a free-running generator or it can be synchronized with external signals. Trigger output pulses for synchronization of external circuits are also available.

The basic generating loop, which is redrawn for greater clarity in Fig. 11-15, consists of two current sources, the ramp capacitor, the Schmitt trigger

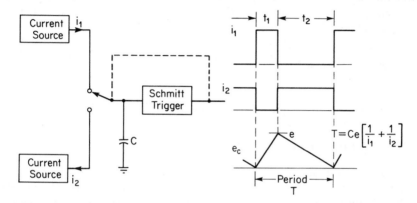

FIGURE 11-15   Simplified current source operation (courtesy Hewlett-Packard Co.).

circuit, and the current-switching circuit (indicated by a simple switch). The two current sources provide a constant current for charging and discharging the ramp capacitor. The ratio of these two currents is determined by the setting of the *symmetry* control which then determines the duty cycle of the output waveform. The *frequency* dial controls the sum of the two currents from the current sources by applying appropriate control voltages to the bases of the current control transistors in the current generators. The size of the ramp capacitor is selected by the *multiplier* switch. These last two controls provide decade switching and vernier control of the frequency of the output.

The *upper* current source, supplying a constant current to the ramp capacitor, charges this capacitor at a constant rate, and the ramp voltage increases linearly. When the positive slope of the ramp voltage reaches the upper limit set by internal circuit components, the Schmitt trigger (a bistable multivibrator) changes state. The trigger circuit output goes negative, reversing the condition of the current control switch, and the capacitor starts discharging. The discharge rate is linear, controlled by the *lower* current source. When the negative ramp reaches a predetermined lower level, the Schmitt trigger switches back to its original state. This now provides a positive trigger circuit output that reverses the condition of the current switch again, cutting off the lower current source and switching on the upper current source. One cycle of operation has now been completed. The entire process, of course, is repetitive and the Schmitt trigger circuit provides negative pulses at a continuous rate.

The output of the Schmitt circuit is passed to the trigger output circuit and to the 50-Ω and 600-Ω amplifiers. The trigger output circuit differentiates the square-wave output from the Schmitt trigger, inverts the resulting pulse, and provides a positive triggering pulse. The 50-Ω amplifier is provided with an output attenuator to allow a vernier control of the signal output voltage. In addition to its free-running mode of operation, the generator can be synchronized or locked in to an external signal. This is accomplished by triggering the Schmitt circuit by an external synchronization pulse.

The unit is powered by an internal supply that provides regulated voltages for all stages of the instrument.

11-3   SIGNAL GENERATORS

11-3.1   Standard Signal Generator

A standard signal generator is often used in the measurement of gain, bandwidth, signal-to-noise (S/N) ratio, standing-wave ratio (SWR), and other circuit properties. It is extensively used in the testing of radio receivers and transmitters.

The standard signal generator is a source of ac energy of accurately known characteristics. The instrument is capable of *modulating* a carrier, or center frequency, which is indicated by a dial setting. Common types of modulating signals are sine wave, square wave, and pulse; the output signal may be either amplitude modulated (AM) or frequency modulated (FM). AM is a common feature of the standard signal generator. When the FM system produces a considerable excursion in frequency at a relatively low cyclical rate, the instrument is known as a sweep-frequency generator (Sec. 11-3.2). The output voltage is read by an output meter and an output attenuator setting.

The elements of a conventional standard signal generator are given in Fig. 11-16. The carrier frequency is generated by a very stable *LC* oscillator, delivering a good sinusoidal waveform and having no appreciable hum or noise modulation. The frequency of oscillation is selected by a frequency range control and a vernier dial setting. The *LC* circuit is designed to give a reasonably constant output over any one frequency range. AM is provided from an internal, fixed-frequency sine-wave generator or from an external

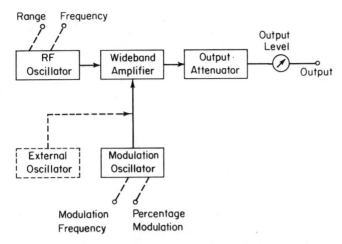

FIGURE 11-16   Elements of a standard signal generator.

source. Modulation takes place in the output amplifier circuit, which delivers the modulated carrier to the output attenuator.

The frequency stability of the basic instrument is limited by the *LC* circuit design of the master oscillator. Since range switching is usually accomplished by selecting appropriate capacitive elements in the oscillator circuit, any change in frequency range upsets the circuit to some extent, and the user must wait until the circuit has stabilized at its new resonant frequency.

Some laboratory-type signal generators use a different approach to frequency generation in order to improve the frequency stability. The functional block diagram of Fig. 11-17 shows a standard signal generator whose master oscillator is optimally designed for the highest frequency range, and frequency dividers are switched in to produce the lower ranges. In this manner, the stability of the top range is imparted to all other ranges. The master oscillator is made insensitive to temperature variations and also to the influence of the following stages by careful circuit design.

The RF oscillator output, after passing through an untuned buffer amplifier (B_1), enters the power amplifier unit. On the highest frequency range (34 MHz–80 MHz), the RF signal passes through an additional buffer (B_2) to the main amplifier (A). For the lower frequency ranges the oscillator signal is applied to a series of frequency dividers and from there through another buffer (B_3) to the power amplifier. The nine 2/1 dividers give a maximum divisor of 512. Therefore the lowest frequency range produced by the cascaded divider chain equals the highest range divided by 512, or 67 kHz– 156 kHz. The buffer amplifiers provide a very high degree of isolation between the master oscillator and the power amplifier, and they practically

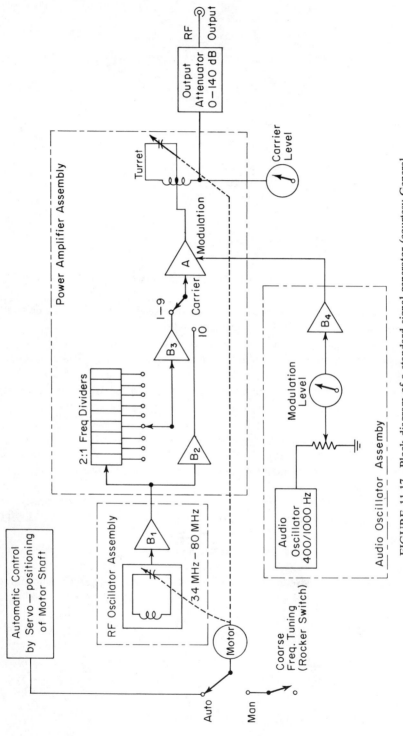

FIGURE 11-17 Block diagram of a standard signal generator (courtesy General Radio Co.).

eliminate all frequency-pulling effects from changes in operating and loading conditions at the output stage. Range-switching effects are also eliminated since the same oscillator is used on all bands.

The master oscillator is tuned by a motor-driven variable capacitor. For fast coarse tuning, a rocker switch on the front panel is pushed which sends the indicator gliding along the slide-rule scale of the main frequency dial at approximately 7 per cent frequency change per second. When the correct location on the main dial is reached, the oscillator can be fine-tuned by means of a large rotary control, with each division corresponding to 0.01 per cent of the main dial setting. A second front panel control ($\Delta F/F$) permits incremental tuning over a limited range and allows very great tuning resolution.

The availability of a motor-driven frequency control presents obvious opportunities for both local and remote automatic tuning, and these are exploited by a programmable automatic frequency control device. With this instrument, one can sweep between adjustable frequency limits and automatically tune to preset frequencies.

The basic modulating function is performed in the power amplifier stage by variation of the base voltage of the power transistor. Two highly stable modulating signals (400 Hz and 1 kHz) are internally generated. The amplitude of the modulating signal can be adjusted by the modulation level control for up to 95 per cent modulation. Any given modulation setting is kept constant over the range of the carrier level control. Provision for application of external modulating signals is also made.

Internal calibration is provided by a 1-MHz crystal oscillator. This reference source is mixed with the master oscillator RF signal and produces a zero beat when the two signals are equal. An external reference signal can be applied to the mixer input, and by using a portion of the crystal oscillator calibration circuitry the instrument functions as a heterodyne frequency meter.

Additional refinements include the autocontrol unit that permits a number of automatic tuning operations. Power supply requirements are rather critical, but the small power consumption of the instrument makes it relatively easy to obtain excellent regulation and stability with very low ripple. The supply voltage for the master oscillator, which is especially sensitive to power variations, is regulated by a temperature-compensated reference circuit.

11-3.2 Sweep-frequency Generator

The sweep-frequency generator is a logical extension of the standard signal generator. It provides a sinusoidal output voltage, usually in the RF range, whose frequency is smoothly and continuously varied over an entire frequency band, usually at a low audio rate. The process of frequency modulation (FM) may be accomplished electronically or mechanically. The mechanical method of varying the frequency of the master oscillator follows from the discussion of the standard signal generator of Sec. 11-3.1 and may be realized by using a motor-driven variable capacitor in the *LC* oscillator circuit. This method is used to advantage in some laboratory instruments and brings the precision and stability of the conventional manually tuned signal generator to the field of swept-frequency measurements. In microwave measurements, a motor-driven mechanically tuned *klystron* may be used to produce a swept RF signal, although more recent developments have led to the construction of electronically tuned oscillators. One such development is the *backward-wave oscillator tube*,* which overcomes the disadvantages of long sweep times and mechanical wear involved with motorized tuning devices.

The elements of a standard sweep-frequency generator are given in Fig. 11-18. The heart of the instrument is the master oscillator whose frequency range can be selected by the range switch. This provides a number of frequency bands, where each band usually covers several octaves. For simplicity, the frequency sweeper is assumed to be a mechanical device that rotates the tuning capacitor in the *LC* master oscillator, causing repetitive sweeps over the complete frequency range. A representative sweep rate would be in the order of 20 sweeps per second. A manual frequency control allows independent manual adjustment of the master oscillator resonant frequency. The frequency sweeper also provides a synchronously varying sweep voltage that can be used to drive the horizontal deflection plates of a CRO or the X axis of an X-Y recorder. Thus the amplitude response of a device, fed by the swept-frequency output of the generator, can be displayed automatically on an oscilloscope or X-Y recorder.

To identify frequencies and bands of special interest, a marker generator provides half-sinusoidal waveforms at any frequency within the sweep range. The marker voltage can be added to the base line of the CRO trace during alternate cycles of the sweep voltage and appears as a superimposed identification mark on the response curve of the device under test.

The automatic level control circuit is basically a closed-loop feedback system that monitors the RF level at some point in the measurement system. This circuit holds the forward power (power delivered to the load) constant with frequency and load impedance variations. A constant power level ideally prevents any source mismatch and also provides a constant readout calibration with frequency.

---

*Hewlett-Packard, Palo Alto, Calif. ("How a Helix Backward-wave Tube Works"), *Application Note* No. AN 12.

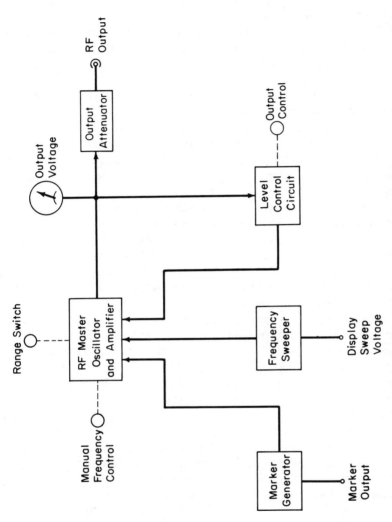

FIGURE 11-18 Elements of a sweep-frequency generator.

11-3.3 Random Noise Generator

The random noise generator is a device that delivers a signal whose instantaneous amplitude is determined at random and is therefore *unpredictable*. True random noise contains no periodic frequency components and has a continuous spectrum. This instrument offers the possibility of using a single measurement as an indicator of performance over a wide frequency band.

Random noise measurements are applied in many measurement fields. In accoustic measurements random noise is used to smooth response curves that might otherwise be difficult to interpret. Random noise in psycho-acoustic measurements has greatly increased the knowledge of the process of hearing. Random noise also best simulates the vibrations that aircraft and rockets are subjected to in flight, and it is commonly used in vibration and fatigue testing of aerospace components and assemblies.

In electrical measurements, noise can be used as the test signal itself. Intermodulation (IM) distortion and crosstalk measurements in communication systems, tests on servo amplifiers, and studies made with analog computers are but a few of the many applications. Noise of known amplitude and known spectral characteristics is most effective for testing various methods of signal detection and recovery in the presence of noise, as in radio, telemetry, radar, and sonar systems.

Random noise generators are available to cover frequencies from near dc to microwave. The method of generating random noise is usually by a semiconductor *noise diode* that delivers frequencies in a band roughly extending from 80 kHz to 220 kHz. The simplified block diagram of a random noise generator for use in the audiofrequency range is shown in Fig. 11-19(a). The output from the noise diode is amplified and heterodyned down to the audio-frequency band in a balanced, symmetrical modulator delivering a symmetrical amplitude distribution.

The output amplifier, as the final stage in the generator, includes a transformer and supplies a floating, single-ended, or balanced output. The filter arrangement that follows the modulator further reduces and controls the bandwidth and supplies an output signal in three spectrum choices: *white* noise, *pink* noise, and *USASI* noise. This is illustrated in Fig. 11-19(b).

White noise is flat from 20 Hz to 25 kHz and has an upper cutoff frequency of 50 kHz with a cutoff slope of -12 dB per octave. Pink noise is so called because of its emphasis (greater amplitude) on the lower frequencies, as in reddish light. Pink noise has a voltage spectrum that is inversely proportional to the square root of the frequency and is used in bandwidth analysis. USASI noise roughly simulates the energy distribution of speech and music frequencies and is used for testing audio amplifiers and loudspeakers.

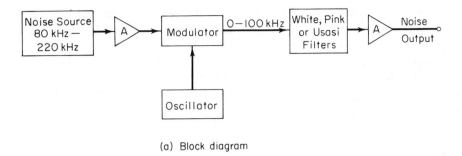

(a) Block diagram

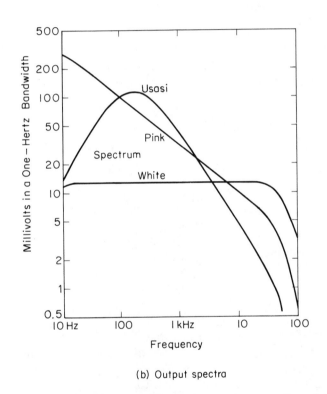

(b) Output spectra

FIGURE 11-19   Random noise generator (courtesy General Radio Co.).

11-4 FUNCTION GENERATORS

A function generator is a versatile instrument that delivers a choice of different waveforms whose frequencies are adjustable over a wide range. The most common output waveforms are the sine, triangular, square, and sawtooth waves. The frequencies of these waveforms may be adjusted from a fraction of a hertz to several hundred kilohertz.

The various outputs of the generator may be available at the same time. For instance, by providing a square wave for linearity measurements in an audio system, a simultaneous sawtooth output may be used to drive the horizontal deflection amplifier of a CRO, providing a visual display of the measurement results. The capability of the function generator to *phase lock* to an external signal source is another useful feature. One function generator may be used to phase lock a second function generator, and the two output signals can be displaced in phase by an adjustable amount. In addition, one generator may be phase locked to a harmonic of the sine wave of another generator. By adjusting the phase and the amplitude of the harmonics, almost any waveform may be generated by the summation of the fundamental frequency generated by the one function generator and the harmonic generated by the other function generator. The function generator can also be phase locked to a frequency standard, and all its output waveforms are then generated with the frequency accuracy and stability of the standard source.

The function generator can supply output waveforms at very low frequencies. Since the low frequency of a simple RC oscillator is limited, a different approach is used in the function generator of Fig. 11-20. This instrument delivers sine, triangular, and square waves with a frequency range of 0.01 Hz to 100 kHz. The frequency control network is governed by the frequency dial on the front panel of the instrument or by an externally applied control voltage. The frequency control voltage regulates two current sources.

The *upper* current source supplies a constant current to the triangle integrator whose output voltage increases linearly with time. The output voltage is given by the well-known relationship

$$e_{\text{out}} = -\frac{1}{C} \int i \, dt \tag{11-28}$$

An increase or a decrease in the current supplied by the upper current source increases or decreases the slope of the output voltage. The voltage comparator multivibrator changes state at a predetermined level on the positive slope of the integrator's output voltage. This change of state cuts off the upper current supply to the integrator and switches on the lower current supply.

The *lower* current source supplies a reverse current to the integrator so that its output decreases linearly with time. When the output voltage reaches a predetermined level on the negative slope of the output waveform, the voltage comparator again switches and cuts off the lower current source while at the same time switching on the upper source again.

The voltage at the output of the integrator has a triangular waveform whose frequency is determined by the magnitude of the current supplied by

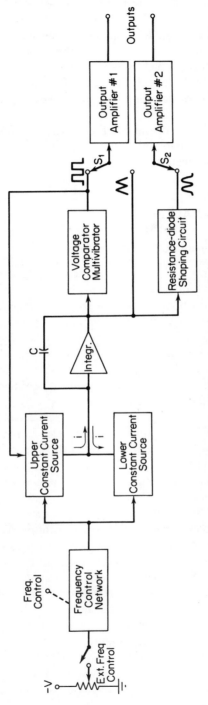

FIGURE 11-20 Basic elements of a function generator.

the constant current sources. The comparator delivers a square-wave output voltage of the same frequency. The third output waveform is derived from the triangular waveform, which is synthesized into a sine wave by a diode resistance network. In this circuit the slope of the triangular wave is altered as its amplitude changes, resulting in a sine wave with less than 1 per cent distortion.

The output circuitry of the function generator consists of two output amplifiers that provide two simultaneous, individually selected outputs of any of the waveform functions.

11-5 WAVE ANALYZERS

11-5.1 Frequency-selective Wave Analyzer

A *wave analyzer* is an instrument designed to measure the relative amplitudes of single-frequency components in a complex or distorted waveform. Basically, the instrument acts as a *frequency-selective voltmeter* which is tuned to the frequency of one signal component while rejecting all the other signal components. Two basic circuit configurations are generally used. For measurements in the audiofrequency range (from 20 Hz to 20 kHz), the analyzer has a filter section with a very narrow passband that can be tuned to the frequency component of interest. An instrument of this type is given in the functional block diagram of Fig. 11-21(a).

The waveform to be analyzed in terms of its separate frequency components is applied to an input attenuator that is set by the *meter range* switch on the front panel. A driver amplifier feeds the attenuated waveform to a high-Q active filter. The filter consists of a cascaded arrangement of RC resonant sections and filter amplifiers. The passband of the total filter section is covered in decade steps over the entire audio range by switching capacitors in the RC sections. Close-tolerance polystyrene capacitors are generally used for selecting the frequency ranges. Precision potentiometers are used to tune the filter to any desired frequency within the selected passband.

A final amplifier stage supplies the selected signal to the meter circuit and to an untuned buffer amplifier. The buffer amplifier can be used to drive a recorder or an electronic counter. The meter is driven by an average-type detector and usually has several voltage ranges as well as a decibel scale.

The bandwidth of the instrument is very narrow, typically about 1 per cent of the selected frequency. Figure 11-21(b) shows a typical attenuation curve of a wave analyzer (General Radio Company, Type 1568-A Wave Analyzer). The initial attenuation rate is approximately 600 dB per octave; the attenuation at one-half and twice the selected frequency is about 75 dB. The filter characteristic also shows that the attenuation is still increasing far

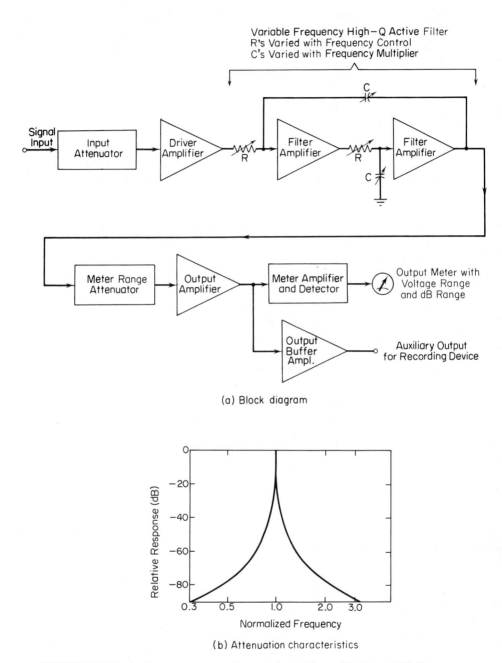

(a) Block diagram

(b) Attenuation characteristics

FIGURE 11-21 Audio-range wave analyzer (adapted from GR Type 1568-A). Characteristics of the active filter show the extremely sharp attenuation at the selected frequency (courtesy General Radio Co.).

from the center frequency, well into the noise level of the instrument itself. The analyzer must have extremely low input distortion, so low, in fact, that it cannot be detected by the analyzer itself.

11-5.2 Heterodyne Wave Analyzer

Measurements in the megahertz range are usually done with another wave analyzer that is particularly suited to the higher frequencies. The input signal to be analyzed is *heterodyned* to a higher intermediate frequency (IF) by an internal local oscillator. Tuning the local oscillator shifts the various signal frequency components into the passband of the IF amplifier. The output of the IF amplifier is rectified and applied to the metering circuit. An instrument that uses the heterodyning principle is often called a *heterodyning tuned voltmeter.*

A wave analyzer using the heterodyning principle is shown in the block diagram of Fig. 11-22. The operating frequency range of this instrument is from 10 kHz to 18 MHz in 18 overlapping bands selected by the frequency range control of the local oscillator. The bandwidth is controlled by an active filter and can be selected at 200, 1,000, and 3,000 Hz.

The input signal enters the instrument through a probe connector that contains a unity-gain isolation amplifier. After appropriate attenuation, the input signal is heterodyned in the mixer stage with the signal from a local oscillator. The output of the mixer forms an intermediate frequency that is uniformly amplified by the 30-MHz IF amplifier. This amplified IF signal is then mixed again with a 30-MHz crystal oscillator signal, which results in information centered on a zero frequency. An active filter with controlled bandwidth and symmetrical slopes of 72 dB per octave then passes the selected component to the meter amplifier and detector circuit. The output from the meter detector can be read off a decibel-calibrated scale or may be applied to a recording device.

11-5.3 Applications

Applications of the wave analyzer are found in the fields of electrical measurements and sound and vibration analysis. For example, *harmonic distortion* of an amplifier can readily be measured, and the contribution of each harmonic to the total distortion figure can be determined. When the passband of the analyzer of Fig. 11-21(a) is tuned to the second harmonic, the fundamental frequency is sufficiently attenuated to reduce its level to substantially less than that of the harmonic. The curve of Fig. 11-21(b) shows that half-frequency attenuation is at least 75 dB. When the third harmonic is selected, the fundamental frequency is attenuated by more than 85 dB. A complete harmonic analysis can be carried out by resolving the individual

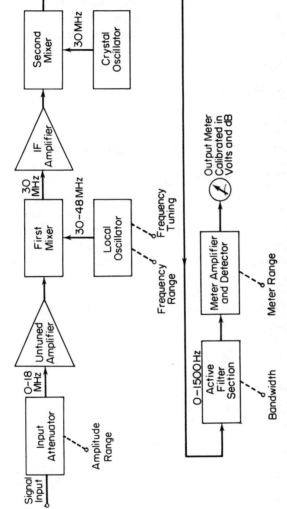

FIGURE 11-22 Functional block diagram of the heterodyning wave analyzer (Adapted from HP Model 312-A, courtesy Hewlett-Packard Co.).

components of a periodic signal and measuring or displaying these components. It is not unusual to be able to separate and display about 50 harmonics.

The wave analyzer is applied industrially in the field of reduction of sound and vibration generated by machines and appliances. The source of the noise or vibration generated by a machine must first be identified before it can be reduced or eliminated. A fine-spectrum analysis with the wave analyzer will show various discrete frequencies and resonances that can then be related to motions within the machine.

11-6 HARMONIC DISTORTION ANALYZERS

11-6.1 Harmonic Distortion

In the ideal case, application of a sinusoidal input signal to an electronic device, such as an amplifier, should result in the generation of a sinusoidal output waveform. Generally, however, the output waveform is not an exact replica of the input waveform because various types of distortion may arise. Distortion may be a result of the inherent nonlinear characteristics of the transistors in the circuit or of the circuit components themselves. Nonlinear behavior of circuit elements introduces harmonics of the fundamental frequency in the output waveform, and the resultant distortion is often referred to as *harmonic distortion* (HD).

A measure of the distortion represented by a particular harmonic is simply the ratio of the amplitude of the harmonic to that of the fundamental frequency, expressed as a percentage. Harmonic distortion is then represented by

$$D_2 = \frac{B_2}{B_1}, \qquad D_3 = \frac{B_3}{B_1}, \qquad D_4 = \frac{B_4}{B_1} \qquad \textbf{(11-29)}$$

where $D_n (n = 2, 3, 4, \ldots)$ represents the distortion of the nth harmonic, B_n represents the amplitude of the nth harmonic and B_1 is the amplitude of the fundamental.

The *total harmonic distortion,* or *distortion factor,* is defined as

$$D = \sqrt{D_2^2 + D_3^2 + D_4^2 + \cdots} \qquad \textbf{(11-30)}$$

Several methods have been devised to measure the harmonic distortion caused either by a single harmonic or by the sum of all the harmonics. Some of the better-known methods are described in the following sections.

11-6.2 Tuned-circuit Harmonic Analyzer

One of the oldest methods of determining the harmonic content of a waveform uses a tuned circuit, as in Fig. 11-23. A series-resonant circuit, consisting of inductor L and capacitor C, is tuned to a specific harmonic

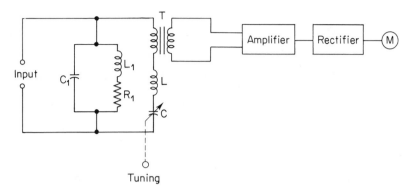

FIGURE 11-23   Functional block diagram of the tuned-circuit harmonic analyzer.

frequency. This harmonic component is transformer-coupled to the input of an amplifier. The output of the amplifier is rectified and applied to a meter circuit. After a reading is obtained on the meter, the resonant circuit is retuned to another harmonic frequency and the next reading is taken, and so on. The parallel-resonant circuit consisting of L_1, R_1, and C_1 provides compensation for the variation in the ac resistance of the series-resonant circuit and also for the variations in the amplifier gain over the frequency range of the instrument.

Although numerous modifications of this basic circuit have been developed, tuned-circuit analyzers generally have two major drawbacks: (1) At low frequencies, very large values for L and C are required and their physical size becomes rather impractical. (2) Harmonics of the signal frequency are often very close in frequency so that it becomes extremely difficult to distinguish between them. Some circuit refinements can lessen this problem and the analyzer does find useful application whenever it is important to measure each harmonic component individually rather than to take a single reading for the total harmonic distortion.

11-6.3   Heterodyne Harmonic Analyzer or Wavemeter

The difficulties of the tuned circuit are overcome in the heterodyne analyzer by using a highly selective, fixed-frequency filter. The simplified block diagram of Fig. 11-24 shows the basic functional sections of the heterodyne harmonic analyzer.

The output of a variable-frequency oscillator is mixed (*heterodyned*) successively with each harmonic of the input signal, and either the sum or the difference frequency is made equal to the frequency of the filter. Since now each harmonic frequency is converted to a constant frequency, it is possible to use highly selective filters of the quartz-crystal type. With this technique, only the constant-frequency signal, corresponding to the particular harmonic being measured, is passed and delivered to a metering circuit. The mixer

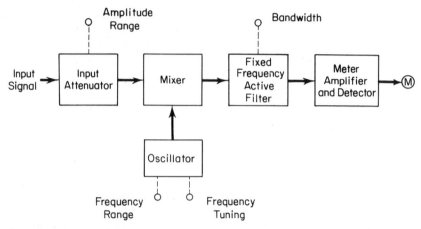

FIGURE 11-24 Block diagram of a heterodyne-type harmonic analyzer or wavemeter.

usually consists of a balanced modulator since it offers a simple means of eliminating the original frequency of the harmonic. The low harmonic distortion generated by the balanced modulator is another advantage over different types of mixers. Excellent selectivity is obtained by using quartz-crystal filters or inverse feedback filters.

On some heterodyne analyzers the meter reading is calibrated directly in terms of voltage; other analyzers compare the harmonics of the impressed signal with a reference voltage, usually by making the reference voltage equal to the amplitude of the fundamental. Direct-reading instruments of the heterodyne type are sometimes known as *frequency-selective voltmeters*. In these instruments the frequency of the input signal is read off a calibrated dial. A low-pass filter in the input circuit excludes the sum of the mixed frequencies and passes only the difference frequency. This voltage is compared to the input signal and read off on a calibrated voltmeter in dBm and volts. The level range for most of these meters is from −90 dBm to +32 dBm.

11-6.4 Fundamental-suppression Harmonic Distortion Analyzer

The fundamental-suppression method of measuring distortion is used when it is important to measure total harmonic distortion (THD) rather than the distortion caused by each component. In this method the input waveform is applied to a network that *suppresses* or *rejects* the fundamental frequency but passes all the harmonic-frequency components for subsequent measurement. This instrument has two major advantages: (1) The harmonic distortion generated within the instrument itself is very small and can be neglected.

(2) The selectivity requirements are not severe because only the fundamental frequency component must be suppressed.

The block diagram of the fundamental suppression HD analyzer is shown in Fig. 11-25. The instrument consists of four major sections: (1) the input circuit with impedance converter, (2) the rejection amplifier, (3) the metering circuit, (4) the power supply. The impedance converter provides a low-noise, high-impedance input circuit, independènt of the signal source impedance placed at the input terminals to the instrument. The rejection amplifier rejects the fundamental frequency of the input signal and passes the remaining frequency components on to the metering circuit where the HD is measured. The metering circuit provides a visual indication of total HD in terms of a percentage of total input voltage.

Two modes of operation are possible: When the function switch is in the "voltmeter" position, the instrument operates as a conventional ac voltmeter, a very convenient feature. In this mode the input signal is applied to the impedance converter circuit through the 1/1 and 100/1 attenuator, which selects the appropriate meter range. The output of the impedance converter then bypasses the rejection amplifier and the signal is applied directly to the metering circuit. The voltmeter section can be used separately for general-purpose voltage and gain measurements.

When the function switch is in the "distortion" postion, the rejection amplifier becomes part of the circuit and distortion measurements are made. In this mode the input signal is applied to a 1-MΩ input attenuator that provides 50-dB attenuation in 10-dB steps, controlled by a front panel switch marked *sensitivity*. When the desired attenuation is selected, the signal is fed to the impedance converter, which is a low-distortion, high input-impedance amplifier circuit whose gain is independent of the source impedance placed at the input terminals. The overall negative feedback in this amplifier results in unity gain and low distortion. Signals having a high source impedance can be measured accurately and the sensitivity selector can be used in the high-impedance positions without distorting the input signal.

The rejection amplifier circuit consists of a preamplifier, a Wien bridge, and a bridge amplifier. The preamplifier receives the signal from the impedance converter and provides additional amplification at extremely low distortion levels. The Wien bridge circuit is used as a rejection filter for the fundamental frequency of the input signal. With the function switch in the "distortion" position, the Wien bridge is connected as an interstage coupling element between the preamplifier and the bridge amplifier. The bridge is tuned to the fundamental frequency of the input signal by setting the *frequency range* selector and is balanced for zero ontput by the coarse and fine *balance* controls. When the bridge is tuned and balanced, the voltage and phase of the fundamental, which appears at the junction of the series reactance and the shunt reactance, are the same as the voltage and phase at

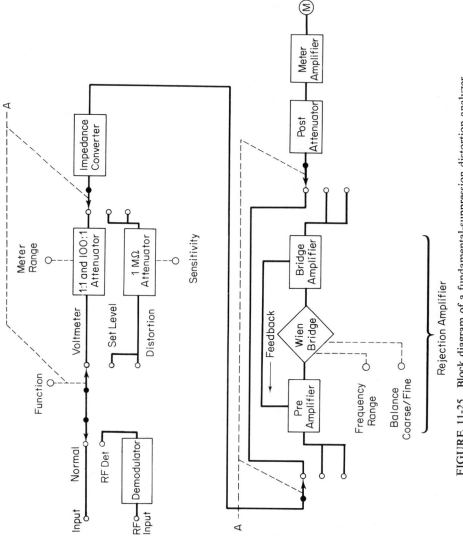

FIGURE 11-25 Block diagram of a fundamental-suppression distortion analyzer (courtesy Hewlett-Packard Co.).

the midpoint of the resistive branch. When these two voltages are equal and in phase, no output signal will appear.

For frequencies other than the fundamental, the Wien bridge offers varying degrees of phase shift and attenuation, and the resulting output voltage is amplified by the bridge amplifier. The output of the bridge amplifier is connected through a post-attenuator to the meter circuit and displayed on the front panel meter. The attenuator limits the signal level to the meter amplifier to 1 mV for full-scale deflection on all ranges. The meter amplifier is a multistage circuit designed for low drift and low noise, and with flat response characteristics. The meter is connected in a bridge-type rectifier and reads the average value of the signal impressed on the circuit. The meter scale is calibrated to the rms value of a sine wave.

The AM detector circuit allows measurement of envelope distortion in AM carriers. The input signal is applied to the demodulator, where the modulating signal is recovered from the RF carrier. The signal is then applied to the impedance converter through the 1-MΩ attenuator and is processed in the same manner as the normal distortion measurements.

The response characteristic of the Wien bridge rejection filter in Fig. 11-25 is modified by negative feedback from the bridge amplifier to the preamplifier. The result of this feedback is shown in the very sharp response curve of Fig. 11-26, causing the rejection of practically all frequency components except the fundamental.

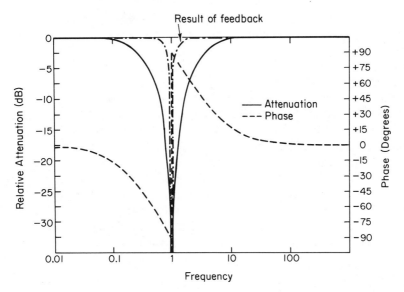

FIGURE 11-26   Rejection characteristic of the Wien bridge, modified by negative feedback.

11-7 SPECTRUM ANALYSIS

11-7.1 Introduction

Spectrum analysis* is defined as the study of the *energy distribution* across the frequency spectrum of a given electrical signal. From this study comes valuable information about bandwidth, effects of various types of modulation, spurious signal generation, and so on—all of it useful in the design and testing of RF and pulse circuitry.

Because of instrumentation capabilities and limitations, spectrum analysis is usually divided into two major categories: (1) audio spectrum analysis and (2) RF spectrum analysis. The RF spectrum analysis, covering the frequency range of 10 MHz to 40 GHz, is the more important, since it includes the vast majority of the communications, industrial instrumentation, navigation, and radar bands.

Originally designed to view the spectrum of a burst of RF energy in radar applications, the spectrum analyzer has developed into a sophisticated instrument capable of graphically presenting amplitude as a function of frequency in a portion of the RF spectrum. The instrument finds application as a tool for measuring attenuation, FM deviation, frequency, and pulse characteristics.

11-7.2 Basic Spectrum Analyzer

The spectrum analyzer is designed to represent, graphically, a plot of amplitude versus frequency of a selected portion of the frequency spectrum under investigation. One of the earliest instruments used during the development of radar pulse techniques involved little more than a simple RF indicator that lacked calibrated controls or broad spectrum coverage but sufficed for the relatively simple task at hand. The modern spectrum analyzer basically consists of a narrow-band *superheterodyne receiver* and a CRO. The receiver is electronically tuned by varying the frequency of the local oscillator. The simplified block diagram of Fig. 11-27 shows the elements of a spectrum analyzer using the swept-frequency design.

A sawtooth generator supplies a sawtooth voltage to the frequency control element of the voltage-tuned local oscillator, which then sweeps through its frequency band at a recurring linear rate. The same sawtooth voltage is simultaneously applied to the horizontal deflection plates of the CRO. The RF signal under investigation is applied to the input of the mixer

---

*Hewlett-Packard, Palo Alto, Calif., *Application Notes* AN 63, AN 63A. General principles and measurement applications are treated in a practical manner.

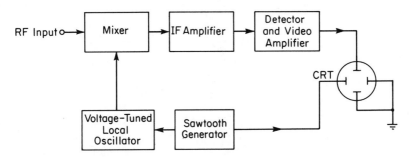

FIGURE 11-27   Elements of a spectrum analyzer of the swept-receiver design.

stage. As the local oscillator is swept through its frequency band by the saw-tooth generator, it will beat with the input signal to produce the required intermediate frequency (IF). An IF component is produced only when the corresponding component is present in the RF input signal. The resulting IF signals are amplified and detected and then applied to the vertical deflection plates of the CRT, producing a display of amplitude versus frequency.

11-7.3   Spectral Displays

To better understand the application of spectrum analysis, it is impor-tant to have a clear understanding of the term *spectral display*.

A CRO is generally used to portray electrical signals with respect to time. For example, pulse risetime, pulsewidth, and repetition rate are read directly on the calibrated X-axis of a CRT, and the observed pattern is a plot of signal amplitude versus time. Such measurements are said to be in the *time domain*. In the spectrum analyzer, signals are broken down into their indivi-dual frequency components and displayed along the X axis of a CRT, which is calibrated in frequency, for a presentation of signal amplitude versus fre-quency. These measurements are then said to be in the *frequency domain*.

Figure 11-28(a) gives a three-dimensional representation of a funda-mental frequency (f_1) and its second harmonic $(2f_1)$ and illustrates the time domain and the frequency domain characteristics. In Fig. 11-28(b) the two signals are portrayed as broken lines; the solid line is the algebraic sum of the instantaneous values of the two signals, as seen on the CRT screen. In Fig. 11-28(c) the two signals are shown in the amplitude-frequency plane and are portrayed on the CRO as two components of the composite signal as the *window* of the spectrum analyzer sweeps across the frequency range of the signal.

It is illustrative to consider the spectra of some common signals and the CRT displays that result when these signals are applied to the spectrum analyzer.

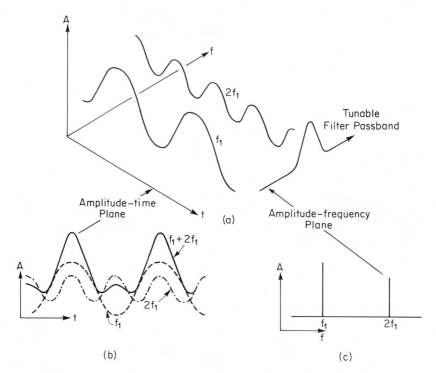

FIGURE 11-28 Three-dimensional presentation of amplitude, frequency, and time. (a) The addition of a fundamental and its second harmonic. (b) View seen in the t–A plane. On an oscilloscope only the composite $f_1 + 2f_1$ would be seen. (c) View seen in the f–A plane. The components of the composite signal are clearly seen (courtesy Hewlett-Packard Co.).

(a) *Continuous-wave (CW) signals* If the analyzer's local oscillator sweeps through a CW input signal slowly, the resulting response on the screen is simply a plot of the IF amplifier passband. A pure CW signal, by definition, has energy only at one frequency and should therefore appear as a single spike on the CRT screen. This will occur provided that the total RF sweep-width, or *spectrum width*, is wide compared to the IF bandwidth in the analyzer.

(b) *Amplitude modulation* When a CW signal of frequency f_c is amplitude modulated by a single tone f_a, sidebands are generated at $f_c + f_a$ and $f_c - f_a$. The analyzer will then display the carrier frequency f_c, flanked by the two sideband frequencies whose amplitudes relative to the carrier depend on the percentage of modulation, as shown in Fig. 11-29.

Provided that the frequency, the spectrum width, and the vertical response of the analyzer are calibrated, the display on the CRT screen provides the following numerical information: (1) carrier frequency, (2) modu-

(a) Time–amplitude plot

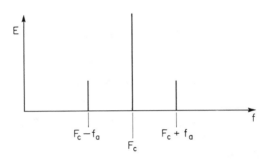

(b) Frequency–amplitude plot

FIGURE 11-29   Single-tone amplitude modulation.

lation frequency, (3) modulation percentage, (4) nonlinear modulation, (5) incidental frequency modulation (as evidenced by jitter of the spectral lines), (6) spurious signal location and strength.

(c) *Frequency modulation*   If a CW signal f_c is frequency modulated at a rate f_r, it will produce an infinite number of sidebands. These are located at intervals of $f_c + nf_r$, where $n = 1, 2, 3, \ldots$. In practice, only the sidebands containing significant power are usually considered. An FM display is shown in Fig. 11-30.

(d) *Pulse modulation*   An ideal rectangular waveform, with zero rise-time and no overshoot or other aberrations, is given in Fig. 11-31(a). This pulse is shown in the time domain, but when its frequency spectrum is

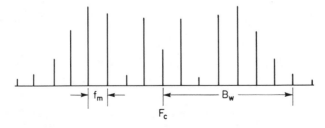

FIGURE 11-30   Amplitude spectrum of single-tone frequency modulation.

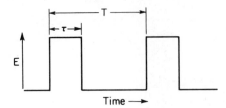

(a) Periodic rectangular pulse train

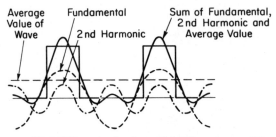

(b) Addition of a fundamental cosine wave and its harmonics to form rectangular pulses

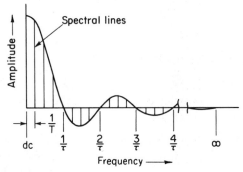

(c) Spectrum of a perfectly rectangular pulse. Amplitudes and phases of an infinite number of harmonics are plotted resulting in smooth envelope.

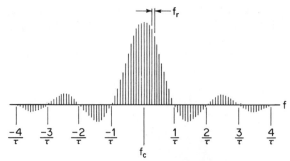

(d) Resultant spectrum of a carrier amplitude modulated with a rectangular pulse.

FIGURE 11-31 Pulse modulation (courtesy Hewlett-Packard Co.).

to be analyzed, it must be broken down into its individual frequency components. This is shown in Fig. 11-31(b), where a constant voltage, a fundamental frequency, and its second harmonic are added algebraically to form a wave that eventually becomes a square wave as more harmonics are added in phase with the fundamental. A spectral plot, in the frequency domain, would have the form given in Fig. 11-31(c), where the amplitudes and phases of an infinite number of harmonics are plotted, resulting in a smooth envelope as shown.

When this pulse is used to amplitude modulate a carrier, the sums and differences of the carrier and *all* the harmonic components contained in the pulse are produced. The harmonic frequencies therefore produce multiple sidebands in exactly the same manner that the modulating signal in amplitude modulation does. These multiple sidebands are generally referred to as *spectral lines* on the analyzer display. There will be twice as many sidebands or spectral lines as there are harmonic frequencies contained in the modulating pulse. Figure 11-31(d) shows the spectral plot resulting from rectangular pulse modulation of a carrier. The individual lines represent the modulation product of the carrier and the modulating pulse frequency with its harmonics. Therefore the lines will be spaced in frequency by an amount equal to the pulse repetition rate of the original pulse waveform. The *main lobe* in the center and the *side lobes* are shown as groups of spectral lines extending above and below the base line. For a perfectly rectangular pulse, the number of side lobes is infinite. The main lobe contains the carrier frequency, represented by the longest line in the center.

The number of application possibilities is as great as the imagination of the user of this instrument. Below is a list of a few representative examples:

(a) Pulsewidth and repetition rate measurements.
(b) Tuning a parametric amplifier.
(c) FM deviation measurement.
(d) RF interference testing.
(e) Antenna pattern measurements.

REFERENCES

1. Doyle, John N., *Pulse Fundamentals.* Englewood Cliffs, N.J.: Prentice-Hall, Inc., 1964.

2. Millman, Jacob, and Herbert Taub, *Pulse, Digital, and Switching Waveforms.* New York: McGraw-Hill Book Company, Inc., 1965.

3. Ryder, John D., *Electronic Fundamentals and Applications.* Englewood Cliffs, N.J.: Prentice-Hall, Inc., 1964.

PROBLEMS

1. Refer to Fig. 11-9 and define the following terms:
   (a) pulsewidth   (b) risetime   (c) overshoot
   (d) ringing   (e) duty cycle   (f) PRR

2. Consider the Wien bridge oscillator of Fig. 11-7 and explain why the frequency of oscillation depends on (a) coupling capacitor C_1; (b) Q_2 base resistor R_7; (c) load resistor R_8; (d) supply voltage V_{CC}.

3. The phase-shift oscillator of Fig. 11-8 uses three RC elements in cascade as the phase-shifting network between output and input. Show that a *two*-element RC network cannot work and explain why a *four*-element RC network is not necessary.

4. Name some of the factors that affect the stability and accuracy of RC oscillators such as the phase-shift or Wien bridge oscillator.

5. What is the major advantage of the RC oscillator over the LC oscillator? Which factors limit the high-frequency oscillation of the RC oscillator?

6. Consider the blocking oscillator of Fig. 11-12 and explain which factor(s) limit the pulse duration. What effect does the ratio of pulsewidth to period have on the transistor?

7. Define the following terms:
   (a) harmonic distortion   (b) total harmonic distortion
   (c) intermodulation distortion

8. The Wien bridge circuit of Fig. 11-7 is slightly altered by connecting inductor L in series with the series RC combination and replacing the parallel RC combination by resistor R_p. Calculate (a) the frequency of oscillation of this circuit; (b) the minimum gain of the two-stage amplifier for a finite value of R_p.

9. Verify Eq. (11-17) for the feedback factor of the phase-shift oscillator of Fig. 11-8 by applying conventional network analysis to the RC feedback circuit. Prove that the phase shift is 180° for $\alpha = 6$ and that at this frequency $\beta = \frac{1}{29}$.

10. Design a phase-shift oscillator to operate at a frequency of 10 kHz. Select a suitable transistor and find the minimum value for the load resistance (R_L in Fig. 11-8) for which the circuit will oscillate. Calculate the RC product required for 10-kHz oscillation and calculate the value for C after choosing a reasonable value of R.

11. For the astable multivibrator of Fig. 11-10 the following component values are specified:
$$R_1 = R_2 = 50 \, k\Omega$$
$$C_1 = C_2 = 0.02 \, \mu F$$
$$R_3 = R_4 = 1 \, k\Omega$$
$$V_{BB} = -10 \, V$$
$$V_{CC} = 10 \, V$$

Calculate (a) frequency of oscillation; (b) amplitude of the output pulse at the collector of Q_2; (c) duty cycle of the output waveform.

12. It is desired to modify the multivibrator of Prob. 11 so that the duty cycle is reduced to 20 per cent while maintaining the original frequency. Indicate which circuit components must be changed and calculate the values of these components.

13. Using a single set of components, draw a circuit arrangement whereby the frequency of an astable multivibrator can be varied over an appreciable range.

14. Draw a circuit arrangement illustrating how an astable multivibrator may be used to display two traces on the screen of a single-beam CRO.

12

Electronic Counters
and Their Application

12-1 ELEMENTS OF THE ELECTRONIC COUNTER

An electronic counter is an instrument designed to measure an unknown frequency or time interval by *comparing* it to a known frequency or time interval. The instrument's logic circuitry is designed to present the result of this comparison measurement in an easy-to-read numerical display. The accuracy of the measurement depends primarily on the stability of the known frequency that is derived from the counter's internal oscillator.

The electronic counter consists of several sections, which can be interconnected in various ways to make different types of measurement. The most important subsystems of the counter are the following:

(a) *Decade counting assembly,* usually with an integrated numerical display, to totalize and display the count
(b) *Signal gate,* to control the duration of the actual count
(c) *Time base,* to supply precise increments of time for frequency or time-interval measurements

Other sections of a general-purpose counter include circuitry for signal shaping and logic control, and binary-coded-decimal output circuitry.

The first part of this chapter deals with the basic building blocks of the electronic counter; the second part is concerned with applications in typical measurement situations.

368

12-2   DECADE COUNTING ASSEMBLY

12-2.1   Bistable Multivibrator

A decade counting assembly is a circuit that produces one single output pulse for every ten input pulses applied to the circuit. Since the base (*radix*) of our numbering system is *ten*, the decade counter is by far the most widely used of all counting assemblies. A counting assembly uses bistable multivibrators or flip-flops. A flip-flop is a circuit with two stable states, capable of remaining in either state indefinitely until triggered by an external signal. This indicates that the flip-flop has a *memory;* it will remember, or store, the last signal it received until a new signal is applied to the circuit.

A well known form of the bistable multivibrator is the emitter-coupled circuit of Fig. 12-1. This circuit forms a symmetrical arrangement of two transistors, with their associated resistive and capacitive components. To explain the operation of the circuit, the assumption is made that initially transistor Q_1 is conducting heavily. (Notice that the circuit has two *pnp*

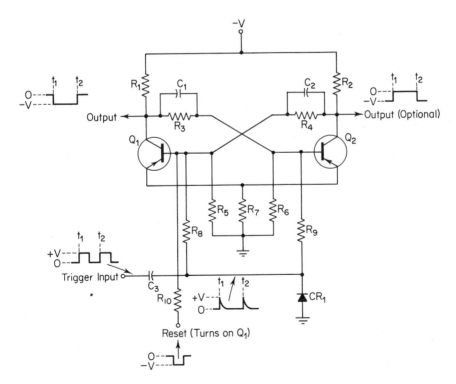

FIGURE 12-1   Emitter-coupled bistable multivibrator or flip-flop. Relevant waveforms are indicated.

transistors using a negative collector supply voltage.) When Q_1 is conducting, its collector voltage is only very slightly negative and is assumed zero for the sake of clarity. The base of Q_2, which is connected to the junction of R_3 and R_6, is then also zero volts. The relatively heavy current through Q_1 causes a voltage drop across R_7 which produces sufficient negative voltage at the emitter of Q_2 to hold it at cutoff. With Q_2 cut off, voltage divider R_2-R_4-R_5 delivers a negative voltage to the base of Q_1 and keeps it conducting. The flip-flop is therefore in a *stable state*, with Q_1 conducting and Q_2 cut off.

This stable state can be reversed by applying a positive voltage to the "trigger input" terminal. Resistors R_8 and R_9 connect the bases of both transistors to this input terminal through capacitor C_3. Now consider the waveforms shown in the diagram of Fig. 12-1. At time $t = t_1$, a positive input voltage is applied to the trigger input terminal, cutting Q_1 off. The Q_1 collector potential goes negative toward the supply voltage. This drives Q_2 into conduction since the base of Q_2 is connected to the R_3-R_6 junction. When Q_2 conducts, its collector voltage becomes almost zero and so does the base of Q_1, allowing Q_1 to remain cut off. The R_1-R_3-R_6 divider delivers a sufficiently large negative voltage to the base of Q_2 to keep it in heavy conduction. The application of the positive trigger pulse has therefore caused the flip-flop to go to its other stable state, with Q_2 conducting and Q_1 cut off. The flip-flop remains in this state until another positive trigger pulse comes along, causing a new transition of states. This next positive pulse, occurring at time $t = t_2$, cuts Q_2 off and starts a sequence of events that ends with Q_1 conducting and Q_2 cut off.

Notice that a positive trigger pulse has no effect on the transistor that is already cut off; it affects only the conducting transistor. The waveforms that accompany the diagram of Fig. 12-1 indicate that *one* output pulse is present at the collector of either Q_1 or Q_2 for every *two* input pulses triggering the flip-flop. This means that *the circuit divides by two*.

Diode CR_1 in the circuit of Fig. 12-1 removes the negative pulse from the differentiated input waveform. Without this diode, the negative pulse would drive the cut-off transistor *on* and the stage would switch from one state to the other, but it would not divide by two. Coupling capacitors C_1 and C_2 assists the circuit in making fast transitions between states; they are called speed-up or *commutating* capacitors. The dc coupling through R_3 and R_4 insures the bistable characteristics of the circuit.

A flip-flop that produces one output pulse after receiving two input pulses is called a *binary* circuit because it is counting in the binary system, by twos. Another feature of the circuit is the "reset" terminal. A negative pulse applied to this terminal reaches the base of Q_1 and forces it to conduct. The flip-flop then returns or resets to its initial condition with Q_1 conducting and Q_2 cut off.

12-2.2 Binary Counter

Consider Fig. 12-2 in which four binaries, represented by logic symbols, are connected in cascade. For convenience and ease of reference, assume that each flip-flop in this cascaded arrangement is identical to the circuit of Fig. 12-1. In the logic symbols, point T represents the input or trigger terminal. The

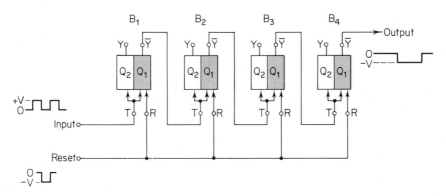

FIGURE 12-2 Four flip-flops in cascade, forming a binary counter. The zero state of each binary is represented by the shaded half of the logic symbol, where transistor Q_1 is conducting.

bases of both transistors are connected to this point so that each successive positive trigger pulse will reverse the state of the binary. The points marked Y and $\bar{Y}$ (read as Y-bar) represent the collectors of the transistors. Comparing the logic symbol with the actual circuit diagram, one sees that the collector of Q_1 has been designated the $\bar{Y}$ output terminal and that the collector of Q_2 has been designated the Y output terminal. The $\bar{Y}$ output terminal of each binary in the chain is connected to the trigger input of the next binary. The positive-going transition at the collector of Q_1 (while switching states from Q_2-conducting to Q_1-conducting) provides the positive input pulse to trigger the next binary. We now arbitrarily say that a binary is in "state 0" when transistor Q_1 conducts and in "state 1" when Q_1 does not conduct. When a binary is in state 1, its output voltage (at the Y terminal) is highly negative; when the binary is in state 0, its output voltage is approximately zero volts. A transition from the 1-state to the 0-state therefore produces a voltage step in the positive-going direction; a transition from the 0-state to the 1-state produces a negative-going output pulse.

In order to start our discussion, a reset pulse, consisting of a negative voltage, is applied to the bases of all the Q_1 transistors, forcing them to conduct. By our earlier convention, all the binaries are now in state 0. Starting from this reference position we observe the waveforms, shown in Fig. 12-3,

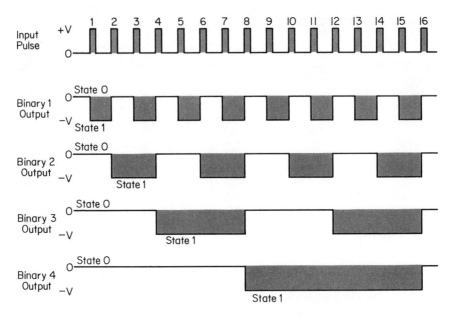

FIGURE 12-3 Waveform chart of a four-binary counter.

which appear at the output of each binary as the result of 16 successive positive trigger pulses applied to the input of the first binary. The first input pulse causes a transition of binary B_1 from state 0 to state 1. As a result of this transition, a negative-going voltage appears at the output of B_1. This negative pulse does *not* cause a transition in the next stage because of the action of diode CR_1.

The second input pulse causes the first binary B_1 to return to its original 0-state and a positive-going voltage appears at the $\bar{Y}$ output terminal. This output pulse triggers the following binary (B_2) and it makes a transition from state 0 to state 1. Its output voltage is again a negative-going pulse that does not affect the next binary (B_3). After two input pulses have been applied, flip-flop B_1 is in state 0 and flip-flop B_2 is in state 1.

The third input pulse causes B_1 to go to its 1-state, causing a negative pulse at its output. A negative pulse into binary B_2 does not affect this flip-flop and it remains in its 1-state. This process can be carried through all of its 16 step and the waveform chart of Fig. 12-3 will be found to be correct.

The operation of the cascaded flip-flops of Fig. 12-2 can be summarized as follows:

(a) Flip-flop B_1 makes a transition at each applied trigger pulse.
(b) Each of the other flip-flops makes a transition only when the preceding flip-flop switches from state 1 to state 0.

Table 12-1

STATES OF THE BINARIES

Number of input pulses	State of binary			
	B_4	B_3	B_2	B_1
0	0	0	0	0
1	0	0	0	1
2	0	0	1	0
3	0	0	1	1
4	0	1	0	0
5	0	1	0	1
6	0	1	1	0
7	0	1	1	1
8	1	0	0	0
9	1	0	0	1
10	1	0	1	0
11	1	0	1	1
12	1	1	0	0
13	1	1	0	1
14	1	1	1	0
15	1	1	1	1
16	0	0	0	0

Table 12-1 lists the states of the four binaries in the chain as a function of the number of input pulses. This table corresponds exactly to the waveform chart of Fig. 12-3. Notice that the order in which the binaries appear in the table is *opposite* to the order in which they are drawn in Fig. 12-2. The ordered array of states 0 and 1 in any row of Table 12-1 is the binary representation of the number of input pulses. We can therefore say that the chain of four flip-flops counts in the binary system of numbers. The chain will reset itself to the 0-state with the application of the sixteenth pulse, and the counting cycle starts again with the seventeenth input trigger. A chain of 4 binaries will count to the number $2^4 = 16$ before it resets. It is clear that the counting range may be extended by adding more binaries: a chain of 5 flip-flops counts to $2^5 = 32$, and in the general case, n binaries count to 2^n.

12-2.3   Decimal Counter

The decimal system, with which we are more familiar, counts to the base of ten. To construct a counter that provides *one* output pulse for every *ten* input triggers we start with a cascaded arrangement of four flip-flops. Feedback is introduced from the later stages to the earlier stages so that the count is *advanced* by six at some time during the first ten counts. This advance of six counts may be made in one or in several steps. In fact, many different circuit arrangements are possible.

The circuit of Fig. 12-4 shows a *scale-of-16* counter (four flip-flops in cascade) which is modified by feedback into a *scale-of-10* counter. This circuit is very common in commercial decade counters and deserves some attention at this point. The feedback connections are shown in the logic diagram of Fig. 12-4(a) by the heavy lines, and the waveform chart is given in Fig. 12-4(b). The waveform chart shows that the count proceeds normally for the first three

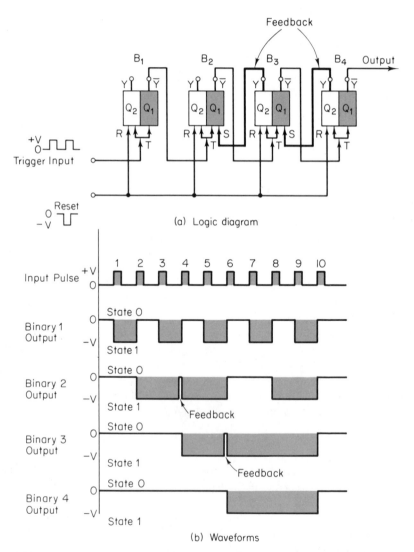

FIGURE 12-4   Four binaries in cascade, modified by feedback into a scale-of-10 counter.

trigger pulses, after which flip-flops B_1 and B_2 are in the 1-state. When the *fourth* input pulse arrives, a series of events takes place: Binary B_1 and binary B_2 go from the 1-state to the 0-state and binary B_3 goes from the 0-state to the 1-state. At this point, the Y-output terminal of binary B_3 gives a positive transition which is coupled to the "set" terminal of binary B_2. This connection is indicated by the heavy line in Fig. 12-4. The set terminal of the flip-flop is connected to the base of the Q_1 transistor and the positive pulse provided by the feedback line from B_3 forces transistor Q_1 of binary B_2 to cutoff. When Q_1 is cut off, Q_2 conducts, and the binary is therefore in the 1-state. In other words, feedback from the Y terminal of B_3 to the S terminal of B_2 causes a transition of B_2 from state 0 to state 1. At this point, the binary chain reads (counting from B_4 to B_1) $0110 = 6$, so that the counter has been advanced by two counts through feedback. The fifth input trigger puts B_1 back to state 1, but it provides no further output pulses for the following binaries that stay in their respective states.

The sixth trigger pulse starts a new sequence of events. First, binary B_1 returns to the 0-state. This transition generates an output pulse that returns binary B_2 to the 0-state. This in turn generates an output pulse that moves binary B_3 to the 0-state. The feedback loop from B_3 to B_2 receives a negative pulse at this instant, which is of no consequence. The transition of binary B_3 to the 0-state produces an output pulse that moves the last binary B_4 to the 1-state. Here again a feedback loop is connected to the Y terminal of B_4, indicated by a heavy line in the diagram. The Y terminal of B_4 is connected to the "set" terminal of binary B_3. The transition of B_4 produces a positive pulse at the Y output which immediately returns binary B_3 to the 1-state. After receiving the sixth input pulse, the situation is then as follows: Binaries B_1 and B_2 are in the 0-state, and binaries B_3 and B_4 are in the 1-state. Again reading from B_4 to B_1, we find that this corresponds to the binary code $1100 = 12$ decimal. The count of the binary chain has therefore been advanced another four steps for a total advance of six counts. Inspection of the waveform chart in Fig. 12-4(b) indicates that the next four pulses simply bring the count up to a total of ten, at which time the binaries all reset to their initial 0-state and the counting process can start all over again.

The net result of this exercise has been that one output pulse has been generated at the $\bar{Y}$ terminal of the last binary after the application of ten input pulses. In other words, the assembly of four binaries now behaves as a decimal, or *decade*, counter. The output pulse of this circuit may serve as a "carry" pulse to a following decade counter assembly (DCA).

12-2.4   Decade Counter with Digital Display

The DCA usually incorporates a digital display system to indicate the state of each individual binary in the chain. A simple indicator that may be used for this purpose is a neon bulb in series with a resistor. This indicator is then connected in the collector circuit of the Y transistor of each binary. When a binary is in the 1-state, the Y transistor will be conducting heavily and the neon lamp lights up. The neon lamps connected to binaries B_1, B_2, B_3, and B_4 are assigned values of 1, 2, 4, and 8, respectively. To determine the count of the DCA, it is necessary only to add the numbers assigned to those neon lamps that are lit.

A much more elegant arrangement involves an electrical readout of the DCA plus a *digital display*. The electrical readout in this case consists of a four-line binary-coded-decimal (BCD) output voltage, where a voltage representing the state of each binary in the DCA is taken from the collector of each of the Y transistors. A binary 1 is then represented by a relatively positive voltage on each line, and a binary 0 is represented by a relatively negative voltage on each line. The ten allowable combinations are summarized in Table 12-2, representing the digits 0 through 9. This binary-coded electrical readout is available for use elsewhere; for example, it may be used for recording on magnetic tape. In most cases, the electrical readout is used to produce a digital display.

Table 12-2

FOUR-LINE CODE TRUTH TABLE

Digit	Four-line code			
	B_4	B_3	B_2	B_1
0	0	0	0	0
1	0	0	0	1
2	0	0	1	0
3	0	0	1	1
4	0	1	1	0
5	0	1	1	1
6	1	1	0	0
7	1	1	0	1
8	1	1	1	0
9	1	1	1	1

A typical application is given in the simplified circuit diagram of Fig. 12-5, which shows the display matrix and the decimal-counter schematics in their basic form. The *display matrix*, consisting of 8 neon lamps and 18 photoconductive elements, is used to convert the binary-coded representation to a digital representation. The schematic diagram indicates that the connection to each numeral in the ten-digit electron tube indicator is made through 3 series-connected photocell elements. A characteristic of the photocell element is that it has a high resistance (several $M\Omega$) when dark and a relatively low

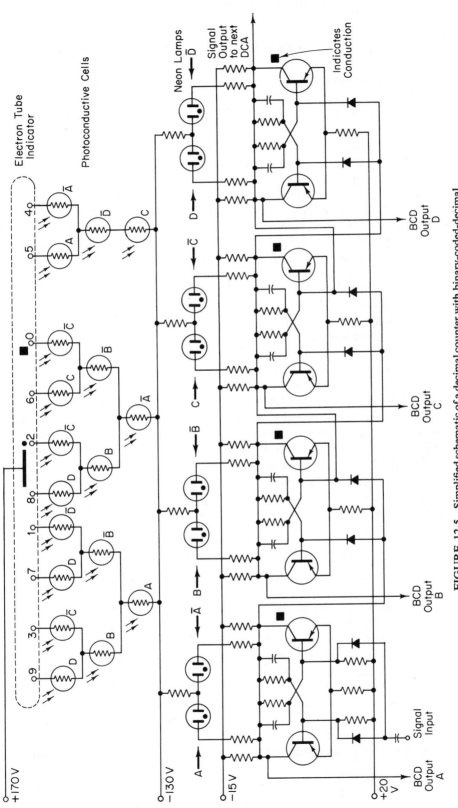

FIGURE 12-5 Simplified schematic of a decimal counter with binary-coded-decimal outputs and electron tube display of the digits (courtesy Hewlett-Packard Co.).

resistance (less than 7,000 Ω) when illuminated. Thus when the 3 photocell elements that constitute a circuit path are illuminated, the resistance in the display path drops to about 20,000 Ω and sufficient current is available to light the display digit.

Illuminating elements for the photocells are neon lamps connected in the collector circuit of each of the eight transistors in the counting circuit. The lamp lights when the transistor conducts. As explained earlier in this section, a four-binary counting circuit has ten states, ten combinations of conducting and nonconducting transistors, each combination corresponding to one digit. There is therefore a pattern of lighted neon lamps for each digit. Assigning a binary weight of 1 when the plain-letter lamp (A, B, C, or D) lights and a weight of 0 when the bar-letter lamp ($\bar{A}$, $\bar{B}$, $\bar{C}$, or $\bar{D}$) lights, the lamp pattern for any digit can be determined from Table 12-2. Figure 12-5 shows the counting circuit with transistors $\bar{A}$, $\bar{B}$, $\bar{C}$, and $\bar{D}$ conducting. The lamps associated with these circuits illuminate the photocell elements in the circuit of the digit 0 display.

It is clear that the decade counting assembly serves a *dual* purpose. It *counts* and *displays* from 0 to 9 input pulses. On the arrival of the tenth input pulse, it resets and displays a zero, but it also produces an output pulse. It therefore *divides* by a factor of ten and may then be called a *decade divider*.

Both functions of the DCA are used in many instrument applications. Consider, for example, the case where several DCAs, each with its own display unit, are connected in cascade. When trigger pulses are applied to the input of the first DCA, its display unit will portray the count of the units from 0 to 9. An output pulse occurs after every tenth input pulse and fires the second DCA, which therefore displays the count of the tens. The third DCA is triggered by every tenth output pulse from the second DCA and portrays the count of the hundreds. The total display is limited only by the number of DCAs available.

12-3 TIME BASE AND ASSOCIATED CIRCUITRY

The time base of an electronic counter invariably consists of a crystal oscillator and a decade divider assembly that reduces the output frequency of the oscillator in steps of ten. The crystal that controls the frequency of the oscillator is usually kept in a temperature-controlled environment (oven) to increase the frequency stability of the instrument. A functional block diagram of a time-base generator for a general-purpose electronic counter is shown in Fig. 12-6.

The time-base generator of Fig. 12-6 contains three major blocks. The *crystal oven* assembly consists of a thermally insulated chamber which contains the piezoelectric crystal whose frequency is to be accurately main-

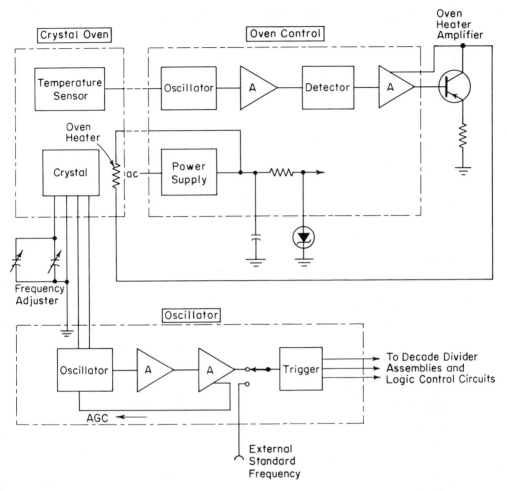

FIGURE 12-6 Functional block diagram showing the crystal oven, oven-control circuit, and oscillator of a general-purpose electronic counter.

tained, an oven heating element, and a thermistor as the temperature-sensing element. This thermistor is connected in a Wheatstone bridge circuit. Temperature variations inside the crystal oven cause small resistance variations in the thermistor and upset the balance of the bridge. The bridge output is connected to the oscillator in the *oven control* assembly and controls the amplitude of its output. The oscillator output is amplified and detected to produce a dc voltage whose magnitude is inversely proportional to the oven temperature. This dc voltage is amplified and applied to the base of the oven heater amplifier whose collector current controls the current through the oven heater and hence the oven temperature.

The *oscillator* assembly includes the oscillator circuit whose frequency is controlled by the crystal in the oven. The oscillator output is amplified and applied to the decade divider assemblies (DDAs) for reduction by factors of ten. The decade dividers operate identically to the decade counter of Sec. 12-2.4, except that the count is not displayed.

12-4   LOGIC CIRCUITS

Logic circuits perform the various switching functions that connect the building blocks of the electronic counter into the configuration needed for the desired counting function (see also Sec. 12-5). The three logic circuits most often used for these switching functions are the OR, AND, and NOT circuits. This section discusses the basic operation of these logic circuits.

12-4.1   OR Gate

The OR gate is a multiple input circuit with a single output, as indicated by the logic symbol in Fig. 12-7(a). The circuit obeys the following definition:

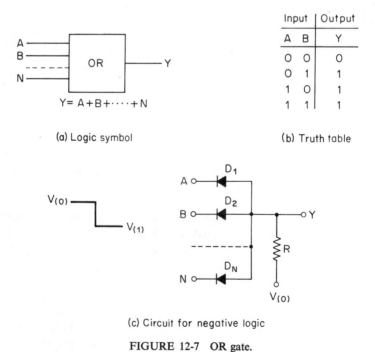

$$Y = A + B + \cdots + N$$

(a) Logic symbol

Input		Output
A	B	Y
0	0	0
0	1	1
1	0	1
1	1	1

(b) Truth table

(c) Circuit for negative logic

FIGURE 12-7   OR gate.

The output of an OR *gate assumes the 1-state if one or more inputs assume the 1-state.* The n inputs to a logic circuit are usually designated by A, B, . . . N and the output by Y. The operation of an OR gate can be described by the Boolean expression $Y = A + B + \ldots + N$, which is to be read as "Y equals A *or* B *or* C . . . *or* N."

It is usual practice to write a truth table for the logic circuit, tabulating all possible input conditions and the corresponding output conditions, as in Fig. 12-7(b). The actual circuitry often contains semiconductor diodes in an arrangement as shown in Fig. 12-7(c). The circuit shown here represents *negative* logic, where the voltage for the 1-state is negative with respect to the voltage for the 0-state. If all the inputs are in the 0-state, the voltage across each diode is equal to $V_{(0)} - V_{(0)}$. In order for a diode to conduct, it must be forward-biased, and in this case none of the diodes will then conduct. Therefore the output voltage equals $V_{(0)}$, and Y is in the 0-state. If input A is changed to the 1-state, diode D_1 will conduct because it is now forward-biased and the output voltage becomes

$$V_{(0)} - [V_{(0)} - V_{(1)}] \approx V_{(1)} \quad \text{(by approximation)}$$

Y is therefore in the 1-state, and each diode, except D_1, is reverse-biased. If two or more inputs are in the 1-state, then the diodes connected to these inputs conduct and all other diodes remain reverse-biased. If for some reason the voltage level $V_{(1)}$ is not identical for all inputs, then the most negative value of $V_{(1)}$ will appear at the output, and all diodes except that one are nonconducting.

12-4.2 AND Gate

The AND gate has two or more inputs and a single output. It operates to satisfy the following definition: *The output of an* AND *gate assumes the 1-state only if all the inputs assume the 1-state.* The IEEE logic symbol for the AND gate is shown in Fig. 12-8(a), together with the Boolean expression $Y = AB \ldots N$, which is to be read as "Y equals A and B . . . and N." The two-input truth table of Fig. 12-8(b) agrees with the Boolean expression.

The diode AND circuit for negative logic is given in Fig. 12-8(c). To understand the operation of the AND circuit, assume that the diodes are ideal (zero forward resistance and infinite reverse resistance) and that the internal resistances of the input voltage sources are zero. If any input to the AND gate is at the 0-level, $V_{(0)}$, the diode connected to that input is *forward*-biased and *conducts*, clamping the output voltage at $V_{(0)}$. Therefore Y is at 0. If all the inputs are in the 1-state at the same time, all the diodes are *reverse*-biased and the output voltage equals $V_{(1)}$, or Y is at 1. The AND gate is often called a *coincidence* gate.

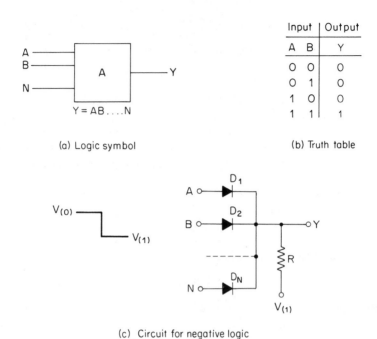

(a) Logic symbol

Y = AB....N

Input		Output
A	B	Y
0	0	0
0	1	0
1	0	0
1	1	1

(b) Truth table

(c) Circuit for negative logic

FIGURE 12-8   AND gate.

12-4.3 NOT Circuit

The NOT or INVERTER circuit has a single input and a single output. It performs the logic function of negation, according to the following definition: *The output of a* NOT *circuit assumes the 1-state only if the input does* NOT *assume the 1-state.* The IEEE logic symbol for the NOT circuit is a little circle drawn at the point where the signal line joins the logic symbol, either at the input or at the output side. This is shown in Fig. 12-9(a), where the Boolean expression $Y = \bar{A}$ is to be read as "Y equals NOT A." The truth table of Fig. 12-9(b) agrees with the Boolean expression.

A transistor circuit for a positive-logic NOT circuit is shown in Fig. 12-9(c). If the input is low, $v_i = V_{(0)}$, the circuit parameters are so chosen that the transistor does not conduct so that its collector voltage is high, $v_o = V_{CC} = V_{(1)}$. If the input is high, $v_i = V_{(1)}$. The circuit parameters are selected so that the transistor saturates and its collector voltage is low, $v_o = V_{EE} = V_{(0)}$. The circuit therefore performs a logic negation in the sense that the output is inverted with respect to the input.

Capacitor C across input resistor R_1 serves to improve the transient response of the circuit.

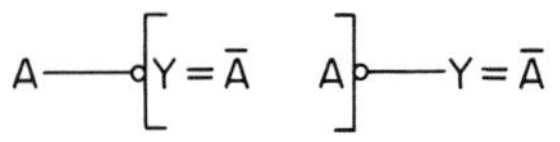

Input	Output
A	Y
0	1
1	0

$$A \longrightarrow \boxed{\,} Y = \bar{A} \qquad A \,\rhd \!\!-\!\!\!- Y = \bar{A}$$

(a) Logic symbol   (b) Truth table

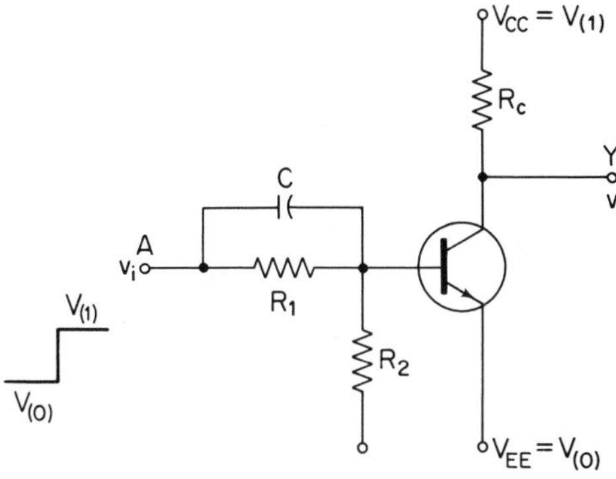

(c) Circuit for negative logic

FIGURE 12-9   NOT circuit.

12-4.4 INHIBIT Circuit

The INHIBIT circuit is a modification of the AND gate, where one of the AND gate terminals is preceded by an *inverter* or NOT circuit, as shown in the circuit diagram of Fig. 12-10(c). This modified AND circuit operates to satisfy the following definition: *The output of an* INHIBIT *gate assumes the 1-state if all the inputs, except the negation input, assume the 1-state.* The IEEE logic symbol and the truth table are given in Fig. 12-10. The truth table is valid for a three-input AND gate with one inhibit terminal (C). The Boolean equation $Y = AB\bar{C}$ is to be read as "Y *equals* A *and* B *and NOT* C," which agrees with the truth table.

The circuit operates as follows: If either of the inputs A or B, or both, are in the 0-state where $V_{(0)} = 0$ V, then at least one of the diodes conducts and the output is clamped to 0 V. If a *coincidence* occurs, so that *both* inputs A and B are in the 1-state [$V_{(1)} = +12$ V], then both diodes D_1 and D_2 are reverse-biased and nonconducting. If at the same time input C is also at the

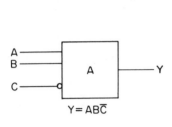

Input			Output
A	B	C	Y
0	0	0	0
0	1	0	0
1	0	0	0
1	1	0	1
0	0	1	0
0	1	1	0
1	0	1	0
1	1	1	0

(a) Logic symbol

(b) Truth table

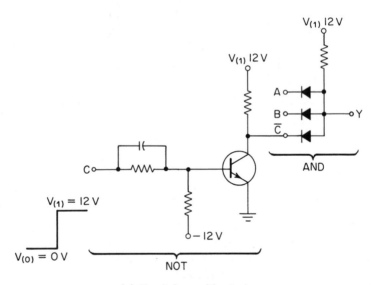

(c) Circuit for positive logic

FIGURE 12-10   INHIBIT circuit.

1-level of $+12$ V, transistor Q in the inverter circuit will be conducting, and the negation terminal $\bar{C}$ will be at the 0-level of 0 V. This forward-biases diode D_3 which clamps the output at 0 V and output terminal Y is in the 0-state. On the other hand, if input terminal C is in the 0-state, $\bar{C}$ is at $+12$ V and diode D_3 is back-biased. The output at the Y terminal will then be at $+12$ V and Y is in the 1-state. The truth table in Fig. 12-10 verifies the anticoincidence conditions of the INHIBIT circuit. An INHIBIT gate with only two input terminals normally passes the input signal. Adding a second input signal of the same logic level prevents the signal from passing through the gate.

12-5   UNIVERSAL COUNTER

The basic building blocks described in the previous sections are connected together to form the universal counter-timer, an example of which is shown in Fig. 12-11. The block diagram of this counter is given in Fig. 12-12.

FIGURE 12-11   Universal counter-timer (courtesy Computer Measurements Co.).

The major circuit blocks can be identified as (a) the time-base generator with decade dividers, (b) the decade counting assembly with digital readout, (c) the input section with amplifiers and attenuators, and (d) the function switch with its associated logic control circuitry.

The function switch is a front panel control that connects the various elements of the counter together in such a way that different kinds of measurement can be performed. These measurement modes usually include period, frequency, time interval, and ratio, as indicated in the block diagram of Fig. 12-12. They are described in detail in Sec. 12-6.

Two important circuit elements in the logic control circuitry are the *main*, or *signal*, *gate* and the *gate control* flip-flop. Their functions are always the same, regardless of the measurement mode selected. The main gate acts as a controlled switch that connects the signal to be counted to the decade counting assembly, while the gate control flip-flop enables and disables the main gate. With the function switch in the "frequency" mode, as shown in the diagram, a control voltage is applied to specific gates in the logic control circuitry. The gates in this circuit allow the input signal to be connected to the "counted-signal" channel of the main gate. The selected output from the time-base dividers is simultaneously gated to the control flip-flop, which enables and disables the main gate. Both control paths are *latched internally* to allow them to operate only in the proper sequence.

With the function switch in the "period" mode, the control voltage is connected to certain gates in the logic control circuitry, which connect the

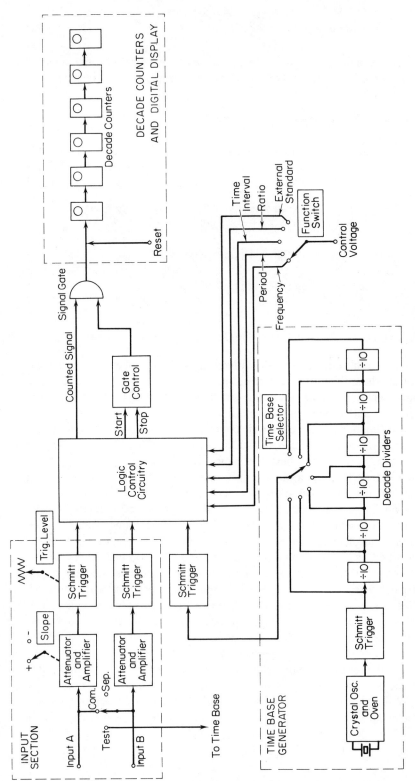

FIGURE 12-12 Simplified block diagram of a universal counter.

time-base signal to the counted-signal channel of the main gate and also the signal input to the gate control for enabling and disabling the main gate. The remaining function switch positions perform similar control functions in the logic control circuitry.

12-6   MEASUREMENT MODES

12-6.1   Frequency Measurement

Frequency can be measured by counting the number of cycles of the unknown signal over a precisely controlled time interval. Figure 12-13 shows the block diagram for a counter in the "frequency" mode of operation.

There are two signals that need to be traced: the *input* signal and the *gating* signal. Both signals are applied to the so-called *signal gate*, or *main gate*, which could be an ordinary two-input AND gate. The input signal, whose frequency is to be measured, is first applied to an amplifier and then to a Schmitt trigger circuit. There it is converted into a square wave whose amplitude is independent of the amplitude of the input waveform. This square-wave signal is then differentiated, so that the signal which arrives at the main gate consists of a series of sharp pulses separated by the period of the original input signal.

The gating signal is derived from the crystal oscillator. In the block diagram of Fig. 12-13, the oscillator, or time-base, frequency is 1 MHz. The time-base output is shaped by a Schmitt trigger circuit, so that positive spikes, 1 μs apart, are applied to the decade dividers. In the example shown, six DDAs are used whose outputs are connected to a time-base selector. This front panel switch allows time intervals from 1 μs to 1 s to be selected. The output from the time-base selector passes through a Schmitt trigger and is applied to the gate control flip-flop. When the first time-base pulse arrives, the gate control flip-flop assumes a state such that an *enable* signal is applied to one terminal of the main gate. Since this is an AND gate, the input signal pulses at the other terminal are allowed to enter the DCAs where they are totalized and displayed. This continues until the second pulse from the DDAs arrives at the gate control flip-flop. The gate control then assumes the other state which removes the enable signal from the main gate. The main gate therefore closes and no further pulses are admitted to the DCAs. The DCA display is now in a state that corresponds to the number of input pulses received during the time interval determined by the time base.

Since frequency can be defined as the number of occurrences of a particular phenomenon in some defined length of time, the counter display corresponds to *frequency*. Usually the time-base selector switch moves the decimal point in the display area, so that the frequency can be read directly in hertz, kilohertz, or megahertz.

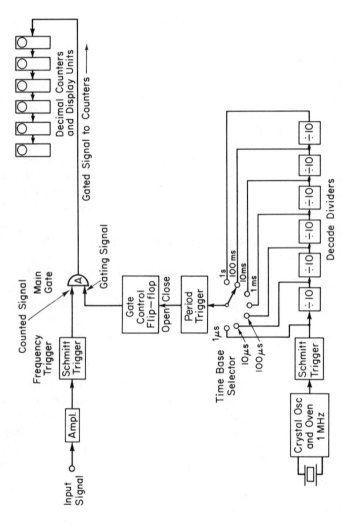

FIGURE 12-13 Block diagram of the electronic counter in the "frequency" mode of operation.

12-6.2 Period Measurement

In some applications it may be desirable to measure the *period* of a signal instead of its frequency. This can be done by rearranging the building blocks used for the frequency measurement so that the counted signal and the gating signal are interchanged. Figure 12-14 shows the block diagram of the counter in the "period" measurement mode. The gating signal is derived from the unknown input signal that now controls the opening and closing of the main gate. The precisely spaced pulses from the crystal oscillator are counted for one period of the unknown frequency. In the example shown in Fig. 12-14 the time base is set to 10 μs (100 kHz time-base frequency), and the number of 100 kHz pulses that occur during one period of the unknown signal is counted and displayed by the DCAs.

The accuracy of the period measurement may be increased greatly by using the "multiple-period average" mode of operation. This type of measurement is similar to the single-period measurement in that the gating signal is derived from the unknown input signal and the counted signal from the time-base oscillator. The basic difference is that the main gate is held open for more than one period of the unknown signal. This is accomplished by passing the unknown signal through one or more DDAs so that the period is extended by a factor of 10, 100, or more.

Figure 12-14 shows the multiple-period average mode of operation as a modification of the single-period measurement by the broken lines in the block diagram. The 1-MHz crystal frequency is divided by 1 DDA to a frequency of 100 kHz (10 μs period). These clock pulses are shaped by the frequency trigger and fed to the main gate to be counted. The input signal whose period is to be measured is amplified, shaped by the period trigger, and fed to 5 DDAs in cascade, counting the input frequency down by a factor of 10^5. This divided signal is now shaped by the multiple-period trigger (another Schmitt trigger circuit) and applied to the gate control flip-flop. The gate control provides the enable and stop pulses for the main gate. Obviously, the main gate will remain open for a greatly increased time interval, in fact an increase of a factor of 10^5. In this case, the DCAs count the number of 10-μs intervals that occur during 100,000 periods of the input signal. The readout logic is so designed that the decimal point will be automatically positioned to display the proper units.

12-6.3 Ratio and Multiple-ratio Measurements

A ratio measurement is, in effect, a period measurement with the *lower* of the two frequencies used as the *gating* signal and the *higher*-frequency signal as the *counted* signal. In other words, the lower-frequency signal takes the place of the time base. The block diagram of Fig. 12-15 applies. The

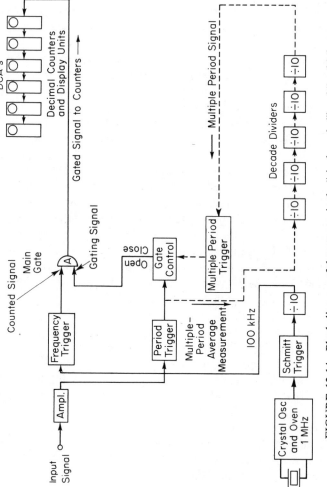

FIGURE 12-14 Block diagram of the counter in the "single period" and "multiple-period average" modes of operation.

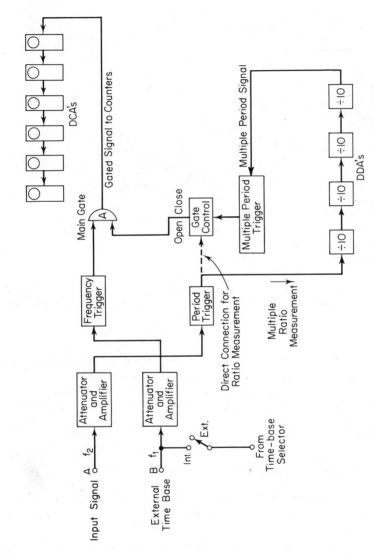

FIGURE 12-15 Block diagram of the counter in the "ratio" and "multiple-ratio" mode of operation.

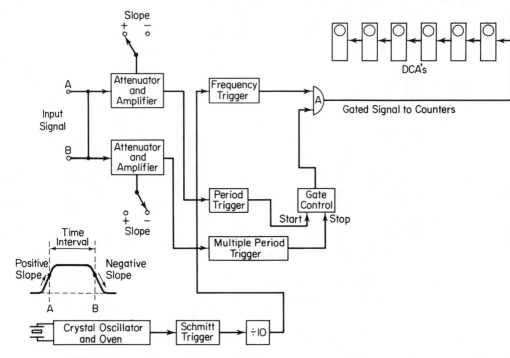

FIGURE 12-16   Block diagram of the counter in the "time-interval" mode of operation.

number of cycles of the higher-frequency signal f_1 which occur during a period of the lower-frequency signal f_2 are counted and displayed in the DCAs. A multiple-ratio measurement extends the number of periods of the lower-frequency signal by a factor of 10, 100, etc. Note that the time-base selector is in the "external" position and that f_1 takes the place of the internal oscillator.

12-6.4   Time-interval Measurement

Time-interval measurements can be made with the same basic blocks as ratio measurements. This measurement is very useful in determining the *pulsewidth* of a certain waveform.

The block diagram for this measurement is given in Fig. 12-16. This configuration shows that the two input terminals A and B are parallelled and that one channel supplies the *enabling* pulse for the main gate and the other channel the *disabling* pulse. The main gate is enabled at a point on the *leading* edge of the input signal waveform and closed at a point on the *trailing* edge of the same waveform. This is facilitated by the "slope selection" feature indicated in the block diagram. The "trigger level" control permits selection of the point on the incoming signal waveform at which the measurement begins and ends.

12-7 MEASUREMENT ERRORS

12-7.1 Gating Error

Frequency and time measurements made by an electronic counter are subject to several inaccuracies inherent in the instrument itself. One very common instrumental error is the *gating error*, which occurs whenever frequency and period measurements are made. For frequency measurement (Sec. 12-6.1) the main gate is opened and closed by the oscillator output pulse. This allows the input signal to pass through the gate and be counted by the DCAs. The gating pulse is not synchronized with the input signal; they are, in fact, two totally unrelated signals.

In Fig. 12-17 the gating interval is indicated by waveform (c). Waveforms (a) and (b) represent the input signal in different phase relationships

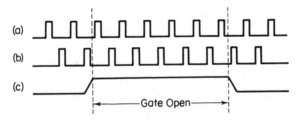

FIGURE 12-17   Gating error.

with respect to the gating signal. Clearly, in one case, six pulses will be counted; in the other case, only five pulses are allowed to pass through the gate. We have therefore a ± 1 count ambiguity in the measurement. In measuring low frequencies, the gating error may have an appreciable effect on the results. Take, for example, the case where a frequency of 10 Hz is to be measured and the gating time equals 1 s (a reasonable assumption). The decade counters would indicate a count of 10 ± 1 count, an inaccuracy of 10 per cent. *Period* measurements are therefore to be preferred over *frequency* measurements at the *lower* frequencies.

The dividing line between frequency and period measurements may be determined as follows: Let

$$f_c = \text{crystal (or clock) frequency of the instrument}$$
$$f_x = \text{frequency of the unknown input signal}$$

In a *period* measurement the number of pulses counted equals

$$N_p = \frac{f_c}{f_x} \qquad (12\text{-}1)$$

In a *frequency* measurement with a 1-s gate time the number of pulses counted is

$$N_f = f_x \qquad (12\text{-}2)$$

The *crossover* frequency (f_o) at which $N_p = N_f$ is

$$\frac{f_c}{f_o} = f_o \quad \text{or} \quad f_o = \sqrt{f_c} \qquad (12\text{-}3)$$

Signals with a frequency *lower* than f_o should therefore be measured in the "period" mode; signals of frequencies *above* f_o should be measured in the "frequency" mode in order to minimize the effect of the ± 1 count gating error. The accuracy degradation at f_o caused by the ± 1 count gating error is $100/\sqrt{f_c}$ per cent.

12-7.2   Time-base Error

Inaccuracies in the time base also cause errors in the measurement. In frequency measurements the time base determines the opening and closing of the signal gate, and it provides the pulses to be counted. Time-base errors consist of oscillator calibration errors, short-term crystal stability errors, and long-term crystal stability errors.

Several methods of *crystal calibration* are in common use. One of the simplest calibration techniques is to zero-beat the crystal oscillator against the standard frequency transmitted by a standards radio station, such as WWV (see Sec. 3-3). This method gives reliable results with accuracy on the order of 1 part in 10^6, which corresponds to 1 cycle of a 1-MHz crystal oscillator. If the zero-beating is done with visual (rather than audible) means, for example, by using a CRO, the calibration accuracy can usually be improved to 1 part in 10^7.

Several very low frequency (VLF) radio stations cover the North American continent with precise signals in the 16–20-kHz range. Low-frequency receivers are available with automatic servo-controlled tuning that can be slaved to the signal of one of these stations. The error between the local crystal oscillator and the incoming signal can then be recorded on a strip-chart recorder. A simplified diagram of this procedure is given in Fig. 12-18. Improved calibration accuracy can be obtained by using VLF stations rather than HF stations because the transmission paths for very low frequencies is shorter than for high-frequency transmissions.

Short-term crystal stability errors are caused by momentary frequency variations due to voltage transients, shock and vibration, cycling of the crystal oven, electrical interference, etc. These errors can be *minimized* by taking frequency measurements over *long* gate times (10 s to 100 s) and multiple-

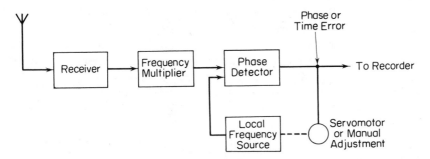

FIGURE 12-18   Calibration of a local frequency source.

period-average measurements. A reasonable figure for short-term stability of a standard crystal-oven combination is on the order of 1 or 2 parts in 10^7.

   Long-term stability errors are the more subtle contributors to the inaccuracy of a frequency or time measurement. Long-term stability is a function of aging and deterioration of the crystal. As the crystal is tempera-ture-cycled and kept in continuous oscillation, internal stresses induced during manufacture are relieved, and minute particles adhering to the surface are shed reducing its thickness. Generally, these phenomena will cause an *increase* in the oscillator frequency.

   A typical curve of frequency change versus time is shown in Fig. 12-19. The *initial* rate of change of crystal frequency may be on the order of 1 part in

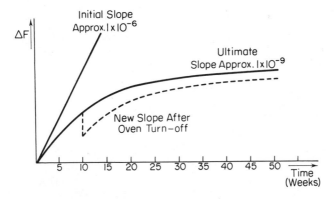

FIGURE 12-19   Frequency change versus time for an oven-controlled crystal.

10^6 per day. This rate will decrease, provided that the crystal is maintained at its operating temperature, normally about 50° to 60°C, with ultimate stabilities of 1 part in 10^9. If, however, the instrument containing the crystal is unplugged from the power source for a period of time sufficient to allow the crystal to cool appreciably, a new slope of aging will ensue when the instrument is put back into operation. It is possible that the actual frequency of oscillation after cool off will vary by several cycles and that the original frequency will not again be reached unless calibration is done.

To show the effect of long-term stability on the absolute accuracy of the measurement, assume that the oscillator was calibrated to within 1 part in 10^9 and that a long-term stability of 1 part in 10^8 per day was reached. Assume further that calibration was done 60 days ago. The guaranteed accuracy at this time is then $1 \times 10^{-9} + 60 \times 10^{-8} = 6.01 \times 10^{-7}$, or 6 parts in 10^7. It can be seen therefore that maximum absolute accuracy can be achieved only if an exact calibration is performed a relatively short time *before* the measurement is taken.

12-7.3 Trigger Level Error

In time-interval and period measurements the signal gate is opened and closed by the input signal. The accuracy with which the gate is opened and closed is a function of the *trigger level error*. In the usual application the input signal is amplified and shaped, and then it is applied to a Schmitt trigger circuit that supplies the gate with its control pulses. Usually the input signal contains a certain amount of unwanted components or noise, which is amplified along with the signal. The time at which triggering of the Schmitt circuit occurs is a function of the input signal amplification and of its signal-to-noise ratio. In general, we can say that trigger time errors are reduced with large signal amplitudes and fast risetimes.

Maximum accuracy can be obtained if the following suggestions are followed:

(a) The effect of the one-count gating error can be minimized by making frequency measurements above $\sqrt{f_c}$ and period measurements below $\sqrt{f_c}$, where f_c is the clock frequency of the counter.
(b) Since long-term stability has a cumulative effect, the accuracy of measurement is mostly a function of the time since the last calibration against a primary or secondary standard.
(c) The accuracy of time measurements is greatly affected by the *slope* of the incoming signal controlling the signal gate. *Large* signal amplitude and *fast* risetime assure maximum accuracy.

12-8 MEASUREMENT APPLICATIONS

12-8.1 Frequency Measurements

The measurement capability of an electronic counter in the "frequency" mode of operation can be extended by using a *heterodyne converter*. This is shown in the block diagram of Fig. 12-20. The input signal is applied to the heterodyne converter, which consists of a reference oscillator and a mixer

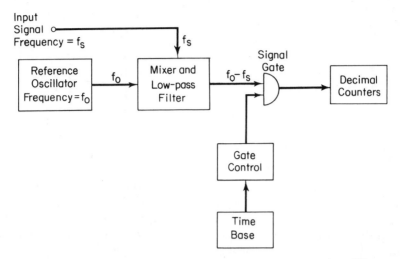

FIGURE 12-20 Heterodyne frequency conversion generates the difference between signal and reference oscillator frequencies.

stage with a low-pass filter. The input signal frequency, f_s, and the reference oscillator frequency, f_o, are applied to the mixer stage, which produces both the sum and the difference of the two frequencies. The low-pass filter, however, allows only the difference frequency to be passed on to the gating circuits of the counter. The counter then counts the frequency $(f_o - f_s)$ or $(f_s - f_o)$, depending on whether the input signal frequency is above or below the reference oscillator frequency. The need to know whether to add or to subtract the counter readout to the reference frequency in order to obtain the frequency of the unknown signal may complicate the operation somewhat, but this method extends the range of a general-purpose counter very effectively. A counter with a time-base frequency of 1 MHz usually has an input frequency range of approximately 5 MHz. The use of the frequency converter extends this range to 500 MHz or higher.

Some of the more sophisticated counters have provision for plug-in units that allow for frequency conversion simply by plugging the appropriate extension into the basic counter frame. The decade divider assemblies (DDAs) in the oscillator circuit of the counter count the time-base frequency of 1 MHz down to 1 Hz, providing a period of 1 s. The advantage of a 1-s time base is that the digital readout of the input frequency is then in cycles per second, a very convenient feature. If a different time base is selected by proper positioning of the front-panel "time-base" control, the decimal point in the digital readout usually is placed in such a position that the readout is again in cycles per second.

It is not necessary to use a time base of 1 s and, in fact, many applications require a different time base. For example, if the winch drum of Fig.

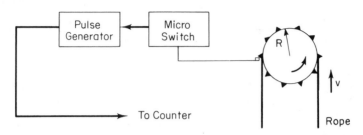

FIGURE 12-21   Winch drum produces 10 pulses per revolution for application to the counter.

12-21 has a circumference of 100 cm, the rope speed (v) in cm per second is 100 times the drum's angular speed (R) in revolutions per second; therefore, $v = 100 R$. The rope speed is read out directly in cm per second if the counter counts 100 pulses per revolution for 1 s. If the rope speed in cm per minute is desired, the counter could be arranged to count 100 pulses per revolution for 60 s, by having 10 cams on the drum.

12-8.2   Time-interval Measurements

In time-interval measurements the signal gate is opened and closed by the *input signal*, allowing the *time-base* frequency to be counted. In the block diagram of Fig. 12-16 the period trigger provides the opening pulse for the main gate while the multiple-period trigger supplies the closing pulse for the main gate. Both pulses are derived from the same input waveform, but one Schmitt trigger responds to the *positive*-going signal while the other Schmitt trigger responds to the *negative*-going waveform. A "trigger level" control allows selection of the point on the incoming waveform, either positive or negative, at which the circuit is triggered. This control can increase noise rejection and decrease the effect of harmonic content on the measurement. The operation of the trigger level control is shown in Fig. 12-22.

One application of time-interval measurements involves determination of pulsewidth and risetime of an unknown waveform, using the *slope selection*

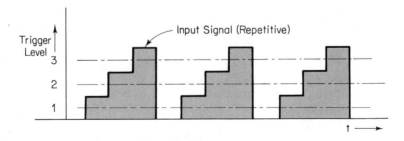

FIGURE 12-22   Trigger-level control.

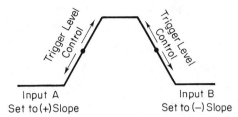

FIGURE 12-23   Slope triggering.   Set to (+) Slope   Set to (−) Slope

feature of the instrument (see Fig. 12-23). The signal gate is opened at a point on the *leading* edge of the input signal by the trigger level control of amplifier *A*. The gate is closed at a point on the *trailing* edge of the input signal by the trigger level control of amplifier *B*. The width of the pulse is registered by the digital readout and depends on the setting of the time-base selector. If the time-base selector was set at 1 μs (1 MHz frequency), the counter reads the time interval directly in μs.

A different application is shown in Fig. 12-24. Here an electronic counter is used to measure the delay times of a relay. The relay functions

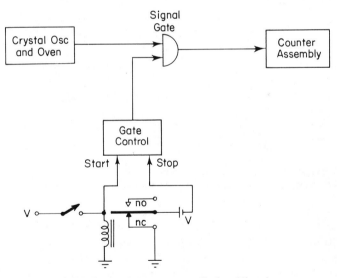

FIGURE 12-24   Measurement of relay delay times.

control the opening and closing of the signal gate and the number of cycles of the time-base generator is counted by the DCAs. The various response times are measured as follows:

Delay time: The gate is opened by the application of coil voltage. The gate is closed by the normally closed contacts, when they open.

Transfer time: The gate is opened by the normally closed contacts, when they open. The gate is closed by the normally open contacts, when they close.

Pick-up time: The gate is opened by the application of coil voltage. The gate is closed by the normally open contacts, when they close.

Drop-out time: The gate is opened by the removal of coil voltage. The gate is closed by the normally open contacts, when they return to their normally open position at the deenergization of the coil.

12-8.3   Digital Voltmeter

The pulse input rate can be made variable to provide analog-to-digital (A/D) conversion. For example, a digital voltmeter can be constructed as shown in Fig. 12-25. An *RC* charging circuit driving a voltage comparator

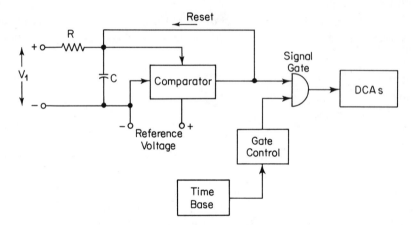

FIGURE 12-25   Digital voltmeter counts the number of charging cycles during the time-base controlled gate interval. Each output pulse resets the capacitor voltage to zero.

produces a pulse output whose *rate* is proportional to the input voltage. With a constant input voltage (V_1), capacitor C charges from 0 V toward the input voltage. When the trigger level is reached, an output pulse is generated. This pulse discharges the capacitor, and the charging cycle starts again. The larger the input voltage V_1, the faster the voltage on the capacitor reaches the trigger voltage and the higher the output pulse rate. The output pulse rate is proportional to the input voltage. The capacitor and the resistor are usually adjusted to provide a convenient scale factor; a commonly used scale factor is 100-Hz full-scale frequency for 100-mV full-scale input voltage. If the gate time is 1 ms, the readout will be directly in millivolts.

12-8.4 Totalizers

Totalizers count and provide a readout of the *total* number of pulses received by the DCAs with no specific gate time being used. They can be used for counting anything, from the number of boxes coming off a production line to pulses from a nuclear particle detector.

Scalers are totalizers with some type of scale factor ahead of the readout. Scalers are particularly useful for converting units. For example, if we obtain one pulse for each egg that rolls down a chute and we want to know how many dozen eggs have rolled by, a scale factor of 12 is applied so that every count in the readout represents 1 dozen eggs.

The same applies to tachometers, where the total number of revolutions is required; the scale factor is the number of pulses from the tachometer generator per revolution. Scaling is easily accomplished by the same technique employed in the generation of time bases, namely, by using *binary* dividers, *decade* dividers, or other types of *feedback* dividers.

An application of the totalizer is the *preset counter* (a specialized scaler) suited for process control. When the total number in the readout reads the same as the preset number (determined by switches), a pulse is generated and the unit stops counting until reset. A contact closure provided at the preset number can be used for machine control. For example, assume that we are winding a coil of wire and we have a pickoff that provides a pulse for every turn wound on the form. If 50 turns are needed, then the contact closure at the preset number can be used to control the winding mechanism and stop it after it has wound the required 50 turns. The same function can be useful in quality-control programs that require a sample from a given number of units. For example, by using a contact closure to drive an ejecting mechanism, every hundredth egg can be taken from the chute and held for inspection. Multiple functions can be performed by using more than one preset number and set of switches. Suppose we want taps on a coil at 10, 20, and 25 turns. By taking 4 preset numbers, we can command the machine to make taps as it reaches the first 3 preset numbers of turns, and to stop at the fourth. The counter would provide momentary contact closures but continue to count until it reached the fourth number.

REFERENCES

1. Millman, Jacob, and Herbert Taub, *Pulse, Digital, and Switching Waveforms.* New York: McGraw-Hill Book Company, Inc., 1965.

2. Doyle, John M., *Pulse Fundamentals.* Englewood Cliffs, N.J.: Prentice-Hall, Inc., 1963.

PROBLEMS

1. Draw a waveform chart illustrating the action in a chain of three binaries.

2. A chain of three binaries is connected as a *reversible* counter and the initial registration is 7. From this initial state draw the waveform chart showing how the state of each binary changes with each of the next three input pulses.

3. Draw the logic block diagram of a chain of four binaries with appropriate feedback loops to reduce the circuit to a 3/1 counter. Illustrate the operation with a waveform chart.

4. Explain how feedback may be used to convert a scale-of-16 counter to a decimal counter. Indicate two possible ways in which the original circuit may be modified.

5. Explain how two separate feedback paths may be used to convert a scale-of-16 counter to a binary counter. Illustrate the operation with a waveform chart.

6. A time-code generator delivers one output pulse per second. Design a circuit in block diagram form which delivers one output pulse per minute (a 60/1 divider).

7. Consider a digital counter-timer of the type illustrated in Fig. 12-11. Draw a simplified block diagram of this counter; explain the function of each block.

8. Consider a digital counter consisting of three basic blocks: a time base, a control gate, and a decade counting assembly. Indicate how these basic blocks are interconnected to perform the following measurements:
   (a) frequency   (b) period
   (c) time interval   (d) total count

9. A digital counter is used to measure a frequency of 2 MHz. The frequency selector of the instrument is set at the kHz position. For what time interval must the gate be open in order to display the correct count?

10. Draw simple circuit diagrams explaining the basic difference in operating principle between a diode OR gate and a diode AND gate.

11. Draw the schematic diagram of a diode OR circuit for use with negative input pulses.

13

Transducers as Input Elements
to Instrumentation Systems

13-1 CLASSIFICATION OF TRANSDUCERS

An electronic instrumentation system consists of a number of components which together are used to perform a measurement and record the result. An instrumentation system generally consists of three major elements: an *input* device, a signal-conditioning or *processing* device, and an *output* device. The input device receives the quantity under measurement and delivers a *proportional* electrical signal to the signal-conditioning device. Here the signal is amplified, filtered, or otherwise modified to a format acceptable to the output device. The output device may be a simple indicating meter, a CRO, or a chart recorder for visual display. It may be a magnetic tape recorder for temporary or permanent storage of the input data, or it may be a digital computer for data manipulation or process control. The kind of system depends on what is to be measured and how the measurement result is to be presented.

The input quantity for most instrumentation systems is *nonelectrical*. In order to use electrical methods and techniques for measurement, manipulation, or control, the nonelectrical quantity is converted into an electrical signal by a device called a *transducer*. One definition states "a transducer is a device which, when actuated by energy in one transmission system, supplies energy in the same form or in another form to a second transmission system." This energy transmission may be electrical, mechanical, chemical, optical (radiant), or thermal.

This broad definition of a transducer includes, for example, devices that convert *mechanical* force or displacement into an electrical signal. These devices form a very large and important group of transducers commonly found in the industrial instrumentation area, and the instrumentation engineer is primarily concerned with this type of energy conversion. Many other physical parameters (such as heat, light intensity, humidity) may also be converted into electrical energy by means of transducers. These transducers provide an output signal when stimulated by a *nonmechanical* input: a thermistor reacts to temperature variations, a photocell to changes in light intensity, an electron beam to magnetic effects, and so on. In all cases, however, the electrical output is measured by standard methods, yielding the magnitude of the input quantity in terms of an *analog* electrical measure.

Transducers may be classified according to their application, method of energy conversion, nature of the output signal, and so on. All these classifications usually result in overlapping areas. A sharp distinction between, and classification of, types of transducers is difficult. Table 13-1 shows a classification of transducers according to the *electrical principles* involved. The first part of the table lists transducers that require external power. These are the *passive* transducers, producing a variation in some electrical parameter, such as resistance, capacitance, and so on, which can be measured as a voltage or

Table 13-1

TYPES OF TRANSDUCERS

Electrical parameter and class of transducer	Principle of operation and nature of device	Typical application
	PASSIVE TRANSDUCERS (EXTERNALLY POWERED)	
Resistance		
Potentiometric device	Positioning of the slider by an external force varies the resistance in a potentiometer or a bridge circuit.	Pressure, displacement
Resistance strain gage	Resistance of a wire or semiconductor is changed by elongation or compression due to externally applied stress.	Force, torque, displacement
Pirani gage or hot-wire meter	Resistance of a heating element is varied by convection cooling of a stream of gas.	Gas flow, gas pressure
Resistance thermometer	Resistance of pure metal wire with a large positive temperature coefficient of resistance varies with temperature.	Temperature, radiant heat
Thermistor	Resistance of certain metal oxides with negative temperature coefficient of resistance varies with temperature.	Temperature
Resistance hygrometer	Resistance of a conductive strip changes with moisture content.	Relative humidity
Photoconductive cell	Resistance of the cell as a circuit element varies with incident light.	Photosensitive relay

Electrical parameter and class of transducer	Principle of operation and nature of device	Typical application
	PASSIVE TRANSDUCERS (EXTERNALLY POWERED) *(cont.)*	
Capacitance		
Variable capacitance pressure gage	Distance between two parallel plates is varied by an externally applied force	Displacement, pressure
Capacitor microphone	Sound pressure varies the capacitance between a fixed plate and a movable diaphragm.	Speech, music, noise
Dielectric gage	Variation in capacitance by changes in the dielectric.	Liquid level, thickness
Inductance		
Magnetic circuit transducer	Self inductance or mutual inductance of ac-excited coil is varied by changes in the magnetic circuit.	Pressure, displacement
Reluctance pickup	Reluctance of the magnetic circuit is varied by changing the position of the iron core of a coil.	Pressure, displacement, vibration, position
Differential transformer	The differential voltage of two secondary windings of a transformer is varied by positioning the magnetic core through an externally applied force.	Pressure, force, displacement, position
Eddy current gage	Inductance of a coil is varied by the proximity of an eddy current plate.	Displacement, thickness
Magnetostriction gage	Magnetic properties are varied by pressure and stress.	Force, pressure, sound
Voltage and current		
Hall effect pickup	A potential difference is generated across a semiconductor plate (germanium) when magnetic flux interacts with an applied current.	Magnetic flux, current
Ionization chamber	Electron flow induced by ionization of gas due to radioactive radiation.	Particle counting, radiation
Photoemissive cell	Electron emission due to incident radiation on photoemissive surface.	Light and radiation
Photomultiplier tube	Secondary electron emission due to incident radiation on photosensitive cathode.	Light and radiation, photosensitive relays
	SELF-GENERATING TRANSDUCERS (NO EXTERNAL POWER)	
Thermocouple and thermopile	An emf is generated across the junction of two dissimilar metals or semiconductors when that junction is heated	Temperature, heat flow, radiation
Moving-coil generator	Motion of a coil in a magnetic field generates a voltage.	Velocity, vibration
Piezoelectric pickup	An emf is generated when an external force is applied to certain crystalline materials, such as quartz.	Sound, vibration, acceleration, pressure changes
Photovoltaic cell	A voltage is generated in a semiconductor junction device when radiant energy stimulates the cell.	Light meter, solar cell

current variation. The second category are transducers of the *self-generating* type, producing an analog voltage or current when stimulated by some physical form of energy. The self-generating transducers do *not* require external power. Although it would be almost impossible to classify all sensors and measurements, the devices listed in Table 13-1 represent a good cross section of commercially available transducers for application in instrumentation engineering. Some of the more common transducers and their application are discussed in the following sections.

13-2 SELECTING A TRANSDUCER

In a measurement system the transducer is the input element with the critical function of transforming some physical quantity to a proportional electrical signal. Selection of the appropriate transducer is therefore the first and perhaps most important step in obtaining accurate results. A number of elementary questions should be asked before a transducer can be selected, for example,

(a) What is the physical quantity to be measured?
(b) Which transducer principle can best be used to measure this quantity?
(c) What accuracy is required for this measurement?

The first question can be answered by determining the *type* and *range* of the measurand. An appropriate answer to the second question requires that the *input* and *output* characterististic of the transducer be compatible with the recording or measurement system. In most cases, these two questions can be answered readily, implying that the proper transducer is selected simply by the addition of an *accuracy* tolerance. In practice, this is rarely possible due to the complexity of the various transducer parameters that affect the accuracy. The accuracy requirements of the total *system* determine the degree to which individual factors contributing to accuracy must be considered. Some of these factors are:

(a) *Fundamental transducer parameters:* type and range of measurand, sensitivity, excitation
(b) *Physical conditions:* mechanical and electrical connections, mounting provisions, corrosion resistance
(c) *Ambient conditions:* nonlinearity effects, hysteresis effects, frequency response, resolution
(d) *Environmental conditions:* temperature effects, acceleration, shock and vibration

(e) *Compatibility of the associated equipment:* zero balance provisions, sensitivity tolerance, impedance matching, insulation resistance

Categories (a) and (b) are basic electrical and mechanical characteristics of the transducer. Transducer accuracy, as an independent component, is contained in categories (c) and (d). Category (e) considers the transducer's compatibility with its associated system equipment.

The total measurement error in a transducer-activated system may be reduced to fall within the required accuracy range by the following techniques:

(a) Using in-place system *calibration* with corrections performed in the data reduction
(b) Simultaneously *monitoring* the environment and correcting the data accordingly
(c) Artificially *controlling* the environment to minimize possible errors

Some *individual* errors are predictable and can be calibrated out of the system. When the entire system is calibrated, these calibration data may then be used to correct the recorded data. *Environmental* errors can be corrected by data reduction if the environmental effects are recorded simultaneously with the actual data. Then the data are corrected by using the known environmental characteristics of the transducers. These two techniques can provide a significant increase in system accuracy.

Another method to improve overall system accuracy is to control *artificially* the environment of the transducer. If the environment of the transducer can be kept unchanged, these errors are reduced to zero. This type of control may require either physically moving the transducer to a more favorable position or providing the required isolation from the environment by a heater enclosure, vibration isolation, or similar means.

13-3   STRAIN GAGES

13-3.1   Gage Factor

The strain gage is an example of a passive tranducer (Table 13-1) that converts a *mechanical* displacement into a change of *resistance*. A strain gage is a thin, wafer-like device that can be attached (bonded) to a variety of materials to measure applied strain. *Metallic* strain gages are manufactured from small-diameter resistance wire, such as Constantan,* or etched from thin foil

---

*A trade name. Constantan is a copper-nickel alloy consisting of 60 per cent copper and 40 per cent nickel.

sheets. The resistance of the wire or metal foil changes with length as the material to which the gage is attached undergoes tension or compression. This change in resistance is proportional to the applied strain and is measured with a specially adapted Wheatstone bridge.

The sensitivity of a strain gage is described in terms of a characteristic called the *gage factor*, K, defined as the unit change in resistance per unit change in length, or

$$\text{gage factor } K = \frac{\Delta R/R}{\Delta l/l} \tag{13-1}$$

where K = gage factor

R = nominal gage resistance

ΔR = change in gage resistance

l = normal specimen length (unstressed condition)

Δl = change in specimen length

The term $\Delta l/l$ in the denominator of Eq. (13-1) is the strain σ, so that Eq. (13-1) can be written as

$$K = \frac{\Delta R/R}{\sigma} \tag{13-2}$$

where σ = the strain in the lateral direction.

The resistance change ΔR of a conductor with length l can be calculated by using the expression for the resistance of a conductor of uniform cross section:

$$R = \rho \frac{\text{length}}{\text{area}} = \frac{\rho \times l}{(\pi/4)d^2} \tag{13-3}$$

where ρ = the specific resistance of the conductor material

l = length of the conductor

d = diameter of the conductor

Tension on the conductor causes an increase Δl in its length and a simultaneous decrease Δd in its diameter. The resistance of the conductor then changes to

$$R_s = \rho \frac{(l + \Delta l)}{(\pi/4)(d - \Delta d)^2} = \rho \frac{l(1 + \Delta l/l)}{(\pi/4)d^2(1 - 2\,\Delta d/d)} \tag{13-4}$$

Equation (13-4) may be simplified by using Poisson's ratio, μ, defined as the ratio of strain in the lateral direction to strain in the axial direction. Therefore

$$\mu = \frac{\Delta d/d}{\Delta l/l} \tag{13-5}$$

Substitution of Eq. (13-5) into Eq. (13-4) yields

$$R_s = \rho \frac{l}{(\pi/4)d^2}\left(\frac{1 + \Delta l/l}{1 - 2\mu\,\Delta l/l}\right) \tag{13-6}$$

which can be simplified to

$$R_s = R + \Delta R = R\left[1 + (1 + 2\mu)\frac{\Delta l}{l}\right] \tag{13-7}$$

The increment of resistance ΔR as compared to the increment of length Δl can then be expressed in terms of the gage factor K where

$$K = \frac{\Delta R/R}{\Delta l/l} = 1 + 2\mu \tag{13-8}$$

Poisson's ratio for most metals lies in the range of 0.25 to 0.35, and the gage factor would then be on the order of 1.5 to 1.7.

For strain-gage applications, a *high sensitivity* is very desirable. A large gage factor means a relatively large resistance change which can be more easily measured than a small resistance change. For Constantan wire, K is about 2, whereas Alloy 479 gives a K value of about 4.

It is interesting to carry out a simple calculation to find out what effect an applied stress has on the resistance change of a strain gage. Hooke's law gives the relationship between stress and strain for a linear stress-strain curve, in terms of the modulus of elasticity of the material under tension. Defining *stress* as the applied force per unit area and *strain* as the elongation of the stressed member per unit length, Hooke's law is written as

$$\sigma = \frac{s}{E} \tag{13-9}$$

where $\sigma =$ strain, $\Delta l/l$ (no units)

       $s =$ stress, kg/cm²

       $E =$ Young's modulus, kg/cm²

Example 13-1: A resistance strain gage with a gage factor of 2 is fastened to a steel member subjected to a stress of 1,050 kg/cm². The modulus of elasticity of steel is approximately 2.1×10^6 kg/cm². Calculate the change in resistance, ΔR, of the strain-gage element due to the applied stress.

SOLUTION: Hooke's law, Eq. (13-9), yields

$$\sigma = \frac{\Delta l}{l} = \frac{s}{E} = \frac{1{,}050}{2.1 \times 10^6} = 5 \times 10^{-4}$$

The sensitivity of the strain gage $k = 2$. Therefore, from Eq. (13-2),

$$\frac{\Delta R}{R} = K\sigma = 2 \times 5 \times 10^{-4} = 10^{-3} \quad \text{or} \quad 0.1\%$$

Example 13-1 illustrates that the relatively high stress of 1,050 kg/cm² results in a resistance change of only 0.1 per cent, a very small change indeed. Actual measurements generally involve resistance changes of very much lower values, and the bridge measuring circuit must be very carefully designed to be able to detect these small changes in resistance.

13-3.2 Metallic Sensing Elements

Metallic strain gages are formed from thin resistance wire or etched from thin sheets of metal foil. Wire gages are generally small in size, are subject to minimal leakage, and can be used in high temperature applications. Foil elements are somewhat larger in size and are more stable than wire gages. They can be used under extreme temperature conditions and under prolonged loading, and they dissipate self-induced heat easily.

Various resistance materials have been developed for use in wire and foil gages. Some of these are described in the following paragraphs.

Constantan is a copper-nickel alloy with a low temperature coefficient. Constantan is commonly found in gages that are used in dynamic strain measurements, where alternating strain levels do not exceed $\pm 1,500$ μcm/cm. Operating temperature limits are from 10°C to 200°C.

Nichrome V is a nickel-chrome alloy used for static strain measurements to 375°C. With temperature compensation, the alloy may be used for static measurements to 650°C and dynamic measurements to 1,000°C.

Dynaloy is a nickel-iron alloy with a high gage factor and a high resistance to fatigue. This material is used in dynamic strain applications when high temperature sensitivity can be tolerated. The temperature range of dynaloy gages is generally limited by the carrier materials and the bonding cement.

Stabiloy is a modified nickel-chrome alloy with a wide temperature compensation range. These gages have excellent stability from cryogenic temperatures to approximately 350°C and good fatigue life.

Platinum-tungsten alloys offer excellent stability and high resistance to fatigue at elevated temperatures. These gages are recommended for static tests to 700°C and dynamic measurements to 850°C. Because the material has a relatively large temperature coefficient, some form of temperature compensation must be used to correct this error.

Semiconductor strain gages are often used in high-output transducers such as load cells. These gages have very high sensitivities, with gage factors from 50 to 200. They are, however, sensitive to temperature fluctuations and often behave in a nonlinear manner.

The size of the finished gage, and the manner in which the wire or foil pattern is arranged, varies with the application. Some bonded gages can be as small as $\frac{1}{8}$ in. by $\frac{1}{8}$ in., although they are generally somewhat larger, and are manufactured to a maximum size of 1 in. long by $\frac{1}{2}$ in. wide. In the usual application, the strain gage is cemented to the structure whose strain is to be measured. The problem of providing a good bonding between the gage and the structure is very difficult. The adhesive material must hold the gage firmly to the structure, yet it must have sufficient elasticity to give under strain without losing its adhesive properties. The adhesive should also be resistant to temperature, humidity, and other environmental conditions.

13-3.3 Gage Configuration

The shape of the sensing element is selected according to the strain to be measured: uniaxial, biaxial, or multidirectional. Uniaxial applications most often use long, narrow sensing elements, as in Fig. 13-1, to maximize

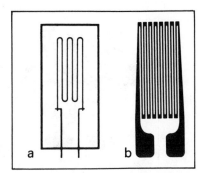

FIGURE 13-1   Uniaxial strain gages. (a) Wire; (b) Foil. (courtesy Gould, Inc., Measurement Systems Division).

the strain sensing material in the direction of interest. End loops are few and short, so that sensitivity to transverse strains is low. Gage length is selected according to the strain field to be investigated. For most strain measurements, the 6 mm gage length offers good performance and easy installation.

Simultaneous measurement of strains in more than one direction can be accomplished by placing single-element gages at the proper locations. However, to simplify this task and provide greater accuracy, multielement, or *rosette*, gages are available.

Two-element rosettes, shown in Fig. 13-2, are often used in force trans-ducers. The gages are wired in a Wheatstone bridge circuit to provide maxi-

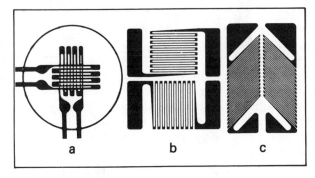

FIGURE 13-2   Two-element rosettes. (a) 90° stacked foil; (b) 90° planar foil; (c) 90° shear planar foil. (courtesy Gould, Inc., Measurement Systems Division).

mum output. For stress analysis, the axial and transverse elements may have different resistances that can be so selected that the combined output is proportional to stress while the output of the axial element alone is proportional to strain. Three-element rosettes are often used to determine the direction and magnitude of principal strains resulting from complex structural loading. The most popular types have 45°- or 60°-angular displacements between the sensing elements, as shown in Fig. 13-3. The 60°-rosettes are

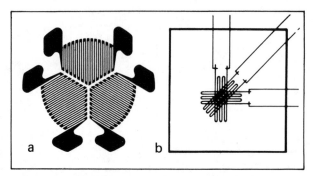

FIGURE 13-3   Three-element rosettes. (a) 60° pianar foil. (b) 45° stacked wire. (courtesy Gould, Inc., Measurement Systems Division).

used when the direction of the principal strains is unknown. The 45°-rosettes provide greater angular resolutions and are normally used when the directions of the principal strains are known.

13-3.4   Unbonded Strain Gage

The unbonded strain gage consists of a stationary frame and an armature that is supported in the center of the frame. The armature can move only in one direction. Its travel in that direction is limited by four filaments of strain-sensitive wire, wound between rigid insulators that are mounted on the frame and on the armature. The filaments are of equal length and arranged as shown in Fig. 13-4(a).

When an external force is applied to the strain gage, the armature moves in the direction indicated. Elements A and D increase in length, whereas elements B and C decrease in length. The resistance change of the four filaments is proportional to their change in length, and this change can be measured with a Wheatstone bridge, as shown in Fig. 13-4(b). The unbalance current, indicated by the current meter, is calibrated to read the magnitude of the displacement of the armature.

The unbonded strain-gage transducer can be constructed in a variety of configurations, depending on the required use. Its principal use is as a *displacement* transducer: A linkage pin can be attached to the armature in order to measure displacement directly. The unit of Fig. 13-4 allows an armature displacement of 0.004 cm on each side of its center position. Using the same construction, the unit will function as a dynamometer, capable of measuring

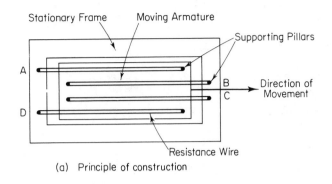

Stationary Frame   Moving Armature

Supporting Pillars

A

B
C   Direction of
Movement

D

Resistance Wire

(a)   Principle of construction

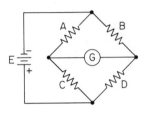

(b)   Wheatstone bridge circuit

FIGURE 13-4   Unbonded strain gage. (a) Principle of construction; (b) Wheatstone bridge circuit.

force. Depending on the number of turns and the diameter of the strain wires, the transducer will measure forces from ± 40 g to ± 2 kg, full-scale.

The transducer becomes a *pressure* pickup when its armature is connected to a metallic bellows or diaphragm. When a bellows is used, force on the end of the bellows is transmitted by a pin to the armature, and the unit functions as a dynamometer. By applying pressure to one side of the bellows and venting the other side to the atmosphere, gage pressures may be read. If the bellows is evacuated and sealed, *absolute pressure* is measured.

Another modification is provided by two pressure connections, one to each side of the bellows or diaphragm, for the measurement of *differential* pressure. Finally, when a weight is fastened to the armature, the transducer becomes an *accelerometer.*

13-4   DISPLACEMENT TRANSDUCERS

13-4.1   Introduction

The concept of converting an applied force into a displacement is basic to many types of transducers. The mechanical elements that are used to convert the applied force into a displacement are called *force-summing* devices.

The force-summing members generally used are the following:

(a) Diaphragm, flat or corrugated
(b) Bellows
(c) Bourdon tube, circular or twisted
(d) Straight tube
(e) Mass cantilever, single or double suspension
(f) Pivot torque

Examples of these force-summing devices are shown in Fig. 13-5. *Pressure* transducers generally use one of the first four types of force-summing members; categories (e) and (f) will be found in *accelerometer* and *vibration* pickups.

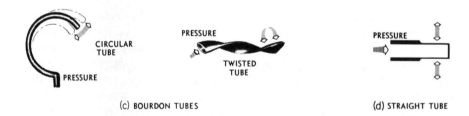

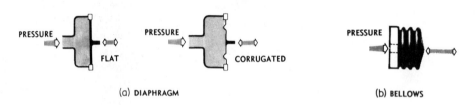

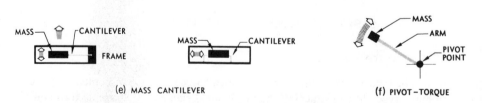

FIGURE 13-5   Force-summing devices (courtesy Statham Instruments, Inc.).

The displacement created by the action of the force-summing device is converted into a change of some *electrical* parameter. The electrical principles most commonly used in the measurement of displacement are

(a) Capacitive
(b) Inductive
(c) Differential transformer
(d) Ionization
(e) Oscillation
(f) Photoelectric
(g) Piezoelectric
(h) Potentiometric
(i) Velocity

These principles are discussed and illustrated in the following sections.

13-4.2   Capacitive Transducer

The capacitance of a parallel-plate capacitor is given by

$$C = \frac{kA\epsilon_0}{d} \quad \text{(farads)} \tag{13-10}$$

where $A =$ the area of each plate, in m^2

$d =$ the plate spacing, in m

$\epsilon_0 = 9.85 \times 10^{-12}$, in F/m

$k =$ dielectric constant

Since the capacitance is inversely proportional to the spacing of the parallel plates, any variation in d causes a corresponding variation in the capacitance. This principle is applied in the capacitive transducer of Fig. 13-6. A force, applied to a diaphragm that functions as one plate of a simple capac-

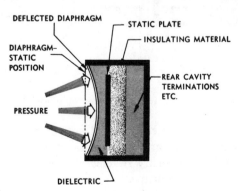

FIGURE 13-6 Capacitive transducer (courtesy Statham Instruments, Inc.).

itor, changes the distance between the diaphragm and the static plate. The resulting change in capacitance could be measured with an ac bridge, but it is usually measured with an oscillator circuit. The transducer, as part of the oscillatory circuit, causes a change in the frequency of the oscillator. This change in frequency is a measure of the magnitude of the applied force.

The capacitive transducer has excellent frequency response and can measure both static and dynamic phenomena. Its disadvantages are sensitivity to temperature variations and possibility of erratic or distorted signals due to long lead length. Also, the receiving instrumentation may be large and complex, and it often includes a second fixed-frequency oscillator for heterodyning purposes. The difference frequency, thus produced, can be read by an appropriate output device such as an electronic counter.

13-4.3 Inductive Transducer

In the inductive transducer the measurement of force is accomplished by the change in the inductance ratio of a pair of coils or by the change of inductance in a single coil. In each case, the ferromagnetic armature is displaced by the force being measured, varying the *reluctance* of the magnetic circuit. Figure 13-7 shows how the air gap is varied by a change in position of the armature. The resulting change in inductance is a measure of the magnitude of the applied force.

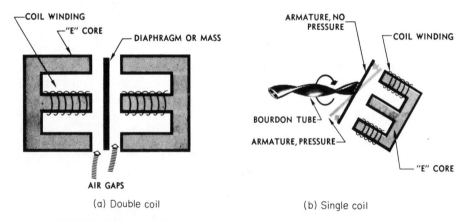

(a) Double coil   (b) Single coil

FIGURE 13-7   Inductive transducers (courtesy Statham Instruments, Inc.).

The coil can be used as a component of an LC oscillator whose frequency then varies with applied force. This type of transducer is used extensively in telemetry systems, with a single coil that modulates the frequency of a local oscillator.

Hysteresis errors of the transducer are almost entirely limited to the mechanical components. When a diaphragm is used as the force-summing member, as shown in Fig. 13-7(a), it may form part of the magnetic circuit. In this arrangement the overall performance of the transducer is somewhat degraded because the desired mechanical characteristics of the diaphragm must be compromised to improve the magnetic performance.

The inductive transducer responds to static and dynamic measurements, and it has continuous resolution and a fairly high output. Its disadvantages are that the frequency response (variation of the applied force) is limited by the construction of the force-summing member. In addition, external magnetic fields may cause erratic performance.

13-4.4 Variable Differential Transformer

The differential transformer transducer measures force in terms of the displacement of the ferromagnetic core of a transformer. The basic construction of the linear variable differential transformer (LVDT) is given in Fig. 13-8. The transformer consists of a single primary winding and two secondary windings which are placed on either side of the primary. The secondaries have an equal number of turns but they are connected in series opposition so that the emfs induced in the coils oppose each other. The position of the movable core determines the flux linkage between the ac-excited primary winding and each of the two secondary windings.

With the core in the center, or *reference*, position, the induced emfs in the secondaries are equal, and since they oppose each other, the output voltage will be 0 V. When an externally applied force moves the core to the left-hand position, more magnetic flux links the left-hand coil than the right-hand coil. The induced emf of the left-hand coil is therefore larger than the induced emf of the right-hand coil. The magnitude of the output voltage is then equal to the *difference* between the two secondary voltages, and it is *in phase* with the voltage of the left-hand coil. Similarly, when the core is forced to move to the right, more flux links the right-hand coil than the left-hand coil and the resulting output voltage is now in phase with the emf of the right-hand coil, while its magnitude again equals the difference between the two induced emfs. Figure 13-8(b) shows the LVDT output voltage as a function of the core position.

The output of the differential transformer may serve as a component in a force-balancing servo system. This is indicated schematically in Fig. 13-9. The output terminals of an *input* transformer and a *balancing* transformer are connected in series opposition. The algebraic sum of the two voltages is fed to an amplifier that drives a two-phase motor. When the two transformers are in their reference positions, the sum of their output voltages is zero and no voltage is delivered to the servo motor. When the core of the input transfor-

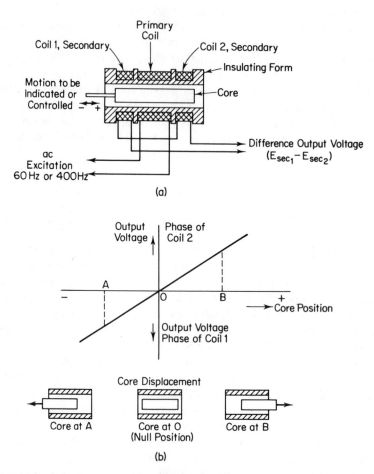

FIGURE 13-8   Linear variable differential transformer (LVDT). (a) The essential components of the LVDT; (b) The relative positions of the core generate the indicated output voltages. The linear characteristics impose limited core movements, which are typically up to 5 mm from the null position.

mer is moved away from its reference position by an externally applied displacement input, an output voltage is delivered to the amplifier and the motor rotates. The motor shaft is mechanically coupled to the core of the balancing transformer. Since the output of the balancing transformer opposes the output of the input transformer, the motor continues to rotate until the outputs of the two transformers are equal. The indicator on the motor shaft is calibrated to read the displacement of the balancing transformer and, indirectly, the displacement of the input transformer.

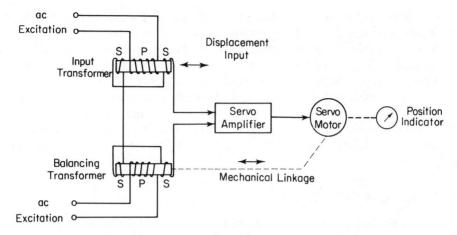

FIGURE 13-9 Displacement measurement using two differential transformers in a closed-loop servo system.

A variation of the moving-core differential transformer is given in Fig. 13-10. Here the primary winding is wound on the center leg of an E core, and the secondary windings are wound on the outer legs of the E core. The armature is rotated by the externally applied force about a pivot point above the center leg of the core. When the armature is displaced from its balance or reference position, the reluctance of the magnetic circuit through one secondary coil is decreased, while, simultaneously, the reluctance of the magnetic

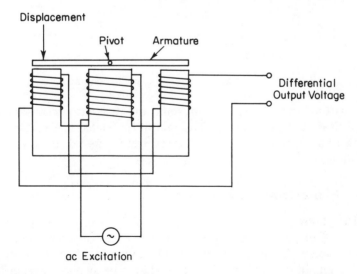

FIGURE 13-10 Differential transformer with an E core and pivoted armature.

circuit through the other secondary coil is increased. The induced emfs in the secondary windings, which are equal in the reference position of the armature, will now be different in magnitude as a result of the applied displacement. The induced emfs again oppose each other, and the transformer operates in the same manner as the moving-core transformer of Fig. 13-8.

The differential transformer provides continuous resolution and shows low hysteresis. Relatively large displacements are required, and the instrument is sensitive to vibration. The receiving instrument must be selected to operate on ac signals, or a demodulator network must be used if a dc output is required.

13-4.5 Oscillation Transducer

This class of transducer uses the force-summing member to change the capacitance or inductance in an LC oscillator circuit. Figure 13-11 shows the basic elements of an LC oscillator whose frequency is affected by a change in the inductance of the coil. The stability of the oscillator must be excellent in order to detect changes in oscillator frequency caused by the externally applied force.

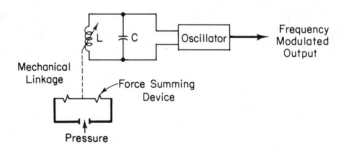

FIGURE 13-11   Basic elements of an oscillation transducer.

This transducer measures both static and dynamic phenomena and is convenient for use in telemetry applications. Its limited temperature range, poor thermal stability, and low accuracy restrict its use to low-accuracy applications.

13-4.6 Photoelectric Transducer

The photoelectric transducer makes use of the properties of a photo-emissive cell or phototube. The phototube is a radiant energy device that controls its electron emission when exposed to incident light. The construction of a phototube is shown in Fig. 13-12(a); its symbol is given in the schematic diagram of Fig. 13-12(b).

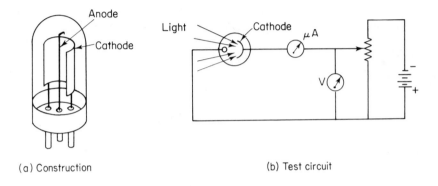

(a) Construction   (b) Test circuit

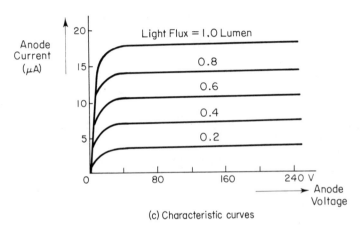

(c) Characteristic curves

FIGURE 13-12   Phototube and its output characteristics.

The large semicircular element is the photosensitive cathode and the thin wire down the center of the tube is the anode. Both elements are placed in a high-vacuum glass envelope. When a constant voltage is applied between the cathode and the anode, the current in the circuit is directly proportional to the amount of light, or light intensity, falling on the cathode. The curves of Fig. 13-12(c) show the anode characteristics of a typical high-vacuum phototube.

Notice that for voltages above approximately 20 V the output current is nearly independent of applied anode voltage but depends entirely on the amount of incident light. The current through the tube is extremely small, usually in the range of a few microamperes. In most cases therefore, the phototube is connected to an amplifier to provide a useful output.

The photoelectric transducer of Fig. 13-13 uses a phototube and a light source separated by a small window whose aperture is controlled by the force-summing member of the pressure transducer. The displacement of the

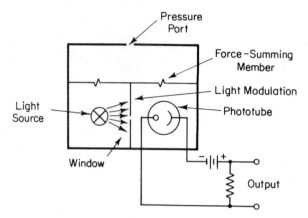

FIGURE 13-13   Elements of a photoelectric transducer.

force-summing member modulates the quantity of incident light on the photo-
senstitive element. According to the curves of Fig. 13-12(c), a change in light
intensity varies the photoemissive properties at a rate approximately linear
with displacement. This transducer can use either a stable source of light or
an ac modulated light.

The advantages of this type of transducer are its high efficiency and its
adaptability to measuring both static and dynamic conditions. The device
may have poor long-term stability, does not respond to high-frequency light
variations, and requires a large displacement of the force-summing member.

13-4.7   Piezoelectric Transducer

Asymmetrical crystalline materials, such as quartz, Rochelle salt, and
barium titanite, produce an emf when they are placed under stress. This
property is used in piezoelectric transducers, where a crystal is placed between
a solid base and the force-summing member, as shown in Fig. 13-14. An

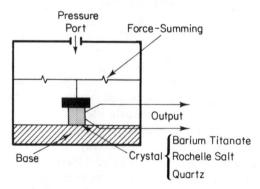

FIGURE 13-14   Elements of a piezoelectric transducer.

ally applied force, entering the transducer through its pressure port, ap, es pressure to the top of a crystal. This produces an emf across the crystal, proportional to the magnitude of the applied pressure.

Since the transducer has a very good high-frequency response, its principal use is in high-frequency accelerometers. In this application its output voltage is typically on the order of 1 to 30 mV per g of acceleration. The device needs no external power source and is therefore self-generating. The principal disadvantage of this transducer is that it cannot measure static conditions. The output voltage is also affected by temperature variations of the crystal.

13-4.8 Potentiometric Transducer

A potentiometric transducer is an electromechanical device containing a resistance element that is contacted by a movable slider. Motion of the slider results in a resistance change that may be linear, logarithmic, exponential, and so on, depending on the manner in which the resistance wire is wound. In some cases, deposited carbon, platinum film, and other techniques are used to provide the resistance element. The basic elements of the potentiometric transducer are given in Fig. 13-15.

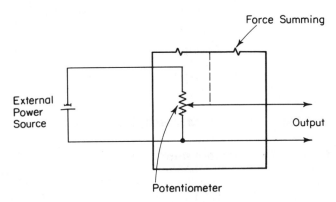

FIGURE 13-15   Principle of the potentiometric transducer.

The potentiometric principle is widely used despite its many limitations. Its electric efficiency is very high and it provides a sufficient output to permit control operations without further amplification. The device may be ac- or dc-excited and can therefore serve a wide range of functions. Because of the mechanical friction of the slider against the resistance element, its life is limited and noise may develop as the element wears out. Large displacements are usually required to move the slider along the entire working surface of the potentiometer.

13-4.9 Velocity Transducer

The velocity transducer essentially consists of a moving coil suspended in the magnetic field of a permanent magnet, as shown in Fig. 13-16. A voltage is generated by the motion of the coil in the field. The output is proportional to the velocity of the coil, and this type of pickup is therefore generally used for the measurement of velocities developed in a linear, sinusoidal, or random manner. Damping is obtained electrically, thus assuring high stability under varying temperature conditions.

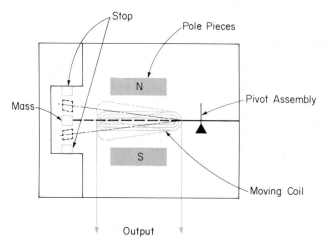

FIGURE 13-16   Elements of a velocity transducer.

13-5 TEMPERATURE MEASUREMENTS

13-5.1 Resistance Thermometers

Resistance-temperature detectors, or resistance thermometers, employ a sensitive element of extremely pure platinum, copper, or nickel wire that provides a definite resistance value at each temperature within its range. The relationship between temperature and resistance of conductors in the temperature range near 0°C can be calculated from the equation

$$R_t = R_{ref}(1 + \alpha\,\Delta t) \qquad (13\text{-}11)$$

where R_t = resistance of the conductor at temperature $t(°C)$

$\quad\quad R_{ref}$ = resistance at the reference temperature, usually 0°C

$\quad\quad \alpha$ = temperature coefficient of resistance

$\quad\quad \Delta t$ = difference between operating and reference temperature

Almost all metallic conductors have a positive temperature coefficient of resistance so that their resistance increases with an increase in temperature. Some materials, such as carbon and germanium, have a negative temperature coefficient of resistance that signifies that the resistance decreases with an increase in temperature. A high value of α is desirable in a temperature-sensing element so that a substantial change in resistance occurs for a relatively small change in temperature. This change in resistance (ΔR) can be measured with a Wheatstone bridge, which may be calibrated to indicate the temperature that caused the resistance change rather than the resistance change itself.

Figure 13-17 shows the variation of resistance with temperature for several commonly used materials. The graph indicates that the resistance of

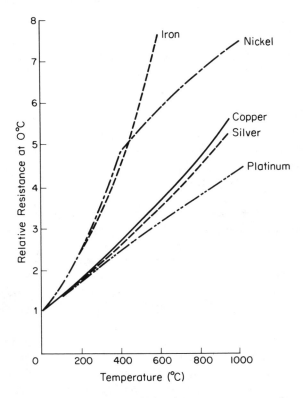

FIGURE 13-17   Relative resistance (R_t/R_{ref}) versus temperature for some pure metals.

platinum and copper increases almost linearly with increasing temperature, while the characteristic for nickel is decidedly nonlinear.

The sensing element of a resistance thermometer is selected according to the intended application. Table 13-2 summarizes the characteristics of the

Table 13-2

RESISTANCE THERMOMETER ELEMENTS

Type	Temperature range	Accuracy	Advantages	Disadvantages
Platinum	$-300°F$ to $+1,500°F$	$\pm 1°F$	Low cost High stability Wide operating range	Relatively slow response time (15 s) Not as linear as copper thermometers
Copper	$-325°F$ to $+250°F$	$\pm 0.5°F$	High linearity High accuracy in ambient temperature range High stability	Limited temperature range (to 250°F)
Nickel	$+32°F$ to $+150°F$	$\pm 0.5°F$	Long life High sensitivity High temperature coefficient	More nonlinear than copper Limited temperature range (to 150°F)

three most commonly used resistance materials. Platinum wire is used for most laboratory work and for industrial measurements of high accuracy. Nickel wire and copper wire are less expensive and easier to manufacture than platinum wire elements, and they are often used in low-range industrial applications.

Resistance thermometers are generally of the probe type for immersion in the medium whose temperature is to be measured or controlled. A typical sensing element for a probe-type thermometer is constructed by coating a small platinum or silver tube with ceramic material, winding the resistance wire over the coated tube, and coating the finished winding again with ceramic. This small assembly is then fired at high temperature to assure annealing of the winding and then it is placed at the tip of the probe. The probe is protected by a sheath to produce the complete sensing element.

Practically all resistance thermometers for industrial applications are mounted in a tube or well to provide protection against mechanical damage and to guard against contamination and eventual failure. Protecting tubes are used at atmospheric pressure; when they are equipped with a pipe-thread bushing, they may be exposed to low or medium pressures. Metal tubes offer adequate protection to the sensing element at temperatures to 2,100°F, although they may become slightly porous at temperatures above 1,500°F and then fail to protect against contamination.

Protecting wells are designed for use in liquids or gases at high pressure such as in pipe lines, steam power plants, pressure tanks, pumping stations, and so on. The use of a protecting well becomes imperative at pressures above 3 at. Protective wells are drilled from solid bar stock, usually carbon steel or stainless steel, and the sensing element is mounted inside. A waterproof junction box with provision for conduit coupling is attached to the top of the well or tube, as shown in Fig. 13-18.

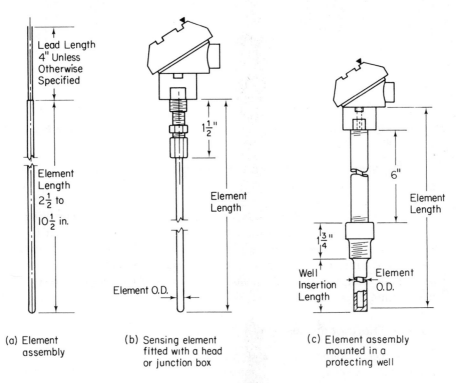

FIGURE 13-18   Resistance thermometers.

A typical bridge circuit with resistance thermometer R_t in the unknown position is shown in Fig. 13-19. The function switch connects three different resistors in the circuit. R_{ref} is a fixed resistor whose resistance is equal to that of the thermometer element at the reference temperature (say, $0°C$). With the function switch in the "REF" position, the zero-adjust resistor is varied until the bridge indicator reads zero. R_{fs} is another fixed resistor whose resistance equals that of the thermometer element for full-scale reading of the current indicator. With the function switch in the "FS" position, the full-scale-adjust resistor is varied until the indicator reads full scale. The function switch is then set to the "MEAS" position, connecting the resistance thermometer R_t in the circuit. When the resistance-temperature characteristic of the thermometer element is linear, the galvanometer indication can be interpolated linearly between the set values of reference temperature and full-scale temperature.

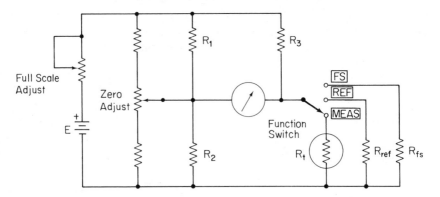

FIGURE 13-19   Bridge circuit with a resistance thermometer as one of the bridge elements.

The Wheatstone bridge has certain disadvantages when it is used to measure the resistance variations of the resistance thermometer. These are the effect of contact resistances of connections to the bridge terminals, heating of the elements by the unbalance current, and heating of the wires connecting the thermometer to the bridge. Slight modifications of the Wheatstone bridge, such as the double slide-wire bridge, eliminate most of these problems. Despite these measurement difficulties, the resistance thermometer method is so accurate that it is one of the standard methods of temperature measurement within the range of $-183°C$ to $630°C$.

13-5.2 Thermocouples

A thermocouple consists of a pair of dissimilar metal wires joined together at one end (*sensing*, or *hot*, junction) and terminated at the other end (*reference*, or *cold*, junction) which is maintained at a known constant temperature (reference temperature). When a temperature difference exists between the sensing junction and the reference junction, an emf is produced that causes a current in the circuit. When the reference junction is terminated by a meter or recording instrument, as in Fig. 13-20, the meter indication will be proportional to the temperature *difference* between the hot junction and the reference junction. This thermoelectric effect, caused by contact potentials at the junctions, is known as the *Seebeck* effect, after the German physicist Thomas Seebeck.

The magnitude of the thermal emf depends on the wire materials used and on the temperature difference between the junctions. Figure 13-21 shows the thermal emfs for some common thermocouple materials. The values shown are based on a reference temperature of $32°F$.

To ensure long life in its operating environment, a thermocouple is protected in an open- or closed-end metal protecting tube or well. To prevent

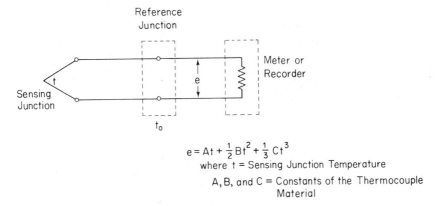

$$e = At + \frac{1}{2}Bt^2 + \frac{1}{3}Ct^3$$
where t = Sensing Junction Temperature
A, B, and C = Constants of the Thermocouple Material

FIGURE 13-20   Basic thermocouple circuit.

contamination of the couple when precious metals are used (platinum and its alloys), the protecting tube is both chemically inert and vacuum tight. Since the thermocouple is usually in a location *remote* from the measuring instrument, connections are made by means of special extension wires called *compensating* wires. Maximum accuracy of measurement is assured when the compensating wires are of the same material as the thermocouple wires.

Thermocouples are available from manufacturers with an NBS certificate of calibration or with a certificate of test based on precision comparison against NBS-certified couples.

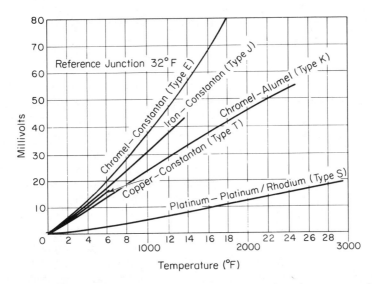

FIGURE 13-21   Thermocouple output voltage as a function of temperature for various thermocouple materials.

The simplest temperature measurement using a thermocouple is by connecting a sensitive millivoltmeter directly across the cold junction. The deflection of the meter is then almost directly proportional to the difference in temperature between the hot junction and the reference junction. This simple instrument has several serious limitations, mainly because the thermocouple can only supply very limited power to drive the meter movement.

The most common method in thermocouple temperature measurements involves using a *potentiometer*. A representative example of this instrument, specifically designed to measure thermocouple voltages, is shown in Fig. 6-7 and its circuit diagram is given in Fig. 6-8 (see Sec. 6-3.2 for a description of this typical instrument).

Many types of *automatic* potentiometers have been developed both for automatic recording of temperatures on chart recorders and for automatic process control. One common system uses a *photoelectric* device, as shown in the simplified schematic diagram of Fig. 13-22. The photocell controls the position of the moving contact of a slide-wire potentiometer. The advantage of this system is that the galvanometer is not subjected to a physical load since it is used only to direct light onto a photocell. The photocell receives its light by reflection from the galvanometer mirror whose angular position is a measure of the unbalance voltage in the potentiometer circuit. The photocell is part of the input circuit to the amplifier, and its resistance controls the input voltage to the amplifier. The amplifier drives a reversible motor that controls the movement of the slider. When the potentiometer is initially balanced, for instance at the reference temperature, there will be a definite amount of light reflected to the photocell. If the light striking the photocell changes because of some change in temperature at the thermocouple junction, the resistance of the photocell changes and affects the input to the amplifier. The amplifier then drives the reversible motor in a direction which tends to restore balance in the potentiometer circuit.

A fairly standard method of dealing with small dc error signals is shown in Fig. 13-23. The thermocouple is again placed in a potentiometer circuit. The unbalance voltage in the potentiometer, caused by temperature variations at the thermocouple hot junction, is applied to a converter. When power is applied to the driving coil of the converter (usually at 60 Hz or 400 Hz), the magnetic armature vibrates in synchronism with the frequency of the coil voltage. The armature alternately connects opposite ends of the transformer primary to the unbalance voltage. The pulsating unbalance voltage in the transformer primary is transferred to the secondary where it is amplified by an ac amplifier and applied to the balancing motor. The polarity of the error signal originating from the potentiometer circuit determines whether the pulse going into the transformer is of one polarity or of the opposite polarity. The polarity of the amplified error signal with respect to the instantaneous line voltage determines the direction of rotation of the balancing motor. The

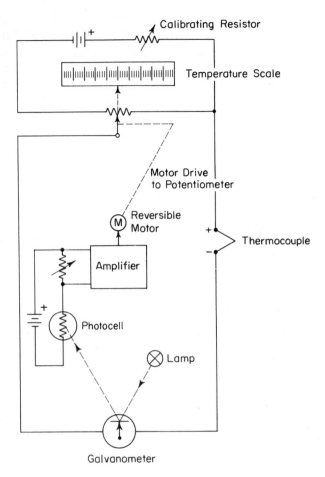

FIGURE 13-22 Automatic balancing potentiometer using a photocell to provide two-directional positioning of the potentiometer.

motor is mechanically coupled to the slider of the potentiometer and drives it in a direction to restore circuit balance.

13-5.3 Thermistor Characteristics

Thermistors, or *thermal resistors*, are semiconductor devices that behave as resistors with a high, usually *negative*, temperature coefficient of resistance. In some cases, the resistance of a thermistor at room temperature may decrease as much as 6 per cent for each 1°C rise in temperature. This high sensitivity to temperature change makes the thermistor extremely well suited to precision temperature *measurement*, *control*, and *compensation*. Thermistors

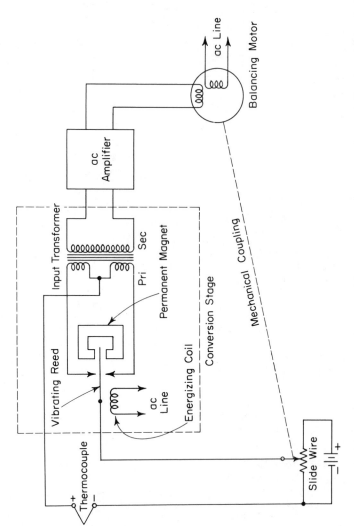

FIGURE 13-23   Automatic balancing potentiometer using a converter to change the dc error signal into ac for amplification and use.

are therefore widely used in such applications, especially in the lower temperature range of $-100°C$ to $300°C$.

Thermistors are composed of a sintered mixture of metallic oxides, such as manganese, nickel, cobalt, copper, iron, and uranium. Their resistances range from $0.5 \, \Omega$ to $75 \, M\Omega$ and they are available in a wide variety of shapes and sizes. Smallest in size are the beads with a diameter of 0.15 mm to 1.25 mm. Beads may be sealed in the tips of solid glass rods to form probes that are somewhat easier to mount than beads. Disks and washers are made by pressing thermistor material under high pressure into flat cylindrical shapes with diameters from 2.5 mm to 25 mm. Washers can be stacked and placed in series or in parallel for increased power dissipation.

Three important characteristics of thermistors make them extremely useful in measurement and control applications: (a) the *resistance-temperature* characteristic, (b) the *voltage-current* characteristic, and (c) the *current-time* characteristic. Representative examples of these characteristic curves are shown in Fig. 13-24.

The resistance-temperature characteristic of Fig. 13-24(a) shows that the thermistor has a very high negative temperature coefficient of resistance, making it an ideal *temperature transducer*. The resistance-versus-temperature variations of the two industrial materials are compared to the characteristics for platinum (a widely used resistance thermometer material). Between the temperatures of $-100°C$ and $+400°C$, the resistance of type A thermistor material changes from 10^7 to 1 Ohm-cm, while the resistance of platinum varies only by a factor of approximately 10 over the same temperature range.

The voltage-current characteristic of Fig. 13-24(b) shows that the voltage drop across a thermistor increases with increasing current until it reaches a peak value beyond which the voltage drop decreases as the current increases. In this portion of the curve the thermistor exhibits a *negative resistance* characteristic. If a very small voltage is applied to the thermistor, the resulting small current does not produce sufficient heat to raise the temperature of the thermistor above ambient. Under this condition, Ohm's law is followed and the current is proportional to the applied voltage. Larger currents, at larger applied voltages, produce enough heat to raise the thermistor above the ambient temperature and its resistance then decreases. As a result, more current is then drawn and the resistance decreases further. The current continues to increase until the heat dissipation of the thermistor equals the power supplied to it. Therefore under any fixed ambient conditions, the resistance of a thermistor is largely a function of the power being dissipated within itself, provided that there is enough power available to raise its temperature above ambient. Under such operating conditions, the temperature of the thermistor may rise $100°C$ or $200°C$, and its resistance may drop to one-thousandth of its value at low current.

This characteristic of *self-heat* provides an entirely new field of uses for

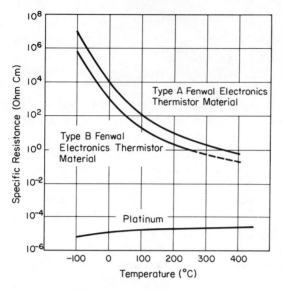

(a) Resistance–temperature characteristic

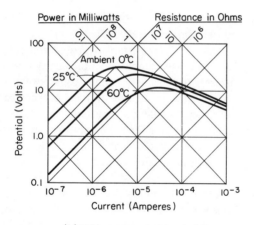

(b) Voltage–current characteristic

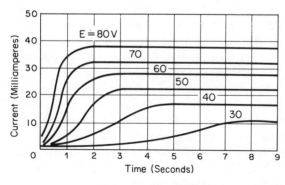

(c) Current–time characteristic

FIGURE 13-24 Character-istic curves of thermistors. (courtesy Fenwall Electronics, Inc.).

434

the thermistor. In the self-heat state the thermistor is sensitive to anything that changes the *rate* at which heat is conducted away from it. It can so be used to measure *flow, pressure, liquid level, composition of gases,* etc. If, however, the rate of heat removal is fixed, the thermistor is sensitive to power input and can be used for *voltage* or *power-level* control.

The current-time characteristic curve of Fig. 13-24(c) indicates the time delay to reach maximum current as a function of the applied voltage. When the self-heating effect just described occurs in a thermistor network, a certain finite time is required for the thermistor to heat and the current to build up to a maximum steady-state value. This time, although fixed for a given set of circuit parameters, may easily be varied by changing the applied voltage or the series resistance of the circuit. This time-current effect provides a simple and accurate means of achieving time delays from milliseconds to many minutes.

13-5.4 Thermistor Applications

Although thermistors are probably best known for their function in the measurement and control of temperature, they can be used in a variety of other applications. A few common applications are described in the following paragraphs.

The thermistor's relatively large resistance change per degree change in temperature (called the *sensitivity*) makes it an obvious choice as a temperature transducer. A typical industrial-type thermistor with a 2,000-Ω resistance at 25°C and a temperature coefficient of 2.9 per cent/°C will exhibit a resistance change of 78 Ω/°C change in temperature. When this thermistor is connected in a simple series circuit consisting of a battery and a micro-ammeter, any variation in temperature causes a change in thermistor resistance and a corresponding change in circuit current. The meter can be calibrated directly in terms of temperature and may be able to resolve temperature variations of 0.1°C. Higher sensitivity is obtained by using the bridge circuit of Fig. 13-25. The 4-kΩ thermistor will readily indicate a temperature change of as little as 0.005°C.

This high sensitivity, together with the relatively high thermistor resistance that may be selected (for example, 100 kΩ), makes the thermistor ideal for *remote* measurement or control, since changes in contact or transmission line resistance due to ambient temperature effects are negligible.

A simple *temperature control* circuit may be constructed by replacing the microammeter in the bridge circuit of Fig. 13-25 with a relay. This is indicated in the typical thermistor temperature control circuit of Fig. 13-26 where a 4-kΩ thermistor is connected in an ac-excited bridge. The unbalance voltage is fed to an ac amplifier whose output drives a relay. The relay contacts are used to control the current in the circuit that generates the heat. These control circuits can be operated to a precision of 0.0001°F.

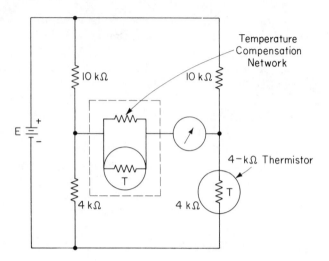

FIGURE 13-25 Temperature measurement with the thermistor in a bridge circuit to improve sensitivity.

Thermistor control systems are inherently sensitive, stable, and fast acting, and they require relatively simple circuitry. The voltage output of the standard thermistor bridge circuit at 25°C will be approximately 18 mV/°C using a 4,000-Ω thermistor in the configuration of Fig. 13-25.

Because thermistors have a negative temperature coefficient of resistance—opposite to the positive coefficient of most electrical conductors and semiconductors—they are widely used to *compensate* for the effects of tem-

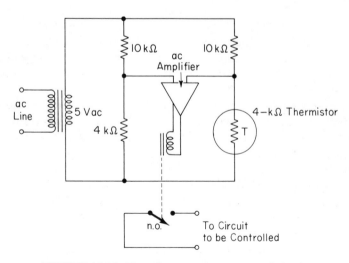

FIGURE 13-26 Thermistor temperature control circuit.

perature on both component and circuit performance. Disk-type thermistors are often used when the maximum temperature does not exceed 125°C. A properly selected thermistor, mounted against or near a circuit element such as a copper meter coil, and experiencing the same ambient temperature changes, can be connected in such a way that the total circuit resistance is constant over a wide range of temperatures. This is shown in the curves of Fig. 13-27, which illustrates the effect of a *compensation network*.

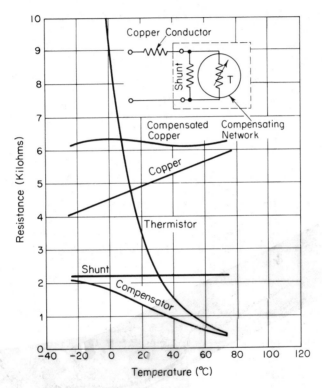

FIGURE 13-27 Temperature compensation of a copper conductor by means of a thermistor network.

The compensator consists of a thermistor, shunted by a resistor. The negative temperature coefficient of this combination equals the positive coefficient of the copper meter coil. The coil resistance of 5,000 Ω at 25°C varies from approximately 4,500 Ω at 0°C to 5,700 Ω at 60°C, representing a change of about ±12 per cent. With a single thermistor compensation network, this variation can be reduced to approximately ±15 Ω, or ±¼ per cent. With double or triple compensation networks, variations can be reduced even further.

In a *thermal conductivity measurement*, two thermistors are connected in adjacent legs of a Wheatstone bridge, as shown in Fig. 13-28. The bridge supply voltage is high enough to raise the thermistors above ambient temperature, typically to about 150°C. One thermistor is mounted in a static area to provide

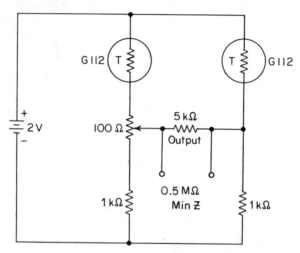

FIGURE 13-28   Thermal conductivity measurement.

temperature compensation while the other is placed in the medium to be measured. Any change in the thermal conductivity of this medium will change the rate at which heat is dissipated from the sensing thermistor, thus changing its temperature. This results in bridge unbalance, which can be calibrated in appropriate units.

In another application, two thermistors are placed in separate cavities in a brass block. With air in both cavities, the bridge is balanced. When the air in one cavity is replaced by pure carbon dioxide, which has a lower conductivity than air, the bridge will be unbalanced because that thermistor becomes hotter and lower in resistance. This amount of unbalance represents 100 per cent CO_2 in the analyzer; 50 per cent CO_2 gives just half the meter reading, and the instrument may be calibrated with a linear scale to read per cent CO_2 in air. Similar calibration may be made for any other mixture of two gases.

If the same bridge uses one thermistor sealed in a cavity in a brass block and the other thermistor mounted in a small pipe, it may be used as a *flow meter*. When no air is flowing through the pipe, the bridge may be balanced. When air flows through the pipe, the thermistor is cooled, and its resistance increases which unbalances the bridge. The amount of cooling is proportional to the rate of flow of the air and the meter may be calibrated in terms of flow in the pipe. Such instruments have been made to measure flow rates as low as 0.001 cm³ per minute.

13-6 PHOTOSENSITIVE DEVICES

13-6.1 Introduction

Photosensitive elements are versatile tools for detecting radiant energy or light. They exceed the sensitivity of the human eye to all the colors of the spectrum and operate even into the ultraviolet and infrared regions.

The photosensitive device has found practical use in many engineering applications. This section deals with the following devices and their applications:

(a) *Vacuum-type phototubes*, used to best advantage in applications requiring the observation of light pulses of short duration, or light modulated at relatively high frequencies.
(b) *Gas-type phototubes*, used in the motion picture industry as sound-on-film sensors.
(c) *Multiplier phototubes*, with tremendous amplifying capability, used extensively in photoelectric measurement and control devices and also as scintillation counters.
(d) *Photoconductive cells*, also known as *photoresistors* or *light-dependent resistors*, find wide use in industrial and laboratory control applications.
(e) *Photovoltaic cells* are semiconductor junction devices used to convert radiation energy into electrical power. A fine example is the *solar cell* used in space engineering applications.

13-6.2 Vacuum Phototube

The phototube was described briefly in Sec. 13-4.6, where an application of the phototube as a pressure transducer was given. The construction details are therefore omitted here.

The photocathode emits electrons when stimulated by incident radiant energy. The most important photocathode now used in vacuum phototubes is the cesium-antimony surface, which is characterized by high sensitivity in the visible spectrum. The type of glass used in the glass envelope determines mainly the sensitivity of the device at other wavelengths. Usually the glass cuts off the transmitted radiation in the ultraviolet region.

Typical voltage-current characteristics are shown in Fig. 13-29(a). When sufficient voltage is applied between the photocathode and the anode, the collected current is almost entirely dependent on the amount of incident light. Vacuum phototubes are characterized by a photocurrent response that is linear over a wide range, so much so that these tubes are frequently used as standards in light-comparison measurements. Figure 13-29(b) shows the linear current-light relationship.

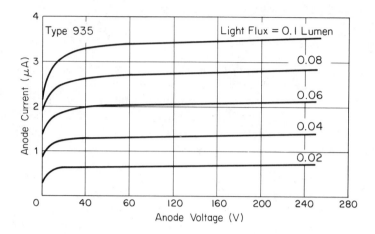

(a) Typical anode characteristics

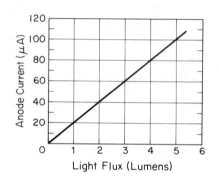

(b) Output current as a function
of light intensity

FIGURE 13-29   Characteristics of a vacuum phototube.

13-6.3   Gas-filled Phototube

The gas-filled phototube has the same general construction as the vacuum phototube, except that the envelope contains inert gas (usually argon) at a very low pressure. Electrons are emitted from the cathode by photoelectric action and accelerate through the gas by the applied voltage at the anode. If the energy of the electrons exceeds the ionization potential of the gas (15.7 V for argon), the collision of an electron and a gas molecule can result in ionization, i.e., the creation of a positive ion and a second electron. As the voltage is further increased above the ionization potential, the current collected by the anode increases because of the higher number of collisions between photoelectrons and gas molecules. If the anode voltage is raised to a very high value, the current becomes uncontrolled; all the gas molecules are then ionized and the tube exhibits a glow discharge. This condition should be avoided because it may permanently damage the phototube. Typical current-voltage characteristics for various light levels are given in Fig. 13-30.

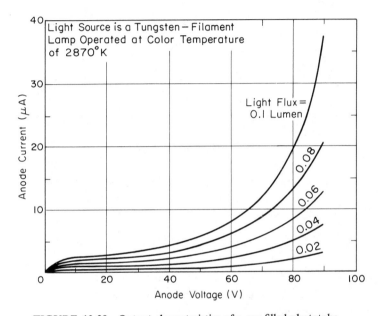

FIGURE 13-30   Output characteristics of a gas-filled phototube.

13-6.4   Multiplier Phototubes

To detect very low light levels, special amplification of the photocurrent is necessary in most applications. The multiplier phototube, or photomultiplier, uses secondary emission to provide current amplification in excess of a factor 10^6 and then becomes a very useful detector for low light levels.

In a photomultiplier the electrons emitted by the photocathode are electrostatically directed toward a secondary emitting surface, called a *dynode*. When the proper operating voltage is applied to the dynode, three to six secondary electrons are emitted for every primary electron striking the dynode. These secondary electrons are focused to a second dynode, where the process is repeated. The original emission from the photocathode is therefore *multiplied* many times.

Figure 13-31 shows a photomultiplier with ten dynodes. The last dynode (10) is followed by the anode that collects the electrons and serves as the signal output electrode in most applications.

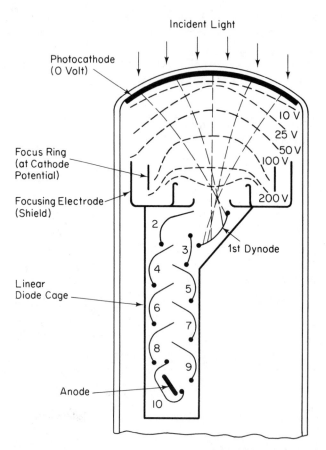

FIGURE 13-31 Linear photomultiplier with Matheson front-end configuration, showing equipotential lines and electron trajectories feeding into a linear diode cage (courtesy Radio Corporation of America).

The linear photomultiplier of Fig. 13-31 (also known as the Matheson tube) has a specially designed focused cage structure with a large effective area for the collection of photoelectrons on the first dynode. The Matheson tube uses a curved cathode and annular rings for electrostatic focusing of the photoelectrons. This construction results in very effective collection of photoelectrons and also in very short transit times (high-frequency response).

The gain of the photomultiplier depends on the number of dynodes and the properties of the dynode material. For a typical ten-dynode tube, such as shown in Fig. 13-31, the gain would be on the order of 10^6 with an applied voltage of 100 V per stage (a 1,000-V supply source would be needed in this case). The spectral response may be controlled by the material of which the cathode and the dynodes are made. The output of the multiplier is linear, similar to the output of the vacuum phototube.

Magnetic fields affect the gain of the photomultiplier because some electrons may be deflected from their normal path between stages and therefore never reach a dynode nor, eventually, the anode. In scintillation-counting applications this effect may be disturbing, and mu-metal magnetic shields are often placed around the photomultiplier tube.

13-6.5 Photoconductive Cells

Photoconductive cells are elements whose conductivity is a function of the incident electromagnetic radiation. Many materials are photoconductive to some degree, but the commercially important ones are cadmium sulfide, germanium, and silicon. The spectral response of the cadmium sulfide cell closely matches that of the human eye, and the cell is therefore often used in applications where human vision is a factor, such as street light control or automatic iris control for cameras.

The essential elements of a photoconductive cell are the ceramic substrate, a layer of photoconductive material, metallic electrodes to connect the device into a circuit, and a moisture-resistant enclosure. A cutaway view of a typical photoconductive cell is shown in Fig. 13-32.

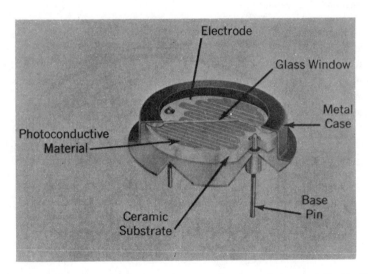

FIGURE 13-32 Cutaway view of a photoconductive cell (courtesy Radio Corporation of America).

A typical application of a practical *on-off* photocell control circuit is given in Fig. 13-33. Resistors R_2, R_3, and R_4 are chosen so that the emitter-to-base bias of Q_2 is sufficiently positive to allow Q_2 to conduct. As a result, the relay in Q_2 collector circuit will be energized. When configuration A is used as the control circuit. the relay is energized when the light on the photocell is

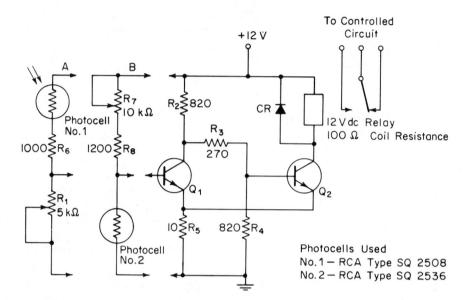

FIGURE 13-33   12-V photocell control circuit (courtesy RCA, Electronic Components and Devices Division).

below a predetermined level. When the photocell is illuminated, the emitter-to-base bias of Q_1 becomes sufficiently positive to allow Q_1 to conduct. Its collector potential becomes considerably less positive, decreasing the bias on Q_2, and Q_2 cuts off, deenergizing the relay. When configuration B is used, the relay will be energized when the light incident on the photocell is above a predetermined level.

Semiconductor junction photocells are used in some applications. The volt-ampere characteristics of a p-n junction may appear as the solid line in Fig. 13-34, but when light is applied to the cell, the curve shifts downward, as shown by the broken line.

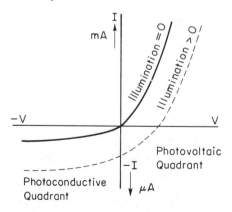

FIGURE 13-34   Current-voltage characteristics of a photojunction diode.

In photoconductive applications the cell is biased in the reverse direction. When the cell is illuminated, the reverse current increases and an output voltage can be developed across an output resistor. This output voltage is then proportional to the amount of incident light. A typical order of magnitude for the increase in output current is approximately 0.7 μA for each 1 footcandle increase in illumination. This increase in photocurrent is linear with an increase in illumination. The time constant of p-n junction photocells is relatively fast, making these devices useful for optical excitation frequencies well above the audio range.

13-6.6 Photovoltaic Cells

Photovoltaic cells may be used in a number of applications. The silicon *solar cell* converts the radiant energy of the sun into electrical power. The solar cell consists of a thin slice of single crystal *p*-type silicon, up to 2 cm square, into which a very thin (0.5 micron) layer of *n*-type material is diffused. The conversion efficiency depends on the spectral content and the intensity of the illumination.

Multiple-unit silicon photovoltaic devices may be used for sensing light in applications such as reading punched cards in the data-processing industry.

Gold-doped germanium cells with controlled spectral response characteristics act as photovoltaic devices in the infrared region of the spectrum and may be used as infrared detectors.

13-7 MAGNETIC MEASUREMENTS

13-7.1 Ballistic Galvanometer

The deflection of a ballistic galvanometer is directly proportional to the electric charge flowing through its coil. Since charge and flux are related by a constant of proportionality, the galvanometer deflection is a measure of the flux so that

$$\phi = K\theta \quad \text{(weber)} \tag{13-12}$$

where ϕ = magnetic flux, in webers

K = constant of proportionality

θ = angular deflection of the galvanometer, in radians

The instrument must be calibrated in the circuit in which the flux measurement is to be made. One calibration technique is described in Sec. 3-5, and the actual flux measurement setup is shown in Fig. 3-8.

To examine the properties of magnetic materials, one single flux measurement is often insufficient. The measurement setup of Fig. 13-35 allows determination of a *hysteresis loop* of a ring sample of magnetic material by

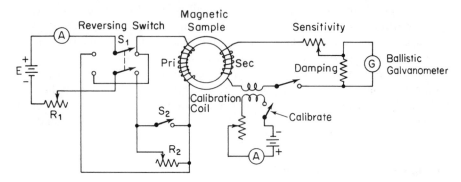

FIGURE 13-35   Circuit using a ballistic galvanometer to determine the hysteresis curve of a magnetic sample.

measuring the flux with a ballistic galvanometer at different values of the magnetizing force. The calibration coil of Fig. 13-35 is used to determine the galvanometer sensitivity K of Eq. (3-9). The hysteresis loop, or BH curve, is plotted by measuring B with the galvanometer for different values of H. In the international system (SI), H is measured in AT/m. The current through the primary winding on the ring sample is controlled by rheostat R_1. Since the number of primary turns and the mean circumference of the sample are known, H can be set to any desired value with R_1.

The hysteresis loop is measured in the following manner: Switch S_2 is initially closed and the current in the primary winding is set by R_1 to the desired value of maximum H. Switch S_1 is reversed several times so that the sample is in a cyclic condition and the deflections of the ballistic galvanometer are read. The average value of the repeated measurements yields the value for maximum flux density B. Switch S_2 is now opened, which shunts R_2 into the current circuit and reduces the magnetizing force by a small amount. The reduction in flux density ΔB is obtained from the galvanometer deflection and the new value of H from the meter reading. Several measurements are made by manipulation of the reversing switch S_1, and the average value of ΔB is found. Now S_2 is closed again and the sample is returned to its initial position of maximum magnetization. Rheostat R_2 is now slightly adjusted to decrease the total magnetization current, and a new set of measurements is made, starting from the initial point of maximum H.

13-7.2   Fluxmeter and Gauss Meter

The *fluxmeter* uses a special moving-coil movement that lacks an internal magnet and pole pieces. This meter is placed in the unknown magnetic field and a current is passed through the meter. The deflection of the fluxmeter depends on the amount of current and the strength of the unknown magnetic field. The amount of current can be controlled with a rheostat and read on another meter whose scale is calibrated in gauss or weber. The amount of current for a standard deflection on the fluxmeter is then directly proportional to the magnetic fieldstrength, and the current reading is a direct indication of the magnetic fieldstrength.

The *gauss meter* operates on a different principle. The torque exerted on a small magnet by magnetic induction is balanced by the restoring torque of a spiral spring. The small magnet is brought into the influence of the unknown magnetic field and rotated for maximum indication of a pointer connected to the restoring spiral spring. The instrument scale is calibrated to read magnetic fieldstrength directly in either gauss or weber.

13-7.3 Magnetic Transducers

Bismuth and *mu-metal* have the property of changing their resistance or impedance if placed in a *transverse* magnetic field. This effect may be used to measure flux density with a conventional Wheatstone bridge. The method is indicated in Fig. 13-36 where two mu-metal wires are placed in the unknown

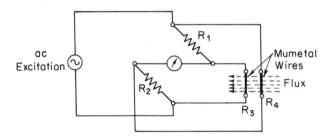

FIGURE 13-36 An ac mu-metal bridge to measure small current magnetic fields.

magnetic field. The impedance of the wires is a function of the strength of the magnetic field and is measured with an ac bridge. This transducer produces an acceptable emf for the input to an instrumentation system and is capable of measuring fieldstrength on the order of milligauss.

A slight variation of the circuit of Fig. 13-36 uses two bismuth spirals in opposite arms of the bridge. The resistance of the bismuth changes with the strength of the magnetic field, and this change is measured with a dc bridge.

The *Hall effect* transducer uses a strip of semiconductor material that is exposed to the unknown magnetic field. When a strip of conductor or semiconductor material carries current in the presence of a transverse magnetic field, an emf is produced between the opposite edges of the semiconductor strip, as shown in Fig. 13-37. Leads 1 and 2 conduct current through a strip of germanium. Leads 3 and 4 are at the same potential when there is no

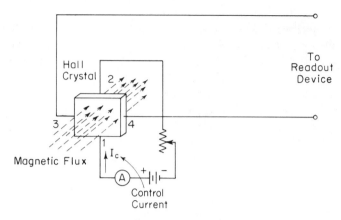

FIGURE 13-37   Circuit for measuring magnetic flux by using the Hall effect of a germanium crystal.

transverse magnetic field passing through the strip. When there is a magnetic flux through the strip, the voltage between leads 3 and 4 is proportional to the product of the current and the fieldstrength. A number of materials exhibit the Hall effect, but in many cases it is so small that it has no practical value. Germanium can be manufactured with a very high Hall coefficient, and germanium probes to measure magnetic flux are used for small flux densities.

REFERENCES

1. Stout, Melville B., *Basic Electrical Measurements*, 2nd ed., chap. 16. Englewood Cliffs, N.J.: Prentice-Hall, Inc., 1960.

2. Bartholomew, Davis, *Electrical Measurements and Instrumentation*, chap. 11. Boston: Allyn and Bacon, Inc., 1963.

3. Partridge, G. R., *Principles of Electronic Measurements*, chap. 13. Englewood Cliffs, N.J.: Prentice-Hall, Inc., 1958.

4. Perry, C. C., and H. R. Lissner, *The Strain Gage Primer*, 2nd ed. New York: McGraw-Hill Book Company, 1962.

5. Lion, Kurt S., *Instrumentation in Scientific Research*. New York: McGraw-Hill Book Company, 1959.

6. Fribance, Austin E., *Industrial Instrumentation Fundamentals*, chaps. 10, 12, 15, 16. New York: McGraw-Hill Book Company, 1962.

7. Minnar, E. J. (ed.), *Instrument Society of America Transducer Compendium*. New York: Plenum Press, 1963.

8. *Capsule Thermistor Course*. Fenwal Electronics, Inc., Framingham, Mass., n. d.

9. *Phototubes and Photocells*, Technical Manual PT-60. Radio Corporation of America, Electronic Components and Devices, Lancaster, Pa., n. d.

10. *Introduction to Transducers for Instrumentation*. Statham Instruments, Inc., Los Angeles, Calif., n. d.

PROBLEMS

1. Name four types of electrical pressure transducer and describe one application of each type.

2. Under what conditions is a "dummy" strain gage used, and what is the function of that gage?

3. What is the difference between a photoemissive, a photoconductive, and a photovoltaic cell? Name one application for each cell.

4. A resistance strain gage with a gage factor of 2.4 is mounted on a steel beam whose modulus of elasticity is 2×10^6 kg/cm^2. The strain gage has an unstrained resistance of 120.0 Ω which increases to 120.1 Ω when the beam is subjected to a stress. Calculate the stress at the point where the strain gage is mounted.

5. The unstrained resistance of each of the four elements of the unbonded strain gage of Fig. 13-4 is 120 Ω. The strain gage has a gage factor of 3 and is subjected to a strain ($\Delta l/l$) of 0.0001. If the indicator is a high-impedance voltmeter, calculate the reading of this voltmeter for a battery voltage of 10 V.

6. The high-impedance voltmeter of Prob. 5 is replaced with a 200-Ω galvanometer with a current sensitivity of 0.5 mm/μA. Calculate the galvanometer indication, in millimeters, for the situation described in Prob. 5.

7. The linear variable differential transformer (LVDT) of Fig. 13-8 produces an output of 2 V rms for a displacement of 50×10^{-6} cm. Calculate the sensitivity of the LVDT in μV/mm. The 2-V output of the LVDT is read on a 5-V voltmeter that has a scale with 100 divisions. The scale can be read to 0.2 division. Calculate the resolution of the instrument in terms of displacement in inches.

14

Analog and Digital Data Acquisition Systems

14-1 INSTRUMENTATION SYSTEMS

Data acquisition systems are used to measure and record signals obtained in basically two ways: (a) Signals originating from *direct measurement* of electrical quantities; these may include dc and ac voltages, frequency, or resistance, and are typically found in such areas as electronic component testing, environmental studies, and quality analysis work. (b) Signals originating from *transducers*, such as strain gages and thermocouples (see Chapter 13).

Instrumentation systems can be categorized into two major classes: analog systems and digital systems. *Analog systems* deal with measurement information in analog form. An analog signal may be defined as a continuous function, such as a plot of voltage versus time, or displacement versus pressure. *Digital systems* handle information in digital form. A digital quantity may consist of a number of discrete and discontinuous pulses whose time relationship contains information about the magnitude or the nature of the quantity.

An *analog data acquisition system* typically consists of some or all of the following elements:

(a) *Transducers* for translating physical parameters into electrical signals.

(b) *Signal conditioners* for amplifying, modifying, or selecting certain portions of these signals.
(c) *Visual display devices* for continuous monitoring of the input signals. These devices may include single- or multichannel CROs, storage CROs, panel meters, numerical displays, and so on.
(d) *Graphic recording instruments* for obtaining permanent records of the input data. These instruments include stylus-and-ink recorders to provide continuous records on paper charts, optical recording systems such as mirror galvanometer recorders, and ultraviolet recorders.
(e) *Magnetic tape instrumentation* for acquiring input data, preserving their original electrical form, and reproducing them at a later date for more detailed analysis.

A *digital data acquisition system* may include some or all of the elements shown in Fig. 14-1. The essential functional operations within a digital sys-

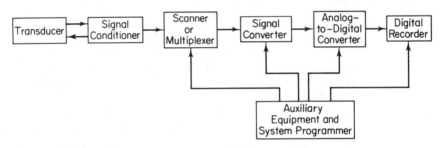

FIGURE 14-1   Elements of a digital data acquisition system.

tem include handling analog signals, making the measurement, converting and handling digital data, and internal programming and control. The function of each of the system elements of Fig. 14-1 is listed below.

(a) *Transducer.*   Translates physical parameters to electrical signals acceptable by the acquisition system. Some typical parameters include temperature, pressure, acceleration, weight displacement, and velocity (see Chapter 13). Electrical quantities, such as voltage, resistance, or frequency, also may be measured directly.
(b) *Signal conditioner.*   Generally includes the supporting circuitry for the transducer. This circuitry may provide excitation power, balancing circuits, and calibration elements. An example of a signal conditioner is a strain-gage bridge balance and power supply unit.
(c) *Scanner, or multiplexer.*   Accepts multiple analog inputs and sequentially connects them to one measuring instrument.

(d) *Signal converter.*   Translates the analog signal to a form acceptable by the analog-to-digital converter. An example of a signal converter is an amplifier for amplifying low-level voltages generated by thermocouples or strain gages.

(e) *Analog-to-digital (A/D) converter.*   Converts the analog voltage to its equivalent digital form. The output of the A/D converter may be displayed visually and is also available as voltage outputs in discrete steps for further processing or recording on a digital recorder.

(f) *Auxiliary equipment.*   This section contains instruments for system programming functions and digital data processing. Typical auxiliary functions include linearizing and limit comparison. These functions may be performed by individual instruments or by a digital computer.

(g) *Digital recorder.*   Records digital information on punched cards, perforated paper tape, magnetic tape, typewritten pages, or a combination of these systems. The digital recorder may be preceded by a coupling unit that translates the digital information to the proper form for entry into the particular digital recorder selected.

Data acquisition systems are used in a large and ever-increasing number of applications in a variety of industrial and scientific areas, such as the biomedical, aerospace, and telemetry industries. The type of data acquisition system, whether analog or digital, depends largely on the intended use of the recorded input data. In general, analog data systems are used when wide bandwidth is required or when lower accuracy can be tolerated. Digital systems are used when the physical process being monitored is slowly varying (narrow bandwidth) and when high accuracy and low per-channel cost is required. Digital systems range in complexity from single-channel dc voltage measuring and recording systems to sophisticated automatic multi-channel systems that measure a large number of input parameters, compare against preset limits or conditions, and perform computations and decisions on the input signal. Digital data acquisition systems are in general more complex than analog systems, both in terms of the instrumentation involved and the volume and complexity of input data they can handle.

Data acquisition systems often use magnetic tape recorders, which are discussed in Sec. 14-2. Digital systems require converters to change analog voltages into discrete digital quantities or numbers. Conversely, digital information may have to be converted back into analog form, such as a voltage or a current, which can then be used as a feedback quantity controlling an industrial process. Conversion techniques are discussed in Secs. 14-3 and 14-4, while scanning or multiplexing equipment is described in Sec. 14-5. Section 14-6 briefly introduces the spatial, or shaft, encoder.

14-2 MAGNETIC TAPE RECORDERS

14-2.1 Elements of a Tape Recorder

A magnetic tape recorder consists of the following basic elements:

(a) The *record head* that responds to an electrical signal and creates a magnetic pattern on a magnetizable medium.
(b) The *magnetic tape*, as the magnetizable medium, that conforms to and retains the magnetic pattern.
(c) The *reproduce head* that detects a magnetic pattern on the tape and converts it into an electrical signal.
(d) *Conditioning devices*, such as amplifiers and filters, to modify the input signal to a format that can be properly recorded on the tape.
(e) A *tape transport mechanism* that moves the tape along the record and reproduce heads at a constant speed.

Three recording methods are specified to meet various industrial requirements: (a) *direct recording*, (b) *frequency modulation*, or *FM*, *recording*, (c) *pulse modulation*, or *PM*, *recording*. Direct recording provides the greatest bandwidth and requires only relatively simple electronic components. FM recording overcomes some of the basic limitations of the direct recording process although sacrificing high-frequency bandwidth. Both direct and FM recording are commonly used in industrial instrumentation systems.

14-2.2 Direct Recording

A recording head is similar to a toroidal transformer with a single winding as shown in Fig. 14-2. The core is made in the form of a closed ring but it has a short *nonmagnetic gap* in it. Signal current in the winding causes magnetic flux that detours around the gap through the magnetic material on the tape. Magnetic tape is simply a ribbon of plastic with tiny particles of magnetic material deposited on it. When the tape is moved across the record head gap, the magnetic tape material is subjected to a flux pattern proportional to the signal current in the head winding. As it leaves the gap, each tiny magnetic particle retains the state of magnetization that was last imposed upon it by the protruding flux. The actual recording therefore takes place at the *trailing* edge of the record head gap.

The magnetic pattern on the tape is *reproduced* by moving the tape across a reproduce head. This head is similar in construction to the record head (in fact, it may be the same head). The magnetic oxide on the moving tape bridges the small nonmagnetic gap in the head, and magnetic lines of flux are shunted through the core. The induced voltage in the head winding

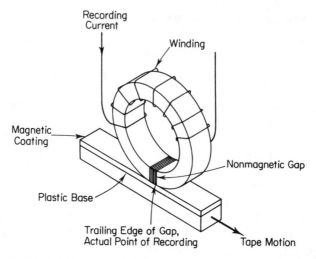

FIGURE 14-2   Simplified diagram of a magnetic recording head.

is proportional to the *rate of change* of flux, and the reproduced signal is therefore the *derivative* of the recorded signal and not the signal itself. Consider the following discussion:

Assuming that the recording current in the head of Fig. 14-2 is sinusoidal, the instantaneous value of this current is given by

$$I_1 = I_m \sin (2\pi f t) \tag{14-1}$$

where I_m = peak value of the recording current

f = frequency of the sinusoidal waveform

The magnetic flux in the recording head is proportional to the recording current, and we can write

$$\begin{aligned} \phi_1 &= k_1 I_1 \\ &= k_1 I_m \sin (2\pi f t) \\ &= \phi_m \sin (2\pi f t) \end{aligned} \tag{14-2}$$

where ϕ_1 = recording flux

k_1 = proportionality constant

f = frequency of the recording current

In the *reproduce* mode the flux through the reproduce head is proportional to the flux deposited on the magnetic tape, and

$$\begin{aligned} \phi_2 &= k_2 \phi_1 \\ &= k_2 \phi_m \sin (2\pi f t) \end{aligned} \tag{14-3}$$

where ϕ_2 = reproduce flux

   k_2 = proportionality constant

The emf induced in the winding of the reproduce head equals

$$e = -N\frac{d\phi_2}{dt}$$
$$= -Nk_2\phi_m 2\pi f \cos(2\pi ft)$$
$$= k_3 f\phi_m \cos(2\pi ft) \qquad (14\text{-}4)$$

Equation (14-4) indicates that the output from the reproduce head is not only proportional to the flux recorded on the tape, but also to the *frequency* of the recording current. The output voltage doubles for every octave rise in frequency of the recording current, and the reproduce head output is therefore subject to a 6-dB/octave rise. The 6-dB/octave rise must be compensated in the reproduce amplifier by a process called *amplitude equalization*, which simply means that the amplifier has an amplitude response that drops at the rate of 6 dB/octave thereby compensating for the 6-dB/octave rise in the reproduce head. This is indicated in Fig. 14-3.

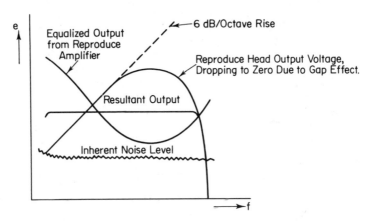

FIGURE 14-3 Amplitude equalization in direct recording.

Furthermore, Eq. (14-4) shows that as the recorded frequency approaches zero, the output voltage decreases and falls below the inherent noise level of the overall recording system. The direct recording process *cannot* be used for recording dc signals.

14-2.3 The Magnetizing Process

It has been assumed that the hysteresis, or *BH*, curve of the magnetic tape material is linear so that the magnetization (*B*) on the tape increases linearly

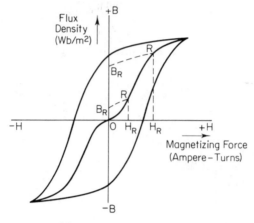

(a) Magnetization curve

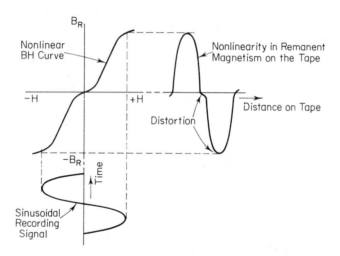

(b) Head-to-tape transfer characteristic

FIGURE 14-4   Nonlinearity in a direct recording process.

with the magnetizing force (H) or number of ampere-turns. Figure 14-4(a) shows a typical BH curve with nonlinear characteristics. Tape magnetization proceeds as follows: As a demagnetized particle on the tape approaches the record head gap, it carries no residual magnetism and is therefore at point O on the BH curve. If it is assumed that one period of the recorded signal is very long compared to the gap length, the particle will pass under the gap area through an essentially constant magnetizing force H_R. This force carries the particle up the BH curve to point R. As the particle leaves the area of the gap, the magnetizing force H_R drops to zero and the particle follows a minor hysteresis loop RB_R, retaining a residual, or remanent, magnetism B_R. Since the *BH* curve is nonlinear, the remanent magnetism of all the particles on the tape follows essentially the same nonlinear pattern, and the entire magnetization process is therefore inherently nonlinear. Distortion in the reproduced signal results unless corrective action is taken. The transfer characteristic of Fig. 14-4(b) illustrates that a sinusoidal recording current deposits a non-linear magnetic pattern on the tape.

Nonlinearity may be greatly reduced by applying an *ac bias current* in series with the recording signal current so that only the linear sections of the magnetization curve are used. Figure 14-5 shows how the recording function

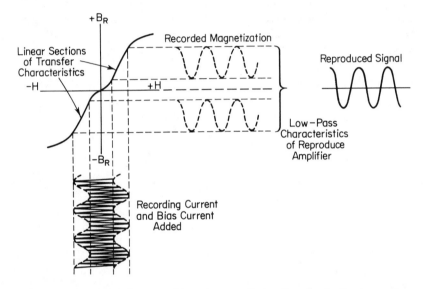

FIGURE 14-5   High-frequency bias reduces nonlinear distortion in direct recording.

is biased into the linear region of the *BH* transfer characteristic simply by *adding* or *mixing* a high-frequency (approximately four times the highest signal frequency) bias signal to the recording signal. The correct amplitude of the bias current depends on the exact *BH* transfer characteristic of the magnetic tape and should be adjusted to reach from center to center of the linear regions, as indicated in Fig. 14-5. Too large bias currents saturate the tape and result in loss of high-frequency response; too low bias currents cause distortion at the low-frequency end. In practice, bias currents are from 1 mA to 20 mA, generally five to thirty times the recording signal current, depending on the tape and head characteristics.

If a sine-wave signal is recorded on the tape, the magnetic intensity of the recorded track varies sinusoidally. The distance on the tape required to record one complete cycle is called the *recorded wavelength* λ, and

$$\lambda = \frac{v}{f} \qquad \textbf{(14-5)}$$

where λ = recorded wavelength

v = tape speed

f = frequency of the recorded signal

A tape recorder with a frequency response specified as 1.2 MHz at a tape speed of 120 ips (inch per second) is capable of a *packing density* of 10,000 cycles per inch. The recorded wavelength of the magnetic pattern on the tape is then 0.0001 in., or $\frac{1}{10}$ mil, which is the limit of the recorder's *resolution*. Both packing density and resolution can be used to describe the response of a recorder, independent of tape speed, and they are more definitive of the recorder's capability than a simple frequency specification at a given speed.

The *high-frequency response* of the direct recording system is limited by several factors, the most important of which is the gap length of the reproduce

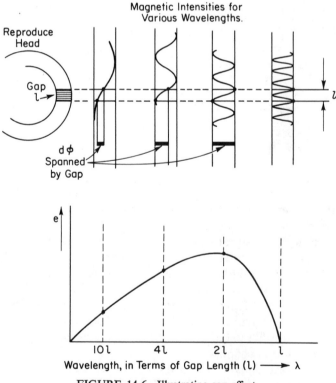

FIGURE 14-6   Illustrating gap effect.

head. As shown in Fig. 14-6, the output voltage of the reproduce head increases with frequency, simply because $d\phi/dt$ increases. This is true up to the point where the dimension of the head gap equals one-half the recorded wavelength. Beyond this point the output voltage decreases rapidly and drops to zero when the gap dimension equals the recorded wavelength. At this point there is no magnetic gradient spanned by the gap and therefore no output voltage. This phenomena is called the *gap effect* and is the most serious restriction on the high-frequency response of a tape recorder.

The *dynamic range*, or signal-to-noise (S/N) ratio, is usually quoted in decibels and is defined as the ratio of the maximum signal to the minimum signal that can be recorded at a certain amount of total harmonic distortion (THD). In general, instrumentation recorders have a S/N ratio of 22 dB to 30 dB at 1 per cent THD and a frequency response down to 400 Hz.

14-2.4 FM Recording

In the FM recording process, the carrier frequency is varied above and below its center value (modulated) in accordance with the *amplitude* of the data signal. The rate at which the carrier frequency deviates from its center value is a function of the *frequency* of the data signal. The amplitude and frequency characteristics that define the data signal are therefore contained in the frequency variations of the FM carrier around its center value. When this modulated FM signal is reproduced (demodulated), the data signal is reconstructed in the FM demodulator by detecting the number and rate of zero crossings.

It is clear that FM recording is extremely sensitive to variations in tape speed (flutter) because tape speed variations introduce apparent modulation of the carrier and are interpreted by the system as unwanted signal (noise). Instability in tape speed therefore reduces the dynamic range of the system.

Since the data signal is contained entirely in the frequency characteristics of the FM carrier, the system is not sensitive to amplitude instability. In the reproduce process residual carrier signals and any out-of-band noise are removed by low-pass filters.

Two important factors in FM recording are deviation ratio and percentage deviation. *Deviation ratio* is defined as the ratio of deviation of the carrier from the center frequency to the signal frequency, or

$$\delta = \frac{\Delta f}{f_m} \tag{14-6}$$

where $\quad \delta$ = deviation ratio

Δf = carrier deviation from center frequency

f_m = data signal frequency

A system with a high deviation ratio generally has a low noise figure. However, Δf is limited by the recorder bandwidth and f_m must be kept high to accommodate all the data signals.

Percentage deviation is defined as the ratio of carrier deviation to center frequency, or

$$m = \frac{\Delta f}{f_0} \tag{14-7}$$

where m = percentage deviation, or modulation index

   Δf = carrier deviation from center frequency

   f_0 = center frequency

With a low percentage deviation, such as 7.5 per cent in FM telemetry subcarrier systems, the effect of flutter becomes noticeable. For example, a 1-per cent deviation caused by flutter in a system with 7.5-per cent deviation corresponds to $1/7.5 \times 100$ per cent = 13.3 per cent noise. The same flutter imposed on a 40-per cent deviation system causes only $1/40 \times 100$ per cent = 2.5 per cent noise signal. Systems with a higher percentage deviation are generally less influenced by tape speed variations.

A well-designed FM carrier recording system will give reasonably good amplitude accuracy, dc response and dc linearity, and low distortion. Frequency response for a given tape speed is greatly reduced, and the record-reproduce electronics is much more complex than for a direct recording system. To keep flutter figures low, the tape transport mechanism must be of good design supplying constant tape speed.

14-2.5   Digital Recording

Digital magnetic tape units are often used as storage devices in digital data processing applications. Digital tape units are generally of two types: *incremental* and *synchronous*. The *incremental* digital recorder is commanded to step ahead (increment) for each digital character to be recorded. The input data may then be at a relatively slow cr even a discontinuous rate. In this way, each character is equally and precisely spaced along the tape.

In the *synchronous* digital recorder, the tape moves at a constant speed (say, 75 cm/s) while a large number of data characters are recorded. The data inputs are at precise rates up to tens of thousands of characters per second. The tape is rapidly brought up to speed, recording takes place, and the tape is brought to a fast stop. In this way a block of characters (a *record*) is written with each character spaced equally along the tape. Blocks of data are usually separated from each other by an erased area on the tape called a *record gap*. The synchronous tape unit starts and stops the tape for each block of data to be recorded.

Characters are represented on magnetic tape by a coded combination of 1-bits in the appropriate tracks across the tape width. The recording technique used in most instrumentation tape recorders is the industry-accepted IBM format of NRZ recording. In this system the tape is magnetically saturated at all times, either in the positive or in the negative direction. The NRZ (nonreturn-to-zero) method uses the *change* in flux direction on the tape to indicate a 1-bit and *no change* in flux direction as a 0-bit. The method is illustrated in Fig. 14-7, where the binary number 1100101 is represented by

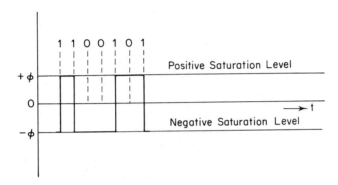

FIGURE 14-7   NRZ (nonreturn-to-zero) recording method for a digital tape recorder. Change in flux direction indicates a 1-bit; no change indicates a 0-bit.

a flux pattern in the NRZ system. Since tape magnetization is independent of frequency and amplitude but relies only on the *polarity* of the recording current, the usual problems of nonlinearity and distortion found in direct and FM recording are nonexistent. The write coils of the tape heads require only sufficient current of the correct polarity to saturate the tape.

Some of the problems encountered in digital tape recording are signal *dropout* and *spurious pulses* (losing or adding data). Signal dropout or loss of pulses becomes serious when the packing density increases (a large number of bits per unit tape length). As a check on dropout errors, most tape systems include a so-called *parity check*. This check involves keeping track of the number of 1-bits initially recorded on the tape by writing a parity check pulse on an extra tape track if the number of ones recorded is even (even parity) or if the number of ones recorded is odd (odd parity). When a dropout occurs, the parity check does not agree with the actual recorded data and a parity error is detected. Some systems use the parity-error system to insert missing bits in the appropriate places in addition to indicating that a parity error has occurred.

Some of the advantages of digital tape recording are high accuracy and relative insensitivity to tape speed, simple electronic conditioning devices, and direct compatibility with computer systems allowing direct transfer of data between tape unit and computer.

14-3 DIGITAL-TO-ANALOG CONVERSION

The problem of converting a digital number to an analog voltage can be solved in several ways. The more sophisticated converters use diodes, transistors, and other elements, but the principle of conversion is best illustrated by the simple passive resistor network of Fig. 14-8.

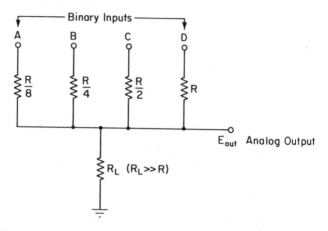

FIGURE 14-8 Basic digital-to-analog converter using a resistive divider network. The binary inputs are at 0 V or +E V, assuming positive logic.

Assume a logic system where a binary 0 is represented by a voltage level of 0 V and a binary 1 by a voltage level of +E V. The binary number, represented by its corresponding combination of voltage levels, is applied to the input terminals of the resistive divider, with the least significant bit (LSB) connected to the terminal marked D. The four input resistors are weighted so that bit 1 (the LSB) has an input resistor of value R, bit 2 has an input resistance of value $R/2$, bit 3 has a resistor of value $R/4$, and so on. The value of R_L, the load resistor, is very large compared to the input resistors. The output voltage E_{out} will be a dc voltage between the values 0 V and +E V, depending on the value of the binary number represented by the four inputs.

The binary number 0001, applied to the input of the converter of Fig. 14-8, applies 0 V to the A, B, and C inputs and +E to the D input. The input resistors act as a voltage divider, connected between 0 and +E V, and consisting of R_D in series with the parallel combination of R_A, R_B, and R_C. The output voltage E_{out} therefore equals $\frac{1}{15}E$ V. If the input equals the binary number 0010, the voltage E_{out} will be $\frac{2}{15}E$ V, and if the input is 0011, the output voltage will be $\frac{3}{15}E$ V. In other words, an increase of one bit at the input to the converter causes an increase in output voltage of $\frac{1}{15}E$ V. When the input reaches its maximum number of 1111, the full output voltage of +E V is obtained. The *digital input* signal is therefore converted, in discrete steps of $\frac{1}{15}E$ V, to an *analog output* voltage.

The *accuracy* of conversion depends on the accuracy of the resistors and the voltage levels of the binary inputs. The resistors are usually carefully selected precision resistors, and the voltage levels of the binary inputs are controlled by a reference supply in order to increase conversion accuracy. The input circuit should also be able to deliver the necessary current without affecting the dc input level.

In a practical circuit, the resistive network—sometimes called a DAC, or digital-to-analog conversion module—is connected to a flip-flop register that *holds* the digital number. Since the divider is simply a passive network, the digital input voltage (the *on* and *off* levels) determines the output voltage. Since digital voltage levels are usually not as precise as required in an analog system, level *amplifiers* may be placed between the flip-flop register and the divider network. These amplifiers switch the inputs to the divider network between ground and a reference voltage supplied by a precision reference supply. The analog output voltage then falls between these two levels.

A practical D/A converter is shown in Fig. 14-9. The basic components

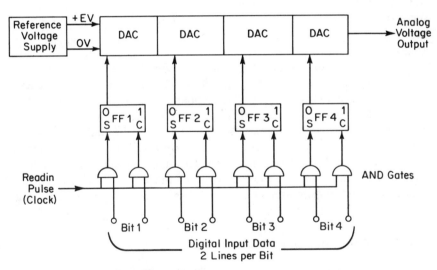

FIGURE 14-9 Digital-to-analog converter.

are recognized as the flip-flop register, the DAC modules which include the level amplifiers, and a reference voltage supply. The digital signals are dropped into the register by a *drop-in* pulse (usually a clock pulse) and are automatically converted by the DAC divider network to the appropriate analog voltage.

It always takes some time for the conversion to be completed after the digital signals are dropped into the register. This *settling* time depends on the number of flip-flops in the register that change state and also on the voltage difference between the original output voltage and the new output voltage. For instance, when the digital input changes from the binary number 0111 to the new binary number 1000, all the flip-flops change their state. The output voltage, however, changes only by $\frac{1}{15}E$ V. *Transients* may occur at the analog output because of variations between transition times of the different flip-flops, and transient current may be drawn from the reference supply. These transients are usually of very short duration (typically, on the order of 2 μs) and can be neglected since the load cannot respond within this time.

14-4 ANALOG-TO-DIGITAL CONVERSION

14-4.1 Introduction

Analog-to-digital conversion is slightly more complex than digital-to-analog conversion, and a number of different methods may be used. Four common conversion methods are described in this section. Of these, the successive-approximation counter is the most widely used A/D converter because it provides excellent performance for a wide range of applications at a reasonable cost.

The *comparator* circuit forms the basis of all A/D converters. This circuit compares an unknown voltage with a reference voltage and indicates which of the two voltages is larger. A comparator is essentially a multistage high-gain differential amplifier, where the state of the output is determined by the *relative polarity* of the two input signals. If, for instance, input signal *A* is greater than input signal *B*, the output voltage is a maximum and the comparator is *on*. If input signal *A* is smaller than input signal *B*, the output voltage is minimum and the comparator is *off*. Since the amplifier has a very high gain, it either saturates or cuts off at relatively low differential input levels, so that it acts as a *binary* device.

14-4.2 Simultaneous A/D Converter

A simple yet effective A/D converter can be built by using several comparator circuits. This is shown in the circuit of Fig. 14-10, where three comparator circuits are used. Each of the three comparators has a *reference input* voltage, derived from a precision reference voltage source. A resistive divider consisting of four equal precision resistors is connected across the reference supply and provides output voltages of $\frac{3}{4}$ V, $\frac{1}{2}$ V, and $\frac{1}{4}$ V, where V is the reference output voltage. The other input terminal of each comparator is driven by the unknown analog voltage.

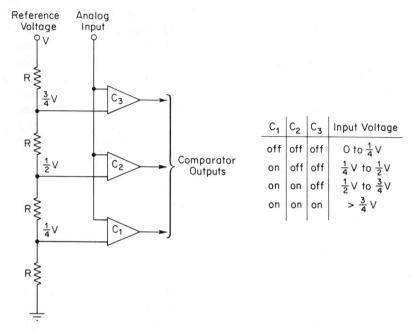

C_1	C_2	C_3	Input Voltage
off	off	off	0 to $\frac{1}{4}$ V
on	off	off	$\frac{1}{4}$ V to $\frac{1}{2}$ V
on	on	off	$\frac{1}{2}$ V to $\frac{3}{4}$ V
on	on	on	> $\frac{3}{4}$ V

FIGURE 14-10   Simultaneous analog-to-digital converter.

In this example the comparator is *on* (providing an output) if the analog voltage is larger than the reference voltage. If none of the comparators is on, the analog input must be smaller than $\frac{1}{4}$ V. If comparator C_1 is on and both C_2 and C_3 are off, the analog voltage must be between $\frac{1}{4}$ V and $\frac{1}{2}$ V. Similarly, if C_1 and C_2 are both on and C_3 is off, the analog voltage must be between $\frac{1}{2}$ V and $\frac{3}{4}$ V; if all the comparators are on, the analog voltage must be greater than $\frac{3}{4}$ V. In total, four different output conditions may exist: from *no* comparators on to *all* comparators on. The analog input voltage can therefore be *resolved* in four equal steps. These four output conditions can be coded to give two binary bits of information. This is shown in the table alongside the diagram of Fig. 14-10. Seven comparators would give three binary bits of information, fifteen comparators would give four bits, etc.

The advantage of the simultaneous system of A/D conversion is its simplicity and speed of operation, especially when low resolution is required. For a high-resolution system (a large number of bits), this method requires so many comparators that the system becomes bulky and very costly.

14-4.3   Counter-type A/D Converter

If the reference voltage to which the analog input is to be compared were *variable*, the number of comparators could be reduced to only *one*. If, for instance, the reference voltage is a linearly increasing voltage (a ramp) that is continuously applied to the comparator input, coincidence of the reference voltage and the unknown voltage could be determined in terms of the time elapsed since the ramp started. But a *digitally controlled variable reference* already exists in the form of the simple D/A converter of Fig. 14-9. This D/A converter can be used to convert a digital number in its DAC register into an analog voltage which can be compared to the unknown analog input by a comparator circuit. If the two voltages are not equal, the digital number in the DAC register is modified and its output is again compared. This is exactly the operation of the circuit of Fig. 14-11.

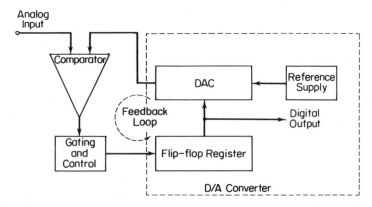

FIGURE 14-11   A/D converter using a DAC to provide the comparison voltage. The contents of the flip-flop register provide the digital output.

The generalized A/D converter of Fig. 14-11 is actually a closed-loop *feedback* system, where the main components are the DAC, the comparator, and some control logic circuitry.

Various methods may be used to *control* the conversion that takes place in the D/A converter. One of the simplest ways is to start the DAC at zero and *count* the number of input pulses required to give an output voltage that equals the analog input.

The *counter-type* A/D converter of Fig. 14-12 contains a D/A converter section consisting of the resistive divider network (DAC), the reference supply, and a six-stage counter which replaces the DAC register of Fig. 14-11. The comparator again receives the unknown analog input for comparison against the generated DAC output voltage. The control circuitry consists of a pulse generator or *clock*, a signal *gate* that steers the clock pulses to the counter, and a *control* flip-flop for starting and stopping the conversion.

When a start signal is given, all the counter flip-flops are cleared and the start-stop flip-flop is reset. This flip-flop provides a gating level (positive logic) to the signal gate, allowing the clock pulses to be applied to the counter

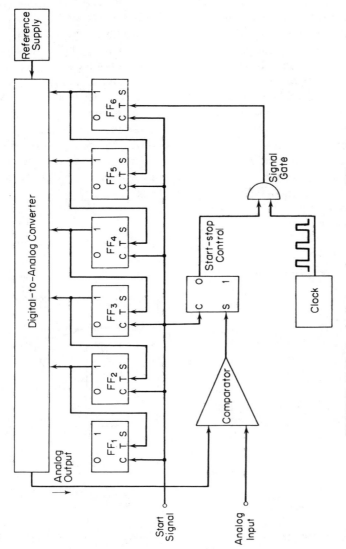

FIGURE 14-12 Logic diagram of a counter-type A/D converter.

register. The clock pulses are propagated through the counter, and the DAC divider output increases in steps toward the top of the reference voltage. When the divider output is equal to the analog input, the comparator switches, delivering an output signal to the start-stop flip-flop. This flip-flop sets and its output drops to zero, blocking the clock pulses at the signal gate. At this instant, the counter stores the number of clock pulses that were required to raise the reference voltage to the level of the analog input voltage. The contents of the counter is then the binary equivalent of the analog input.

The *conversion time* is measured from the moment a request is given to the moment a digital output is available. For the counter-type A/D converter the conversion time depends on the magnitude of the analog voltage and is therefore not constant. If the input signal is variable, it is also important to know when the input signal had the value given by the digital output. The uncertainty of this time measure is called the *aperture* time (sometimes also called *window* or *sample time*). The aperture occurs at the end of the conversion, as shown in the waveform diagram of Fig. 14-13.

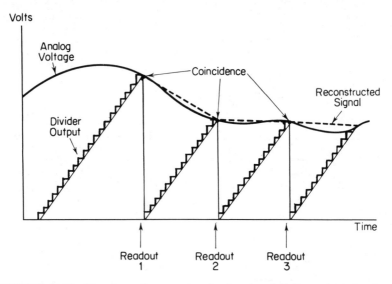

FIGURE 14-13   Waveform diagram of analog input and divider output showing points of coincidence when readout occurs.

The resolution of the counter-type converter is improved by adding extra bits (extra counter stages). This addition can be done at small extra cost. The conversion time, however, increases rapidly with the number of bits used, since an N-bit converter needs time for 2^N counts to accumulate. Higher resolution is therefore obtained at the cost of increased conversion time.

One method of decreasing the conversion time is to divide the counter into sections. For instance, a 10-bit converter could be divided into two sections of five bits each. At the start of the conversion, the least significant section of the counter is preset to all ones and counts are inserted only into the most significant section. When the comparator indicates that the analog input level has been exceeded, the least significant section of the counter is cleared, reducing the DAC divider output. Pulses are then inserted into the least significant section until the correct value is reached. The maximum number of steps required to complete a conversion is 2^5 for the most significant counter and 2^5 for the least significant counter, giving a total of 2^6 steps. This is a maximum of 64 counts (2^6) versus 1,024 counts (2^{10}) for the standard counter.

The section counter technique is frequently used in digital voltmeters, where the output is to be in decimal notation. Each section of the counter then represents a sectional digit.

14-4.4   Continuous A/D Converter

The big disadvantage of the counter converter is that the entire comparison process starts from the beginning each time a coincidence has been detected by the comparator. This means low resolution and low speed.

A slight modification of the counter method involves replacing the simple counter with a *reversible counter*, or *up-down counter*. This allows the converter to continuously follow the analog input voltage whatever the direction in which this voltage changes. Once the converter starts running, the digital equivalent of the input voltage can be sampled at any time, and an extremely rapid readout is possible.

The simplified logic block diagram of Fig. 14-14 represents the continuous converter. The illustration contains four basic parts: (a) the up-down counter, (b) the D/A converter, (c) the comparator, (d) the synchronization and control logic.

An ordinary binary counter counts in the forward direction (*up*) when the trigger input of the succeeding binary is connected to the 1 output of the preceding binary. The count will proceed in the reverse direction (*down*) if the coupling is made instead to the 0 output of the preceding binary. The two methods may be used simultaneously to produce an up-down counter. In Fig. 14-14 additional AND gates are used in the trigger circuits of the binaries to make sure that counts are accumulated only at the desired moment; *synchronization* of the count and the comparison must take place.

The D/A converter section is identical to the basic resistive divider of Fig. 14-8; the reference supply provides the required precision voltage for accurate conversion. The 0 outputs of the binaries are connected directly to the DAC, but it should be understood that appropriate level conversion takes place between the binaries and the DAC input terminals.

The comparator again compares the analog input voltage with the DAC output voltage, providing two possible output voltage levels. When the analog input voltage is larger than the feedback voltage (DAC output), the

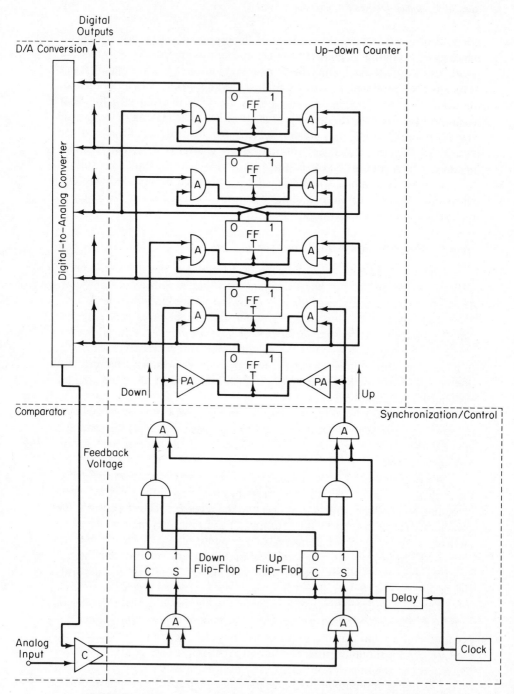

FIGURE 14-14 Simplified logic block diagram of the continuous converter.

appropriate comparator output terminal is connected to the set input of the *up* flip-flop via a gating element. Similarly, when the analog input voltage is smaller than the feedback voltage, the comparator provides an output voltage at its other terminal which is then connected, via a gate, to the set terminal of the *down* flip-flop. The actual transfer of the comparator output signals to the up and down flip-flops is controlled by gating pulses from the clock, which controls the synchronization of the entire measurement cycle. The outputs of the up and down flip-flops are *exclusive OR'd* together to make sure that no count takes place when both flip-flops are set (a safety precaution).

At the start of a measuring cycle, when all the flip-flops are cleared, the clock generates a pulse that samples the comparator output. If the analog input is larger than the feedback voltage (as it usually is when a measurement first starts), the *up* flip-flop is set. The delayed clock pulse is then allowed to trigger the first binary, while at the same time, it conditions the trigger gate of the succeeding binary. The 0 output of the first binary is connected to the DAC, and the resulting DAC output is compared to the analog input by the comparator. If the analog input is still larger, the next clock pulse sets the *up* flip-flop again, allowing the delayed clock pulse to trigger the first binary back to its original state and also to trigger the next binary. The count then has advanced by one and the corresponding DAC output is used for comparison against the analog input. The procedure repeats until the feedback voltage equals the analog input, at which time the comparator output is zero and the count is stopped.

If the analog input changes to a lower value, the next clock pulse detects this at the comparator output and sets the *down* flip-flop. Now the delayed clock pulse is allowed to enter the binary counter at the trigger input of the first binary, but the count is carried from stage to stage at the 0-output side of the binaries of Fig. 14-14, so that the contents of the counter is reduced by one. The DAC output then also drops the appropriate amount and the next comparison determines whether the *up* flip-flop or the *down* flip-flop will be set. The counter therefore *continuously follows* the analog input voltage.

The waveform diagrams of Fig. 14-15 illustrate the action of the continuous converter. The *aperture* is the time for the last step. The assumption is made that the analog input voltage does not change more than ± 1 LSB (the smallest increment of the DAC) between conversion steps. To meet this requirement, the maximum rate of change of the input voltage must not exceed the maximum rate of change of the converter.

14-4.5 Successive-approximation A/D Converter

The successive-approximation A/D converter compares the analog input to a DAC reference voltage that is repeatedly divided in half. The process is illustrated in Fig. 14-16, where a four-digit binary number (1000), repre-

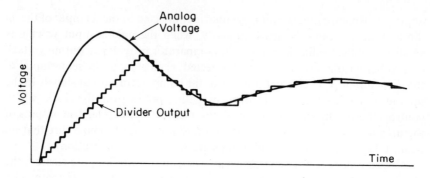

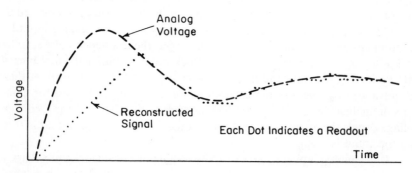

FIGURE 14-15   Waveform diagrams illustrating the action of the continuous A/D converter.

senting the full reference supply voltage V, is divided in half (binary number 100), corresponding to $\frac{1}{2}$ V. A comparison between this reference voltage ($\frac{1}{2}$ V) and the analog input is made. If the result of this comparison shows that this first approximation was too small ($\frac{1}{2}$ V is smaller than the analog input), then the next comparison will be made against $\frac{3}{4}$ V (binary number 110). If the comparison showed that the first approximation was too large ($\frac{1}{2}$ V larger than the analog input), then the next comparison will be made against $\frac{1}{4}$ V (binary number 010). After four successive approximations, the digital number is resolved. A six-digit number will be resolved in six successive approximations. This compares very favorably with the sixty-four (2^6) comparisons needed with a conventional counter-type converter.

The successive-approximation method is a little more elaborate than the previous methods since it requires a special *control register* to gate pulses to the first bit, then to the second bit, and so on. The additional cost of the control register, however, is small, and the converter can handle continuous and

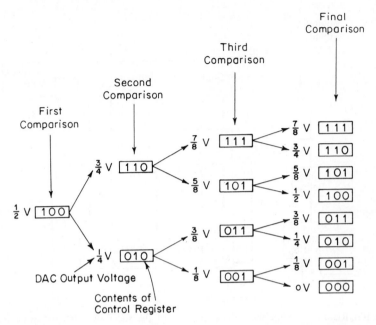

FIGURE 14-16   Operation of the successive-approximation A/D converter.

discontinuous signals with large and small resolutions at moderate speed and moderate cost.

The generalized block diagram of Fig. 14-17 shows the basic successive approximation converter. The converter uses a digital control register with gateable 1 and 0 inputs, a digital-to-analog converter with reference supply, a comparison circuit, a control timing loop, and a distribution register. The distribution register is like a ring counter with a single 1 circulating in it to determine which step is taking place.

At the beginning of the conversion cycle, both the control register and the distribution register are set with a 1 in the *most significant bit* (MSB) and a 0 in all bits of less significance. The distribution register therefore registers that the cycle has started and that the process is in its first phase. The control register, which now reads 1000 . . . , causes an output voltage at the digital-to-analog converter section of *one-half* of the reference supply. At the same instant, a pulse enters the timing delay chain. By the time the D/A converter and the comparator have settled, this delayed pulse is gated with the comparator output. When the *next* most significant bit is set in the control register by the action of the timing chain, the most significant bit either remains in the 1-state or it is reset to the 0-state, depending on the comparator output. The single 1 in the distribution register is shifted to the next position and keeps track of the number of comparisons made.

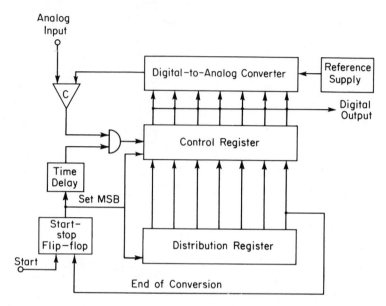

FIGURE 14-17 Simplified block diagram of the successive-approximation A/D converter.

This procedure repeats, following the diagram of Fig. 14-16, until the final approximation has been corrected and the distribution register indicates the end of the conversion. Synchronization is not required in this system because the comparator controls only one flip-flop at a time.

For a successive-approximation converter, the digital output corresponds to some value that the analog input had during the conversion. Thus the aperture time is equal to the total conversion time. This is illustrated in the waveform reconstruction of Fig. 14-18. Aperture time of this converter can be reduced by using redundancy techniques or by using a *sample-and-hold* circuit.

14-4.6 Sample-and-hold Circuit

A sample-and-hold circuit is used with an A/D converter when it is necessary to convert a high-frequency signal that is varying too rapidly to allow an accurate conversion. The sample-and-hold circuit is basically an operational amplifier that charges a capacitor during the *sample* mode and retains the charge of the capacitor during the *hold* mode. The sample-and-hold circuit can be represented by the simple switch and capacitor of Fig. 14-19.

When the switch is first closed, the capacitor charges to the value of the input voltage and then follows the input (assuming a low driving source

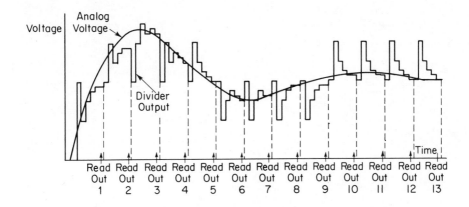

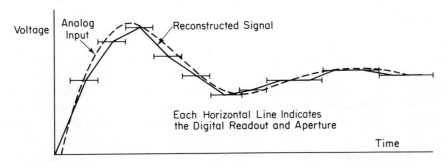

FIGURE 14-18   Waveform diagrams illustrating the operation of the successive-approximation A/D converter.

impedance). When the switch is opened, the capacitor holds the voltage that it had at the time the switch was opened (assuming a high-impedance load).

The *acquisition time* of the sample-and-hold is the time required for the capacitor to charge up to the value of the input signal after the switch is first shorted. The *aperture time* is the time required for the switch to change state and the uncertainty in the time that this change of state occurs. The *holding time* is the length of time the circuit can hold the charge without dropping more than a specified percentage of its initial value.

It is possible to build a sample-and-hold circuit exactly as shown in Fig. 14-19. Often, however, the circuit is built with fast-acting transistor switches and an operational amplifier to increase the available driving current into the capacitor or to isolate the capacitor from an external load on the output. However the sample-and-hold circuit is built, it always acts as the simple switch and capacitor shown.

An actual sample-and-hold circuit is shown in the schematic diagram of

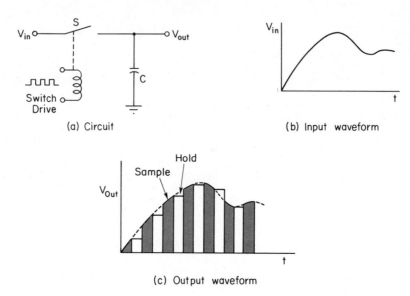

(a) Circuit

(b) Input waveform

(c) Output waveform

FIGURE 14-19   Operation of the sample-and-hold circuit.

Fig. 14-20. The sample pulse operates switches 1 and 3; the hold pulse operates switches 2 and 4. The sample-and-hold control pulses are complementary. In the sample mode the hold capacitor is charged up by the operational amplifier. In the hold mode the capacitor is switched into the feedback loop, while the input resistor R_i and the feedback resistor R_f are switched to ground. Since the input to the amplifier remains within a few μV to ground (except during switching), the input impedance is 10 kΩ in both the sample and the hold modes.

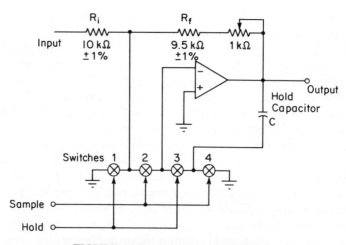

FIGURE 14-20   Sample-and-hold circuit.

14-5 MULTIPLEXING

14-5.1 Digital-to-analog Multiplexing

It is often necessary or desirable to combine, or *multiplex,* a number of analog signals into a single digital channel or, conversely, a single digital channel into a number of analog channels. Both digital signals and analog voltages can be multiplexed.

In digital-to-analog conversion a very common application of multiplexing is found in computer technology, where digital information, arriving sequentially from the computer, is distributed to a number of analog devices, such as a CRO, a pen recorder, an analog tape recorder, and so on. There are two ways to accomplish multiplexing: The first method uses a *separate* D/A converter for each channel, as shown in Fig. 14-21. The second method uses

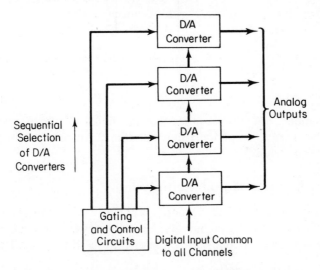

FIGURE 14-21   D/A multiplexer using several converters.

one *single* D/A converter, together with a set of analog multiplexing switches and sample-and-hold circuits on each analog channel, as shown in Fig. 14-22.

In the system of Fig. 14-21 the digital information is applied simultaneously to all channels, and channel selection is made by gating clock pulses to the appropriate output channels. One D/A converter is required per channel, so that the initial cost may be somewhat higher than the second system, but the advantage is that the analog information is available at the DAC output for an indefinite period of time (as long as the contents of the DAC flip-flop register are gated to the DAC).

The second method, illustrated in Fig. 14-22, uses only one D/A converter and is therefore slightly lower in initial cost. The multiple sample-and-

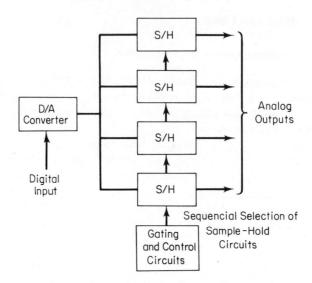

FIGURE 14-22 D/A multiplexer using *one* converter and several sample-and-hold circuits.

hold technique, however, requires that the signal on the sample-and-hold circuits be renewed at periodic intervals (the capacitors do not hold their charge indefinitely).

14-5.2 Analog-to-digital Multiplexing

In analog-to-digital conversion it is convenient to multiplex the analog inputs rather than the digital outputs. A possible system is given in Fig. 14-23, where switches, either solid-state or relays, are used to connect the analog

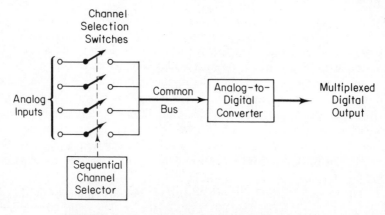

FIGURE 14-23 Multiplexed A/D conversion system.

inputs to a *common bus*. This bus then goes into a single A/D converter that is used for all channels.

The analog inputs are switched *sequentially* to the bus by the channel selector control circuitry. If simultaneous time samples from all channels are required, a sample-and-hold circuit may be used ahead of each multiplexer switch. In this manner, all channels would be sampled simultaneously and then switched to the converter sequentially.

It is also possible to multiplex by using a separate comparator for each analog channel. This system is shown in Fig. 14-24, where it is used with a

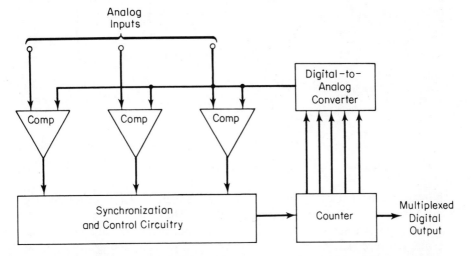

FIGURE 14-24   Counter-type A/D converter with multiplexed input.

counter-type A/D converter. The input of each comparator is connected to the output of the DAC. The other input to each comparator is connected to the separate analog input channels. Synchronization and control circuitry is required to operate the counter and sample the comparators. At the start of the multiplexing process, the counter is cleared and count pulses are applied to the counter. The D/A converter translates the counter output and provides an analog output voltage, which is fed to all the comparators. When one of the comparators indicates that the D/A output is greater than the input voltage on that channel, the contents of the counter are read out. Counting is then resumed until the next signal is received, when the correct comparator is identified and the counter contents read out again.

14-6 SPATIAL ENCODERS

A *spatial encoder* is a mechanical converter that translates the angular position of a shaft into a digital number. It is therefore an analog-to-digital converter, where the analog quantity is *nonelectrical*. An important application of this type of encoder is in the self-balancing potentiometer (Sec. 6-8), where the encoder is mounted on the shaft of the self-balancing motor and therefore *reads* the balancing emf applied to the measuring circuit.

The encoder consists of a cylindrical disk with the coding patterns arranged in concentric rings on one side of the disk. This is shown in Fig. 14-25. The patterns are alternating segments of conducting (black) and

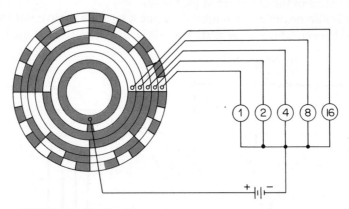

FIGURE 14-25   Spatial encoder using a binary counting system.

nonconducting (white) material. The nonconductive areas are formed by depositing a thin layer of insulating material on the conductive disk. The disk is contacted by a set of brushes, one for each ring, arranged radially from the center outward. The number of segments on the concentric rings decrease, in a binary count (32-16-8-4-2), from a total of 32 (16 conductive and 16 non-conductive) on the outside ring to 2 on the inside ring. With a battery and a lamp matrix connected as shown in Fig. 14-25, each angular position of the driving shaft would have a different combination of brush contacts bearing on the conductive surface areas. The lamp matrix would light in a binary-coded pattern.

Since the outer ring, or *commutator,* of this disk encoder has 32 distinct areas, the resolution would be $\frac{1}{32} \times 360° = 11\frac{1}{4}°$ (1 digit). The resolution can be improved by increasing the number of commutators, thereby decreasing the angle.

Spatial encoders can be made to produce any desired digital number system. The encoder shown in Fig. 14-26 is a binary-coded decimal converter where the eight coded commutators produce a readout from 0 to 99. In the position shown, the commutator segments make contact with the brushes connected to readout circuits 20, 4, and 1, giving a total indication of 25 on the readout panel. If the disk were to be rotated clockwise for one division of the outer ring, the commutators would make contact with readout circuits 20, 4, and 2, for a totalized display of 26.

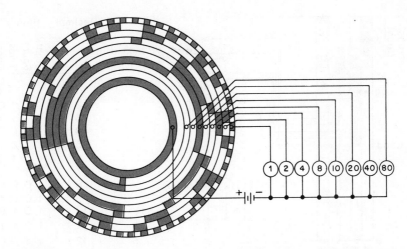

FIGURE 14-26   Binary-coded-decimal disk converter capable of a readout from 0 to 99. Since the outer commutator is divided into 100 segments, the angular position of the disk can be resolved to 3.6°.

Suppose now that the disk is rotated for only half a division from its original position in Fig. 14-26. The possibility exists that the brush on the outer ring would still make contact with the commutator segment before the brush on the second ring leaves its commutator area. The readout circuit would then indicate a reading of 20, 4, 2, and 1, for a total reading of 27, whereas the correct reading should be either 25 or 26. The possibility of ambiguous readings with this type of encoder has led to the development of several other systems to overcome the problem. One of these systems uses an extra outer commutator, with the conducting segments placed in regions where ambiguities may occur. When the brush on this outer ring makes contact, it automatically adds the proper increment of driving voltage to move the disk to a region of nonambiguity.

One common method in the binary number system uses two sets of brushes, arranged in a V-shaped pattern. This type of encoder is called the *V-brush binary encoder*. Its operation is based on the natural progression of the digits in the binary number system. Note that when the digit in the "unit" column changes from a 0 to a 1, none of the digits in the other columns changes. Furthermore, when the digit in the unit column changes from a 1 to a 0, the digit in the twos column changes. Similarly, when the digit in the twos column changes from a 0 to a 1, the digits in the more significant columns are unchanged, and when the digit in the twos column goes from a 0 to a 1, the digit in the fours column changes. This relationship is true for a change of digits in any column.

This logic is shown in diagrammatic form in Fig. 14-27, which represents a section of a binary-coded disk converter. Each of the five commutators

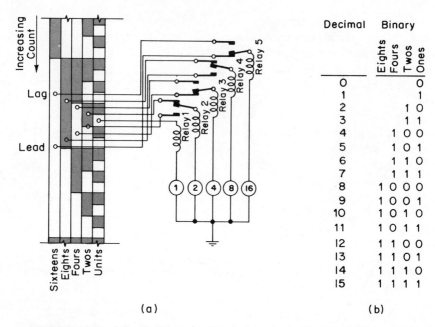

Decimal	Binary
	Eights Fours Twos Ones
0	0
1	1
2	1 0
3	1 1
4	1 0 0
5	1 0 1
6	1 1 0
7	1 1 1
8	1 0 0 0
9	1 0 0 1
10	1 0 1 0
11	1 0 1 1
12	1 1 0 0
13	1 1 0 1
14	1 1 1 0
15	1 1 1 1

(a)   (b)

FIGURE 14-27   V-brush binary encoder counting in the binary number system.

represents a digit in a five-digit binary number. The right-hand commutator represents the least significant digit, corresponding to the units column of the binary table in Fig. 14-27(b). The next commutator to the left represents the next least significant digit, corresponding to the twos column, and so on. The shaded portions of the commutator are connected to the battery circuit, similar to the connections of Fig. 14-26. In Fig. 14-27 one brush is placed on the least-significant-digit, or units, commutator. On the next commutator, two brushes are placed: One brush is *leading* the brush on the units commutator by half a division (half a count); the other brush is *lagging* the brush on the units commutator by half a count. The next two brushes, placed on the fours commutator, are displaced by half a count with respect to the two brushes on the previous commutator. All the other commutators have a leading and a lagging brush, displaced by the same half-count distance.

When the single brush on the units commutator reads a 0, as is indicated in Fig. 14-27, the lagging brush of the twos commutator reads correctly, while the leading brush may give an ambiguous reading. When the disk advances one division, the brush on the units container reads a 1, but now the leading brush on the next commutator reads correctly, whereas the lagging brush may give the ambiguous reading. Similarly, when the output of the twos commutator is a 0, the lagging brush on the next commutator reads correctly; and when the output of the twos commutator is a 1, the leading brush on the next commutator reads correctly.

It follows then that the single brush on the first ring could be used to control some logic circuitry that selects the leading or the lagging brush on the other commutating segments. Such a logic system can be provided by electromechanical relays, such as shown in Fig. 14-27, or by transistor circuitry.

If the brush on the units ring reads a 0, relay 1 is not energized, and the lagging brush on the twos ring is connected to relay No. 2 and indicator No. 2. When the brush on the units ring reads a 1, relay No. 1 is energized, removing the lagging brush and connecting the leading brush of the twos ring into the No. 2 display circuit. Similarly, when the brush on the twos commutator reads a 1, it would energize the relay in circuit No. 2, connecting circuit No. 3 to the leading brush on the fours commutator, and so on. In other words, automatic selection of the proper brushes prevents ambiguities in the readings.

REFERENCES

1. Thomas, Harry E., and Carole A. Clark, *Handbook of Electronic Instruments and Measurement Techniques*, chap. 6. Englewood Cliffs, N.J.: Prentice-Hall, Inc., 1967.

2. Bartholomew, Davis, *Electrical Measurements and Instrumentation*, chap. 7. Boston: Allyn and Bacon, Inc., 1963.

3. Ryder, John D., *Electronic Fundamentals and Applications*, 3rd ed., chaps. 14, 15. Englewood Cliffs, N.J.: Prentice-Hall, Inc., 1964.

4. Magnetic Tape Recording Handbook. Hewlett-Packard Application Note No. 89, 1967.

5. How to Use Shaft Encoders. Datex Division, Conrac Corporation, 1965.

6. Logic Handbook. Digital Equipment Corporation, Maynard, Mass., 1967.

Appendix

ABBREVIATIONS, SYMBOLS, AND PREFIXES*

The use of symbols, prefixes, and abbreviations follows the recommendations of the International Electrotechnical Commission, the American National Standards Institute, Inc., the Institute of Electrical and Electronics Engineers, and other scientific and engineering organizations. Where there is not agreement among these groups, the usage favored by the majority is chosen.

Abbreviations and Symbols

a	atto (10^{-18})		F	farad, Faraday
A	ampere		°F	degrees Fahrenheit
Å	angstrom		f	frequency, femto (10^{-15})
ac	alternating current		fm	frequency modulation
afc	automatic frequency control		FOB	free on board
am	amplitude modulation			
			G	conductance, giga (10^9)
ANSI	American National Standards Institute, Inc.		g	gram, gravitational constant
APS	American Physical Society		g_m	transconductance
ASA	Acoustical Society of America		H	henry
ASTM	American Society for Testing and Materials		h	hour, Planck's constant, hecto (10^2)
avc	automatic volume control		hf	high frequency
avg	average		h_f	forward current-transfer ratio
B	susceptance		h_i	short-circuit input impedance
bar	bar (10^5N/m²)			
BCD	binary-coded decimal		h_o	open-circuit output admittance
c	speed of light, centi (10^{-2})		h_r	reverse voltage-transfer ratio
C	capacitance, coulomb		Hz	hertz (cycle per second)
°C	degrees Celsius (Centigrade)		HTL	hearing threshold level
cd	candela			
CIF	cost, insurance, freight		I	current
CML	current-mode logic		IC	integrated circuit
COD	cash on delivery		ID	inside diameter
cw	continuous wave		IEC	International Electrotechnical Commission
d	deci (10^{-1})		IEEE	Institute of Electrical and Electronics Engineers
D	dissipation factor		if	intermediate frequency
da	deka (10)		in.	inch
dB	decibel		ISA	Instrument Society of America
dBm	decibel referred to one milliwatt		ISO	International Standards Organization
dc	direct current			
DCTL	direct-coupled transistor logic		j	$\sqrt{-1}$
dia	diameter		J	joule
DTL	diode-transistor logic			
DUT	device under test		k	kilo (10^3)
			°K	degrees Kelvin
e	electronic charge		l	liter (10^{-3} m³)
E	voltage		L	inductance
EIA	Electronic Industries Association		lb	pound
emf	electromotive force		LC	inductance-capacitance

*Courtesy General Radio Company, Concord, Mass.

lm	lumen		s	second, series (as L_s)
log	logarithm		shf	super-high frequency
lx	lux		sq	square
m	meter, milli (10^{-3})		sync	synchronous, synchronizing
M	mega (10^6)		T	period, Tesla, tera (10^{12})
max	maximum		t	time
mbar	millibar		TTL	transistor-transistor logic
mil	0.001 inch		TSA	times series analysis
min	minimum, minute			
mo	month		uhf	ultra-high frequency
n	nano (10^{-9})		v	velocity
N	newton		V	volt
			VA	volt ampere
oz	ounce		vhf	very-high frequency
			vlf	very-low frequency
p	page, parallel (as L_P), pico (10^{-12})		W	watt
P	poise (10^{-5}N · s/m^2)		Wb	Weber
PF	power factor		wt	weight
ppm	parts per million			
pps	pulses per second		X	reactance
pk-pk	peak-to-peak		Y	admittance
PRF	pulse repetition frequency		yr	year
Q	quality factor (storage factor)		Z	impedance
			α	short-circuit forward current-transfer ratio (common base)
R	resistance			
®	registered trademark		β	short-circuit forward current-transfer ratio (common emitter)
rad	radian			
RC	resistance-capacitance		Γ	reflection coefficient
RCTL	resistor-capacitor-transistor logic		Δ	increment
re	referred to		δ	loss angle
rf	radio frequency		θ	phase angle
RH	relative humidity		λ	wavelength
rms	root-mean-square		μ	micro (10^{-6})
			Ω	ohm
rpm	revolutions per minute		℧	mho
RTL	resistor-transistor logic		ω	angular velocity ($2\pi f$)

Prefixes

Orders of magnitude from 10^{-18} to 10^{12} are designated by the following prefixes:

Order	Prefix	Symbol
10^{12}	tera	T
10^9	giga	G
10^6	mega	M
10^3	kilo	k
10^2	hecto	h
10	deka	da
10^{-1}	deci	d
10^{-2}	centi	c
10^{-3}	milli	m
10^{-6}	micro	μ
10^{-9}	nano	n
10^{-12}	pico	p
10^{-15}	femto	f
10^{-18}	atto	a

DECIBEL CONVERSION TABLES

In communications systems the ratio between any two amounts of electric or acoustic power is usually expressed in units on a logarithmic scale. The decibel ($\frac{1}{10}$th of the bel) on the briggsian or base-10 scale and the neper on the napierian or base-e scale are in almost universal use for this purpose.

Since voltage and current are related to power by impedance, both the decibel and the neper can be used to express voltage and current ratios, if care is taken to account for the impedances associated with them. In a similar manner the corresponding acoustical quantities can be compared.

From Tables 1 and 2 on the following pages conversions can be made in either direction between the number of decibels and the corresponding power, voltage, and current ratios. Both tables can also be used for nepers by application of a conversion factor.

Decibel　The number of decibels N_{dB} corresponding to the ratio between two amounts of power P_1 and P_2 is

$$N_{dB} = 10 \log_{10} \frac{P_1}{P_2}$$

When two voltages E_1 and E_2 or two currents I_1 and I_2 operate in identical impedances,

$$N_{dB} = 20 \log_{10} \frac{E_1}{E_2} \quad \text{and} \quad N_{dB} = 20 \log_{10} \frac{I_1}{I_2}$$

If E_1 and E_2 and I_1 and I_2 operate in unequal impedances,

$$N_{dB} = 20 \log_{10} \frac{E_1}{E_2} + 10 \log_{10} \frac{Z_2}{Z_1} + 10 \log_{10} \frac{k_1}{k_2}$$

and

$$N_{dB} = 20 \log_{10} \frac{I_1}{I_2} + 10 \log_{10} \frac{Z_1}{Z_2} + 10 \log_{10} \frac{k_1}{k_2}$$

where Z_1 and Z_2 are the absolute magnitudes of the corresponding impedances and k_1 and k_2 are the values of power factor for the impedances. E_1, E_2, I_1, and I_2 are also the absolute magnitudes of the corresponding quantities. Note that Table 1 and Table 2 can be used to evaluate the impedance and power factor terms, since both are similar to the expression for power ratio.

Neper　The number of nepers N_{nep} corresponding to a power ratio P_1/P_2 is

$$N_{nep} = \frac{1}{2} \log_e \frac{P_1}{P_2}$$

For voltage ratios E_1/E_2 or current ratios I_1/I_2 working in identical impedances,

488

$$N_{\text{nep}} = \log_e \frac{E_1}{E_2} \quad \text{and} \quad N_{\text{nep}} = \log_e \frac{I_1}{I_2}$$

Relations Between Decibels and Nepers

Multiply decibels by 0.1151 to find nepers. Multiply nepers by 8.686 to find decibels.

TO FIND VALUES OUTSIDE THE RANGE OF CONVERSION TABLES

Decibels to Voltage and Power Ratios: Table 1, p. 491

Number of decibels positive (+) Subtract +20 decibels successively from the given number of decibels until the remainder falls within range of Table 1. To find the voltage ratio, multiply the corresponding value from the righthand voltage-ratio column by 10 for each time 20 dB was subtracted. To find the power ratio, multiply the corresponding value from the right-hand power-ratio column by 100 for each time 20 dB was subtracted.

Example: Given: 49.2 dB

$$49.2 \text{ dB} - 20 \text{ dB} - 20 \text{ dB} = 9.2 \text{ dB}$$

Voltage ratio: 9.2 dB $\longrightarrow$ 2.884

$$2.884 \times 10 \times 10 = 288.4$$

Power ratio: 9.2 dB $\longrightarrow$ 8.318

$$8.318 \times 100 \times 100 = 83180$$

Number of decibels negative (−) Add +20 decibels successively to the given number of decibels until the sum falls within the range of Table 1. For the voltage ratio, divide the value from the left-hand voltage-ratio column by 10 for each time 20 dB was added. For the power ratio, divide the value from the left-hand power-ratio column by 100 for each time 20 dB was added.

Example: Given: −49.2 dB

$$-49.2 \text{ dB} + 20 \text{ dB} + 20 \text{ dB} = -9.2 \text{ dB}$$

Voltage ratio: −9.2 dB $\longrightarrow$ 0.3467

$$0.3467 \times \tfrac{1}{10} \times \tfrac{1}{10} = 0.003467$$

Power ratio: −9.2 dB $\longrightarrow$ 0.1202

$$0.1202 \times \tfrac{1}{100} \times \tfrac{1}{100} = 0.00001202$$

489

Voltage Ratios to Decibels: Table 2, p. 493

For ratios smaller than those in Table 2 Multiply the given ratio by 10 successively until the product can be found in Table 2. From the number of decibels thus found, subtract $+20$ decibels for each time the ratio was multiplied by 10.

Example: Given: Voltage ratio $= 0.0131$

$$0.0131 \times 10 \times 10 = 1.31$$

From Table 2: $1.31 \longrightarrow 2.345$ dB

$$2.345 \text{ dB} - 20 \text{ dB} - 20 \text{ dB} = -37.655 \text{ dB}$$

For ratios greater than those in Table 2 Divide the given ratio by 10 successively until the remainder can be found in Table 2. To the number of decibels thus found, add $+20$ dB for each time the ratio was divided by 10.

Example: Given: Voltage ratio $= 712$

$$712 \times \tfrac{1}{10} \times \tfrac{1}{10} = 7.12$$

From Table 2: $7.12 \longrightarrow 17.050$ dB

$$17.050 \text{ dB} + 20 \text{ dB} + 20 \text{ dB} = 57.050 \text{ dB}$$

Table 1

CONVERSION OF DECIBELS TO POWER AND VOLTAGE
(OR CURRENT) RATIOS

To account for the sign of the decibel:
(a) For positive (+) dB values use the two right-hand columns. The voltage and power ratios are greater than unity.
(b) For negative (−) dB values use the two left-hand columns. The voltage and power ratios are less than unity.

−dB+

Voltage Ratio	Power Ratio	dB	Voltage Ratio	Power Ratio
1.0000	1.0000	0	1.000	1.000
.9886	.9772	.1	1.012	1.023
.9772	.9550	.2	1.023	1.047
.9661	.9333	.3	1.035	1.072
.9550	.9120	.4	1.047	1.096
.9441	.8913	.5	1.059	1.122
.9333	.8710	.6	1.072	1.148
.9226	.8511	.7	1.084	1.175
.9120	.8318	.8	1.096	1.202
.9016	.8128	.9	1.109	1.230
.8913	.7943	1.0	1.122	1.259
.8810	.7762	1.1	1.135	1.288
.8710	.7586	1.2	1.148	1.318
.8610	.7413	1.3	1.161	1.349
.8511	.7244	1.4	1.175	1.380
.8414	.7079	1.5	1.189	1.413
.8318	.6918	1.6	1.202	1.445
.8222	.6761	1.7	1.216	1.479
.8128	.6607	1.8	1.230	1.514
.8035	.6457	1.9	1.245	1.549
.7943	.6310	2.0	1.259	1.585
.7852	.6166	2.1	1.274	1.622
.7762	.6026	2.2	1.288	1.660
.7674	.5888	2.3	1.303	1.698
.7586	.5754	2.4	1.318	1.738
.7499	.5623	2.5	1.334	1.778
.7413	.5495	2.6	1.349	1.820
.7328	.5370	2.7	1.365	1.862
.7244	.5248	2.8	1.380	1.905
.7161	.5129	2.9	1.396	1.950
.7079	.5012	3.0	1.413	1.995
.6998	.4898	3.1	1.429	2.042
.6918	.4786	3.2	1.445	2.089
.6839	.4677	3.3	1.462	2.138
.6761	.4571	3.4	1.479	2.188
.6683	.4467	3.5	1.496	2.239
.6607	.4365	3.6	1.514	2.291
.6531	.4266	3.7	1.531	2.344
.6457	.4169	3.8	1.549	2.399
.6383	.4074	3.9	1.567	2.455
.6310	.3981	4.0	1.585	2.512
.6237	.3890	4.1	1.603	2.570
.6166	.3802	4.2	1.622	2.630
.6095	.3715	4.3	1.641	2.692
.6026	.3631	4.4	1.660	2.754
.5957	.3548	4.5	1.679	2.818
.5888	.3467	4.6	1.698	2.884
.5821	.3388	4.7	1.718	2.951
.5754	.3311	4.8	1.738	3.020
.5689	.3236	4.9	1.758	3.090
.5623	3162	5.0	1.778	3.162
.5559	.3090	5.1	1.799	3.236
.5495	.3020	5.2	1.820	3.311
.5433	.2951	5.3	1.841	3.388
.5370	.2884	5.4	1.862	3.467
.5309	.2818	5.5	1.884	3.548
.5248	.2754	5.6	1.905	3.631
.5188	.2692	5.7	1.928	3.715
.5129	.2630	5.8	1.950	3.802
.5070	.2570	5.9	1.972	3.890
.5012	.2512	6.0	1.995	3.981
.4955	.2455	6.1	2.018	4.074
.4898	.2399	6.2	2.042	4.169
.4842	.2344	6.3	2.065	4.266
.4786	.2291	6.4	2.089	4.365
.4732	.2239	6.5	2.113	4.467
.4677	.2188	6.6	2.138	4.571
.4624	.2138	6.7	2.163	4.677
.4571	.2089	6.8	2.188	4.786
.4519	.2042	6.9	2.213	4.898
.4467	.1995	7.0	2.239	5.012
.4416	.1950	7.1	2.265	5.129
.4365	.1905	7.2	2.291	5.248
.4315	.1862	7.3	2.317	5.370
.4266	.1820	7.4	2.344	5.495
.4217	.1778	7.5	2.371	5.623
.4169	.1738	7.6	2.399	5.754
.4121	.1698	7.7	2.427	5.888
.4074	.1660	7.8	2.455	6.026
.4027	.1622	7.9	2.483	6.166
.3981	.1585	8.0	2.512	6.310
.3936	.1549	8.1	2.541	6.457
.3890	.1514	8.2	2.570	6.607
.3846	.1479	8.3	2.600	6.761
.3802	.1445	8.4	2.630	6.918
.3758	.1413	8.5	2.661	7.079
.3715	.1380	8.6	2.692	7.244
.3673	.1349	8.7	2.723	7.413
.3631	.1318	8.8	2.754	7.586
.3589	.1288	8.9	2.786	7.762
.3548	.1259	9.0	2.818	7.943
.3508	.1230	9.1	2.851	8.128
.3467	.1202	9.2	2.884	8.318
.3428	.1175	9.3	2.917	8.511
.3388	.1148	9.4	2.951	8.710
.3350	.1122	9.5	2.985	8.913
.3311	.1096	9.6	3.020	9.120
.3273	.1072	9.7	3.055	9.333
.3236	.1047	9.8	3.090	9.550
.3199	.1023	9.9	3.126	9.772

Table 1

CONVERSION OF DECIBELS TO POWER AND VOLTAGE
(OR CURRENT) RATIOS (*cont.*)

dB ← →

Voltage Ratio	Power Ratio	dB	Voltage Ratio	Power Ratio	Voltage Ratio	Power Ratio	dB	Voltage Ratio	Power Ratio
.3162	.1000	10.0	3.162	10.000	.1585	.02512	16.0	6.310	39.81
.3126	.09772	10.1	3.199	10.23	.1567	.02455	16.1	6.383	40.74
.3090	.09550	10.2	3.236	10.47	.1549	.02399	16.2	6.457	41.69
.3055	.09333	10.3	3.273	10.72	.1531	.02344	16.3	6.531	42.66
.3020	.09120	10.4	3.311	10.96	.1514	.02291	16.4	6.607	43.65
.2985	.08913	10.5	3.350	11.22	.1496	.02239	16.5	6.683	44.67
.2951	.08710	10.6	3.388	11.48	.1479	.02188	16.6	6.761	45.71
.2917	.08511	10.7	3.428	11.75	.1462	.02138	16.7	6.839	46.77
.2884	.08318	10.8	3.467	12.02	.1445	.02089	16.8	6.918	47.86
.2851	.08128	10.9	3.508	12.30	.1429	.02042	16.9	6.998	48.98
.2818	.07943	11.0	3.548	12.59	.1413	.01995	17.0	7.079	50.12
.2786	.07762	11.1	3.589	12.88	.1396	.01950	17.1	7.161	51.29
.2754	.07586	11.2	3.631	13.18	.1380	.01905	17.2	7.244	52.48
.2723	.07413	11.3	3.673	13.49	.1365	.01862	17.3	7.328	53.70
.2692	.07244	11.4	3.715	13.80	.1349	.01820	17.4	7.413	54.95
.2661	.07079	11.5	3.758	14.13	.1334	.01778	17.5	7.499	56.23
.2630	.06918	11.6	3.802	14.45	.1318	.01738	17.6	7.586	57.54
.2600	.06761	11.7	3.846	14.79	.1303	.01698	17.7	7.674	58.88
.2570	.06607	11.8	3.890	15.14	.1288	.01660	17.8	7.762	60.26
.2541	.06457	11.9	3.936	15.49	.1274	.01622	17.9	7.852	61.66
2512	.06310	12.0	3.981	15.85	.1259	.01585	18.0	7.943	63.10
.2483	.06166	12.1	4.027	16.22	.1245	.01549	18.1	8.035	64.57
.2455	.06026	12.2	4.074	16.60	.1230	.01514	18.2	8.128	66.07
.2427	.05888	12.3	4.121	16.98	.1216	.01479	18.3	8.222	67.61
.2399	.05754	12.4	4.169	17.38	.1202	.01445	18.4	8.318	69.18
.2371	.05623	12.5	4.217	17.78	.1189	.01413	18.5	8.414	70.79
.2344	.05495	12.6	4.266	18.20	.1175	.01380	18.6	8.511	72.44
.2317	.05370	12.7	4.315	18.62	.1161	.01349	18.7	8.610	74.13
.2291	.05248	12.8	4.365	19.05	.1148	.01318	18.8	8.710	75.86
.2265	.05129	12.9	4.416	19.50	.1135	.01288	18.9	8.811	77.62
.2239	.05012	13.0	4.467	19.95	.1122	.01259	19.0	8.913	79.43
.2213	.04898	13.1	4.519	20.42	.1109	.01230	19.1	9.016	81.28
.2188	.04786	13.2	4.571	20.89	.1096	.01202	19.2	9.120	83.18
.2163	.04677	13.3	4.624	21.38	.1084	.01175	19.3	9.226	85.11
.2138	.04571	13.4	4.677	21.88	.1072	.01148	19.4	9.333	87.10
.2113	.04467	13.5	4.732	22.39	.1059	.01122	19.5	9.441	89.13
.2089	.04365	13.6	4.786	22.91	.1047	.01096	19.6	9.550	91.20
.2065	.04266	13.7	4.842	23.44	.1035	.01072	19.7	9.661	93.33
.2042	.04169	13.8	4.898	23.99	.1023	.01047	19.8	9.772	95.50
.2018	.04074	13.9	4.955	24.55	.1012	.01023	19.9	9.886	97.72
.1995	.03981	14.0	5.012	25.12	.1000	.01000	20.0	10.000	100.00
.1972	.03890	14.1	5.070	25.70					
.1950	.03802	14.2	5.129	26.30					
.1928	.03715	14.3	5.188	26.92					
.1905	.03631	14.4	5.248	27.54					
.1884	.03548	14.5	5.309	28.18					
.1862	.03467	14.6	5.370	28.84					
.1841	.03388	14.7	5.433	29.51					
.1820	.03311	14.8	5.495	30.20					
.1799	.03236	14.9	5.559	30.90					
.1778	.03162	15.0	5.623	31.62					
.1758	.03090	15.1	5.689	32.36					
.1738	.03020	15.2	5.754	33.11					
.1718	.02951	15.3	5.821	33.88					
.1698	.02884	15.4	5.888	34.67					
.1679	.02818	15.5	5.957	35.48					
.1660	.02754	15.6	6.026	36.31					
.1641	.02692	15.7	6.095	37.15					
.1622	.02630	15.8	6.166	38.02					
.1603	.02570	15.9	6.237	38.90					

dB ← →

Voltage Ratio	Power Ratio	dB	Voltage Ratio	Power Ratio
3.162×10^{-1}	10^{-1}	10	3.162	10
10^{-1}	10^{-2}	20	10	10^{2}
3.162×10^{-2}	10^{-3}	30	3.162×10	10^{3}
10^{-2}	10^{-4}	40	10^{2}	10^{4}
3.162×10^{-3}	10^{-5}	50	3.162×10^{2}	10^{5}
10^{-3}	10^{-6}	60	10^{3}	10^{6}
3.162×10^{-4}	10^{-7}	70	3.162×10^{3}	10^{7}
10^{-4}	10^{-8}	80	10^{4}	10^{8}
3.162×10^{-5}	10^{-9}	90	3.162×10^{4}	10^{9}
10^{-5}	10^{-10}	100	10^{5}	10^{10}

To convert decibel ratios outside the range of this table, use the method illustrated on page 489 of the introduction to these tables.

Table 2

CONVERSION OF VOLTAGE (OR CURRENT) AND POWER RATIOS
TO DECIBELS

To find the number of decibels corresponding to a given power ratio—Assume the given power ratio to be a voltage ratio and find the corresponding number of decibels from the table. The desired result is exactly one-half of the number of decibels thus found.

Example: *Given:* a power ratio of 3.41.

Find: 3.41 in the table:

$$3.41 \longrightarrow 10.655 \text{ dB (voltage)}$$

$$10.655 \text{ dB} \times \tfrac{1}{2} = 5.328 \text{ dB (power)}$$

Voltage Ratio	.00	.01	.02	.03	.04	.05	.06	.07	.08	.09
1.0	.000	.086	.172	.257	.341	.424	.506	.588	.668	.749
1.1	.828	.906	.984	1.062	1.138	1.214	1.289	1.364	1.438	1.511
1.2	1.584	1.656	1.727	1.798	1.868	1.938	2.007	2.076	2.144	2.212
1.3	2.279	2.345	2.411	2.477	2.542	2.607	2.671	2.734	2.798	2.860
1.4	2.923	2.984	3.046	3.107	3.167	3.227	3.287	3.346	3.405	3.464
1.5	3.522	3.580	3.637	3.694	3.750	3.807	3.862	3.918	3.973	4.028
1.6	4.082	4.137	4.190	4.244	4.297	4.350	4.402	4.454	4.506	4.558
1.7	4.609	4.660	4.711	4.761	4.811	4.861	4.910	4.959	5.008	5.057
1.8	5.105	5.154	5.201	5.249	5.296	5.343	5.390	5.437	5.483	5.529
1.9	5.575	5.621	5.666	5.711	5.756	5.801	5.845	5.889	5.933	5.977
2.0	6.021	6.064	6.107	6.150	6.193	6.235	6.277	6.319	6.361	6.403
2.1	6.444	6.486	6.527	6.568	6.608	6.649	6.689	6.729	6.769	6.809
2.2	6.848	6.888	6.927	6.966	7.008	7.044	7.082	7.121	7.159	7.197
2.3	7.235	7.272	7.310	7.347	7.384	7.421	7.458	7.495	7.532	7.568
2.4	7.604	7.640	7.676	7.712	7.748	7.783	7.819	7.854	7.889	7.924
2.5	7.959	7.993	8.028	8.062	8.097	8.131	8.165	8.199	8.232	8.266
2.6	8.299	8.333	8.366	8.399	8.432	8.465	8.498	8,530	8.563	8.595
2.7	8.627	8.659	8.691	8.723	8.755	8.787	8.818	8.850	8.881	8.912
2.8	8.943	8.974	9.005	9.036	9.066	9.097	9.127	9.158	9.188	9.218
2.9	9.248	9.278	9.308	9.337	9.367	9.396	9.426	9.455	9.484	9.513
3.0	9.542	9.571	9.600	9.629	9.657	9.686	9.714	9.743	9.771	9.799
3.1	9.827	9.855	9.883	9.911	9.939	9.966	9.994	10.021	10.049	10.076
3.2	10.103	10.130	10.157	10.184	10.211	10.238	10.264	10.291	10.317	10.344
3.3	10.370	10.397	10.423	10.449	10.475	10.501	10.527	10.553	10.578	10.604
3.4	10.630	10.655	10.681	10.706	10.731	10.756	10.782	10.807	10.832	10.857
3.5	10.881	10.906	10.931	10.955	10.980	11.005	11.029	11.053	11.078	11.102
3.6	11.126	11.150	11.174	11.198	11.222	11.246	11.270	11.293	11.317	11.341
3.7	11.364	11.387	11.411	11.434	11.457	11.481	11.504	11.527	11.550	11.573
3.8	11.596	11.618	11.641	11.664	11.687	11.709	11.732	11.754	11.777	11.799
3.9	11.821	11.844	11.866	11.888	11.910	11.932	11.954	11.976	11.998	12.019
4.0	12.041	12.063	12.085	12.106	12.128	12.149	12.171	12.192	12.213	12.234
4.1	12.256	12.277	12.298	12.319	12.340	12.361	12.382	12.403	12.424	12.444
4.2	12.465	12.486	12.506	12.527	12.547	12.568	12.588	12.609	12.629	12.649
4.3	12.669	12.690	12.710	12.730	12.750	12.770	12.790	12.810	12.829	12.849
4.4	12.869	12.889	12.908	12.928	12.948	12.967	12.987	13.006	13.026	13.045
4.5	13.064	13.084	13.103	13.122	13.141	13.160	13.179	13.198	13.217	13.236
4.6	13.255	13.274	13.293	13.312	13.330	13.349	13.368	13.386	13.405	13.423
4.7	13.442	13.460	13.479	13.497	13.516	13.534	13.552	13.570	13.589	13.607
4.8	13.625	13.643	13.661	13.679	13.697	13.715	13.733	13.751	13.768	13.786
4.9	13.804	13.822	13.839	13.857	13.875	13.892	13.910	13.927	13.945	13.962
5.0	13.979	13.997	14.014	14.031	14.049	14.066	14.083	14.100	14.117	14.134
5.1	14.151	14.168	14.185	14.202	14.219	14.236	14.253	14.270	14.287	14.303
5.2	14.320	14.337	14.353	14.370	14.387	14.403	14.420	14.436	14.453	14.469
5.3	14.486	14.502	14.518	14.535	14.551	14.567	14.583	14.599	14.616	14.632
5.4	14.648	14.664	14.680	14.696	14.712	14.728	14.744	14.760	14.776	14.791
5.5	14.807	14.823	14.839	14.855	14.870	14.886	14.902	14.917	14.933	14.948
5.6	14.964	14.979	14.995	15.010	15.026	15.041	15.056	15.072	15.087	15.102
5.7	15.117	15.133	15.148	15.163	15.178	15.193	15.208	15.224	15.239	15.254
5.8	15.269	15.284	15.298	15.313	15.328	15.343	15.358	15.373	15.388	15.402
5.9	15.417	15.432	15.446	15.461	15.476	15.490	15.505	15.519	15.534	15.549

Table 2

CONVERSION OF VOLTAGE (OR CURRENT) AND POWER RATIOS
TO DECIBELS (*cont.*)

Voltage Ratio	.00	.01	.02	.03	.04	.05	.06	.07	.08	.09
6.0	**15.563**	**15.577**	**15.592**	**15.606**	**15.621**	**15.635**	**15.649**	**15.664**	**15.678**	**15.692**
6.1	15.707	15.721	15.735	15.749	15.763	15.778	15.792	15.806	15.820	15.834
6.2	15.848	15.862	15.876	15.890	15.904	15.918	15.931	15.945	15.959	15.973
6.3	15.987	16.001	16.014	16.028	16.042	16.055	16.069	16.083	16.096	16.110
6.4	16.124	16.137	16.151	16.164	16.178	16.191	16.205	16.218	16.232	16.245
6.5	16.258	16.272	16.285	16.298	16.312	16.325	16.338	16.351	16.365	16.378
6.6	16.391	16.404	16.417	16.430	16.443	16.456	16.468	16.483	16.496	16.509
6.7	16.521	16.534	16.547	16.560	16.57ɔ	16.586	16.598	16.612	16.625	16.637
6.8	16.650	16.663	16.676	16.688	16.701	16.714	16.726	16.739	16.752	16.764
6.9	16.777	16.790	16.802	16.815	16.827	16.840	16.852	16.865	16.877	16.890
7.0	**16.902**	**16.914**	**16.927**	**16.939**	**16.951**	**16.964**	**16.976**	**16.988**	**17.001**	**17.013**
7.1	17.025	17.037	17.050	17.062	17.074	17.086	17.098	17.110	17.122	17.135
7.2	17.147	17.159	17.171	17.183	17.195	17.207	17.219	17.231	17.243	17.255
7.3	17.266	17.278	17.290	17.302	17.314	17.326	17.338	17.349	17.361	17.373
7.4	17.385	17.396	17.408	17.420	17.431	17.443	17.455	17.466	17.478	17.490
7.5	17.501	17.513	17.524	17.536	17.547	17.559	17.570	17.582	17.593	17.605
7.6	17.616	17.628	17.639	17.650	17.662	17.673	17.685	17.696	17.707	17.719
7.7	17.730	17.741	17.752	17.764	17.775	17.786	17.797	17.808	17.820	17.831
7.8	17.842	17.853	17.864	17.875	17.886	17.897	17.908	17.919	17.931	17.942
7.9	17.953	17.964	17.975	17.985	17.996	18.007	18.018	18.029	18.040	18.051
8.0	**18.062**	**18.073**	**18.083**	**18.094**	**18.105**	**18.116**	**18.127**	**18.137**	**18.148**	**18.159**
8.1	18.170	18.180	18.191	18.202	18.212	18.223	18.234	18.244	18.255	18.266
8.2	18.276	18.287	18.297	18.308	18.319	18.329	18.340	18.350	18.361	18.371
8.3	18.382	18.392	18.402	18.413	18.423	18.434	18.444	18.455	18.465	18.475
8.4	18.486	18.496	18.506	18.517	18.527	18.537	18.547	18.558	18.568	18.578
8.5	18.588	18.599	18.609	18.619	18.629	18.639	18.649	18.660	18.670	18.680
8.6	18.690	18.700	18.710	18.720	18.730	18.740	18.750	18.760	18.770	18.780
8.7	18.790	18.800	18.810	18.820	18.830	18.840	18.850	18.860	18.870	18.880
8.8	18.890	18.900	18.909	18.919	18.929	18.939	18.949	18.958	18.968	18.978
8.9	18.988	18.998	19.007	19.017	19.027	19.036	19.046	19.056	19.066	19.075
9.0	**19.085**	**19.094**	**19.104**	**19.114**	**19.123**	**19.133**	**19.143**	**19.152**	**19.162**	**19.171**
9.1	19.181	19.190	19.200	19.209	19.219	19.228	19.238	19.247	19.257	19.226
9.2	19.276	19.285	19.295	19.304	19.313	19.323	19.332	19.342	19.351	19.360
9.3	19.370	19.379	19.388	19.398	19.407	19.416	19.426	19.435	19.444	19.453
9.4	19.463	19.472	19.481	19.490	19.499	19.509	19.518	19.527	19.536	19.545
9.5	19.554	19.564	19.573	19.582	19.591	19.600	19.609	19.618	19.627	19.636
9.6	19.645	19.654	19.664	19.673	19.682	19.691	19.700	19.709	19.718	19.726
9.7	19.735	19.744	19.753	19.762	19.771	19.780	19.789	19.798	19.807	19.816
9.8	19.825	19.833	19.842	19.851	19.860	19.869	19.878	19.886	19.895	19.904
9.9	19.913	19.921	19.930	19.939	19.948	19.956	19.965	19.974	19.983	19.991

Voltage Ratio	0	1	2	3	4	5	6	7	8	9
10	**20.000**	**20.828**	**21.584**	**22.279**	**22.923**	**23.522**	**24.082**	**24.609**	**25.105**	**25.575**
20	26.021	26.444	26.848	27.235	27.604	27.959	28.299	28.627	28.943	29.248
30	29.542	29.827	30.103	30.370	30.630	30.881	31.126	31.364	31.596	31.821
40	32.041	32.256	32.465	32.669	32.869	33.064	33.255	33.442	33.625	33.804
50	33.979	34.151	34.320	34.486	34.648	34.807	34.964	35.117	35.269	35.417
60	35.563	35.707	35.848	35.987	36.124	36.258	36.391	36.521	36.650	36.777
70	36.902	37.025	37.147	37.266	37.385	37.501	37.616	37.730	37.842	37.953
80	38.062	38.170	38.276	38.382	38.486	38.588	38.690	38.790	38.890	38.988
90	39.085	39.181	39.276	39.370	39.463	39.554	39.645	39.735	39.825	39.913
100	**40.000**	—	—	—	—	—	—	—	—	—

To convert voltage and power ratios outside the range of this table, use the method illustrated on page 490 of the introduction to these tables.

Index